MOLECULAR
BIOLOGY
INTELLIGENCE
UNIT

HEAT SHOCK PROTEINS AND CYTOPROTECTION:
ATP-DEPRIVED MAMMALIAN CELLS

Alexander E. Kabakov, Ph.D.
Vladimir L. Gabai, Ph.D.

Medical Radiology Research Center
Russian Academy of Medical Sciences
Obninsk, Russia

CHAPMAN & HALL
I(T)P An International Thomson Publishing Company

New York • Albany • Bonn • Boston • Cincinnati • Detroit • London • Madrid • Melbourne •
Mexico City • Pacific Grove • Paris • San Francisco • Singapore • Tokyo • Toronto • Washington

R.G. LANDES COMPANY
AUSTIN

MOLECULAR BIOLOGY INTELLIGENCE UNIT
HEAT SHOCK PROTEINS AND CYTOPROTECTION: ATP-DEPRIVED MAMMALIAN CELLS

R.G. LANDES COMPANY
Austin, Texas, U.S.A.

U.S. and Canada Copyright © 1997 R.G. Landes Company and Chapman & Hall

Please address all inquiries to the Publishers:
R.G. Landes Company, 810 S. Church Street, Georgetown, Texas, U.S.A. 78626
Phone: 512/ 863 7762; FAX: 512/ 863 0081

North American distributor:
Chapman & Hall, 115 Fifth Avenue, New York, New York, U.S.A. 10003

CHAPMAN & HALL

U.S. and Canada ISBN: 0-412-13231-1

While the authors, editors and publisher believe that drug selection and dosage and the specifications and usage of equipment and devices, as set forth in this book, are in accord with current recommendations and practice at the time of publication, they make no warranty, expressed or implied, with respect to material described in this book. In view of the ongoing research, equipment development, changes in governmental regulations and the rapid accumulation of information relating to the biomedical sciences, the reader is urged to carefully review and evaluate the information provided herein.

Library of Congress Cataloging-in-Publication Data
Kabakov, Alexander E., 1960-
    Heat shock proteins and cytoprotection: ATP-deprived mammalian cells/
Alexander E. Kabakov, Vladimir L. Gabai
      p. cm. — (Molecular biology intelligence unit)
Includes bibliographical references and index.
ISBN 0-57059-413-9 (alk. paper)
    1. Heat shock proteins—Physiological effect. 2. Adenosine triphosphate. 3. Molecular chaperones. 4. Energy metabolism.   I. Gabai, Vladimir L., 1959- . II. Title.
III. Series.
  QP552.H43K33 1996
  572'.475—dc21
                                                  96-39322
                                                    CIP

PUBLISHER'S NOTE

R.G. Landes Company publishes six book series: *Medical Intelligence Unit, Molecular Biology Intelligence Unit, Neuroscience Intelligence Unit, Tissue Engineering Intelligence Unit, Biotechnology Intelligence Unit* and *Environmental Intelligence Unit.* The authors of our books are acknowledged leaders in their fields and the topics are unique. Almost without exception, no other similar books exist on these topics.

Our goal is to publish books in important and rapidly changing areas of bioscience and environment for sophisticated researchers and clinicians. To achieve this goal, we have accelerated our publishing program to conform to the fast pace in which information grows in bioscience. Most of our books are published within 90 to 120 days of receipt of the manuscript. We would like to thank our readers for their continuing interest and welcome any comments or suggestions they may have for future books.

Shyamali Ghosh
Publications Director
R.G. Landes Company

CONTENTS

INTRODUCTION

Discovery of the stress response mechanism is, undoubtedly, the outstanding achievement in the field of molecular biology of the cell. Indeed, the amazing ability of cells to react rapidly and adequately to stresses has engaged scientists' attention for decades. An interest in this problem among molecular biologists arose in 1962 after brief publication of F. Ritossa's description of the formation of new chromosomal puffs (i.e., specific gene activity) in *Drosophila* salivary gland following thermal or chemical exposures. Later the products of such stress-induced gene transcription were designated as heat-shock proteins (HSPs) or stress proteins and further reclamation of this "terra incognita" developed in two directions: (1) Studying the structure and properties of HSPs and (2) Unraveling the molecular mechanism of the heat-shock response. By the middle of the 1990s, considerable progress had been reached in both areas. At the present time, we know how a cell senses elevated temperature and which biochemical events in the heat-shocked cell lead to the transcriptional response and stimulation of HSP synthesis. Likewise, it has been proven that the expression of HSPs is an autoregulatory process in all living organisms from bacteria to man.

Perhaps one of the most advanced trends was the characterization of HSPs as molecular chaperones. In an unstressed cell the constitutively expressed HSPs regulate protein folding, protein translocation across membranes, assembly and disassembly of protein oligomers, degradation of aged proteins, etc. Furthermore, being molecular chaperones, HSPs are able to prevent undesirable inter- and intraprotein interactions, thus preserving unfolded or immature polypeptides from aggregation and misfolding. Since heating causes unfolding and aggregation of cellular proteins, no one doubts the same chaperoning function of HSPs also defines the thermoresistance of cells. Obviously, the role of the heat shock response lies in the stimulation of HSP synthesis, the most persuasive evidence showed that an increase in the level of cellular HSPs confers thermotolerance. An explanation of the phenomenon of acquired (stress-induced) thermotolerance was based on the following two assertions: (1) Heating damages cellular proteins that trigger heat shock gene expression followed by synthesis and accumulation of HSPs; (2) Excess HSPs protect cellular proteins from unfolding and aggregation under repeated thermal exposure and facilitate refolding, disaggregation and/or degradation of heat-denatured proteins during cell recovery. This explains how a cell can survive extreme heating and restore its vital functions afterward. Moreover, both postulates are supported by the following observations: (1) the heat shock response is induced following an appearance inside a cell of abnormal (aggregated or misfolded) proteins; and (2) cells overexpressing HSPs exhibit more thermostability of major subcellular systems and metabolic processes.

Thermal stress, ultraviolet irradiation, changes in pH of cell environment, and treatments with oxidants, detergents, ethanol, ions of transition metals, amino acid analogs, etc. can induce HSP synthesis and thermotolerance in the subjected cells. Evidently, the majority of these exposures as well as heat shock are "proteotoxic", i.e., they damage cellular proteins causing unfolding and aggregation. Thus, it is easy to imagine that, on the one hand, abnormal proteins within stressed cells promote the heat shock transcription response and HSP synthesis while, on the other hand, induced HSPs protect cellular proteins and can help an affected cell to overcome consequences of proteotoxic stresses. Thus, we have an understanding of how HSP expression may be triggered under stress and why a stressed cell requires this triggering.

What is not yet clear, however, is the link between HSP expression and fluctuations in the levels of cellular ATP. It was already known that the temporary suppression of ATP generation in mitochondria activates transcription of heat shock genes. This problem has the same history as the heat shock story, since the first Ritossa papers (1962-1964) reported that 2,4-dinitrophenol, an uncoupler of oxidative phosphorylation, as well as thermal exposure, induces puffs in *Drosophila* salivary gland chromosomes. Likewise, in vitro inhibition of cell respiration by mitochondrial poisons and transient anoxia per se are powerful inducers of the heat shock transcriptional response. This creates the impression that a cell undergoing short-term energy deprivation responds directly to the activation of heat shock genes. For many years, such a stress response was the subject of study for molecular pathologists rather than classical biochemists, since it occurs in mammalian tissues following various pathological states such as hypoxia, ischemic insult or hypoglycemia. Actually, numerous works performed in vivo and on isolated reperfused organs demonstrated that transient ischemia (i.e., an inadequate supply of oxygen and nutrients to a tissue because of limited blood flow) stimulates HSP synthesis in mammalian cells. In the beginning of the 1990s, several research groups described the protective effect of previous heat shock or preconditioning ischemia on the survival of the ischemic (energy-deprived) cells. Most of the research shows that the elevated cell viability under ischemic-like conditions is associated with the induced expression of HSPs, but no real mechanism of HSP-mediated cytoprotection from ischemia has been established.

Using an in vitro model of ischemia, Benjamin and Williams demonstrated that the drop in ATP level leads to an immediate activation of the heat shock transcription factor (HSF), which results in the HSP mRNA synthesis following the metabolic stress. Benjamin and Williams also found that constitutive overexpression of the 70 kDa HSP (HSP70), a major chaperone of eukaryotes, protects mouse cells from injury during ATP depletion. Taken together the results suggest that the HSP expression is a universal adaptive device allowing a cell to survive not only heating but also energetically unfavorable (ATP-depleting) conditions. Hence, the possibility to induce a tolerance to ischemia by preconditioning treatments may be a basis for optimism among clinicians, since it is a novel approach to the problems of myocardial infarction, brain stroke and other complications caused by a failure in the blood supply. Significant resistance to myocardial ischemia has already been demonstrated in transgene mice overexpressing inducible HSP70.

With that in mind, both the induction of the heat shock response by transient ATP depletion and the HSP-mediated cytoprotection under energy starvation warrant an explanation because too many questions remain unanswered. The main questions are: (1) Do HSPs themselves sense the changes in ATP generation? (2) Why does the transient drop in cellular ATP activate HSF, thus triggering transcription of heat shock genes? (3) Can the ATP depletion per se be considered the proteotoxic factor for intracellular proteins? (4) How can the chaperone properties of HSPs compensate for ATP deficiency or maintain the viability of ATP-depleted cells? (5) What are the major cellular targets for the protective action of HSPs during ATP deprivation? (6) How can HSP70 rescue a cell devoid of ATP, if HSP70 itself possesses ATP-ase activity and therefore requires ATP for the chaperoning function? It is obvious that all these questions are interrelated.

In this book we would like to present an overview of the studies on the role of cell energetics in the mechanism of the stress response induction. In addition, we would like to provide the most interesting facts, speculations and hypotheses concerning the involvement of HSPs in cytoprotection under ATP depletion. We hope that our book will aid in clarification of the above questions, which could be especially helpful for scientists working at the crossroads of molecular biology and medicine.

PREFACE

In the present book we intend to tell about an intriguing phenomenon, namely how mammalian cells become tolerant to deprivation of oxygen and nutrient substrates. Of course, having read this story our readers will not learn how to do without air and food but hopefully they will get an insight into molecular mechanisms of cellular adaptation to anoxia and starvation. Since such an adaptation also represents a great interest for medicine, our book should be equally recommended to both molecular biologists and clinicians.

Adenosine-5'-triphosphate (ATP), being the immediate source of a cell's energy, is constantly consumed for maintenance of ionic gradients, biosynthesis, cell motility, macromolecule transport etc. At the same time, ATP is permanently reproduced via glycolysis and oxidative phosphorylation if substrates and oxygen are available. Starvation and anoxia as well as blockade of the ATP-generating pathways by specific inhibitors result in rapid depletion of cellular ATP. Sustained ATP deprivation kills cells. This is exactly what happens in tissues upon ischemic insult when many of cells remain without glucose and oxygen owing to limitation of arterial blood flow. Ischemic depletion of ATP evokes mass cell injury and death in the affected organ—the typical examples are myocardial infarction and brain stroke. As a cause of fatal human pathologies, ischemia is successfully studied on various models with experimental animals and cell cultures. In this book, we review in detail molecular processes provoked by ischemia or ischemia-like treatments in mammalian cells.

Cellular response to ischemic stress or transient ATP depletion includes expression of certain genes encoding so called "heat shock proteins" (HSPs). This term arose because these proteins were first discovered as heat-inducible ones. HSPs perform functions of molecular chaperones, preserving unfolded and immature cellular proteins from aggregation and misfolding. Interaction of major HSPs (HSP100, HSP70, HSP60) with protein substrates occurs in an ATP-dependent manner. In heat-shocked cells HSP synthesis is dramatically enhanced, which serves for cytoprotection, since accumulated HSPs are able to minimize heat-induced aggregation of cellular proteins and to facilitate post-stress cell recovery.

Recently it was established that excess HSPs also rescue mammalian cells from lethal injury during ATP-depleting ischemic stress. In turn, transient drop in cellular ATP as well as ischemic episodes causes HSP accumulation in surviving cells. These phenomena indicate the existence of a special adaptive mechanism which is based on (I) enhancement of HSP expression in response to periods of energy starvation and (II) HSP-mediated cytoprotection during ATP deprivation. This intriguing issue of molecular and cellular biology is closely related to such actual medical problems as protection of human tissues against ischemic injury (e.g., alleviation of myocardial infarction and brain stroke) or tumor resistance to some kinds of anticancer therapy. If you wish to know:

- how mammalian cells sense ATP decrease and why it promotes transcription of heat shock genes;
- how overexpressed HSPs can elevate survival of ATP-deprived mammalian cells;
- which HSPs are involved in the cytoprotection and what their targets are;
- what current and future clinical applications of HSP-mediated cytoprotection against ischemic injury may be,

you could find this book worth reading.

Acknowledgments

First of all, we would like to thank our publisher, Ronald G. Landes, for initiating this writing. We greatly appreciate all works executed by Landes Bioscience Publishers on design, advertising and distribution of our book.

Likewise, we express gratitude to many of our colleagues and assistants for their practical and theoretical contribution in this project. Among them should be recognized O. Bensaude, K.R. Budagova, E.N. Denishik, M.-F. Dubois, H.H. Kampinga, S.A. Loktionova, N.H. Lukashova, Yu.M. Makarova, Ya.V. Malutina, A.A. Michels, A.O. Molotkov, A.F. Mosin, V.A. Mosina, V.T. Nguyen, B.S. Polla, G.J.J. Stege, W.J. Welch and I.A. Zamulaeva.

Moreover, we wish to thank personally Olivier Bensaude—collaboration with him has strongly advanced us in the understanding of how intracellular proteins respond to ATP depletion. We are also very grateful to Harm H. Kampinga for his assistance in discussion of some points at issue.

Finally, we would like to acknowledge those researchers who provided antibodies for our investigation; they are: R.S. Adelstein, A.M. Belkin, O. Bensaude, M. Gaestel, H.H. Kampinga, W.J. Welch, and A.C. Wikstrom.

Fulfillment of this project was in part supported by *Association pour la recherche sur le cancer* (France) and the Russian Foundation for Basic Research (Grants 94-04-13338 and 95-04-11316).

CHAPTER 1

HEAT SHOCK PROTEINS AND THE REGULATION OF HEAT SHOCK GENE EXPRESSION IN EUKARYOTES

Our first chapter is devoted to the general description of stress proteins and peculiarities of their expression in eukaryotic cells. We also introduce readers to the modern views of the problem of "negative regulation" of heat shock gene transcription to facilitate an understanding of the subsequent sections of this book.

1.1. THE FAMILY OF STRESS PROTEINS AND MOLECULAR CHAPERONES

Although the heat shock-induced formation of new chromosomal puffs (see Fig. 1.1) was described for the first time as early as 1962 by F. Ritossa,[1] the biological meaning of this reaction remained enigmatic. In 1973, Tissieres working at Caltech used metabolic incorporation of radioactive precursors in polypeptides to analyze the effect of heat shock on protein synthesis in *Drosophila* salivary gland. The data of his experiments were earth-shaking: heat shock activated the synthesis of only a few polypeptides and strongly inhibited the synthesis of most others.[2] Research on the isolation of "heat shock" mRNA, its hybridization to the polytene chromosomes and its translation in vitro have provided convincing evidence that the polypeptides, whose synthesis was enhanced following thermal stress, are translated on the RNA transcripts from the induced puffs.[3,4] Now it is generally accepted that genes which are expressed during stress exposure encode special proteins possessing cytoprotective properties. These inducible proteins can be united in a family of so called "heat shock proteins" (HSPs) or "stress proteins." Another widely referred to term is "molecular chaperones" because the main function of most HSPs appears to be the saving of cellular proteins and immature polypeptides from "dangerous liaisons," i.e., from inappropriate intra- and intermolecular bonds leading to protein misfolding and aggregation.

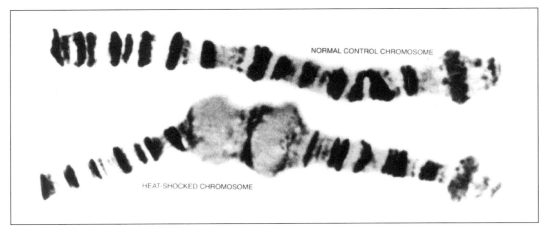

Fig. 1.1. Heat shock response (puffs) in Drosophila *polytene chromosome. The same puffs can be induced by inhibitors of energy metabolism (see ref. 1). Reproduced with permission from Welch WJ, Scientific American 1993; 268: 56-62.*

Most who have worked in the field of cellular biochemistry are acquainted with the problem of protein aggregation. Indeed, the extremely high concentration of protein (100-150 mg/ml) is located in the microvolume of a cell. Moreover, many of the uncompleted polypeptide chains are permanently present on ribosomes and, in addition, many intracellular proteins carry hydrophobic domains. All these factors might result in spontaneous protein aggregation in vivo even without any stress and the risk of the aggregation certainly increases with temperature, since hyperthermia induces mass denaturation of thermolabile cellular proteins, thus unmasking their hydrophobic sites. Apart from heat shock, many other "proteotoxic" stresses affect the folding of cellular proteins and may cause aggregation.

To minimize the problem of protein misfolding and aggregation in vivo a family of special proteins, namely HSPs and molecular chaperones, exists. Most of the members of this family are universally conserved proteins represented by a spectrum of functional homologues in all cellular compartments and organelles. The properties of molecular chaperones enable them to control protein folding in vivo and therefore to regulate such cellular processes as protein maturation, protein sorting and

translocation across membranes, protein degradation, subunit assembly in oligomeric complexes, etc. Of course, not all molecular chaperones belong to the HSP class of proteins and the chaperoning actions are not the sole function of HSPs. The purpose of this section is to review briefly the major classes of HSPs, to characterize the main properties of HSPs as molecular chaperones and to emphasize their role in the protection of eukaryotic cells from stresses.

1.1.1. THE HSP100 CLASS OF STRESS PROTEINS

The family comprises highly conserved stress proteins with MW 104-110 kDa, which are found in both prokaryotes and eukaryotes. Each member of the family is an ATPase and usually contains two ATP-binding domains or more rarely only one. These proteins are heat-inducible but constitutively expressed forms also exist.[5,6]

HSP104 is required for induction of thermotolerance in yeast[7] and it is able to protect yeast cells from high concentrations of ethanol but not from the toxic effect of cadmium.[8] The biochemical activity of HSP100s providing cell survival under extreme conditions is not fully clear. At least some members of this family are involved in proteolysis and seem to promote degradation of damaged proteins.[6] Perhaps the activity of HSP100s is functionally interrelated with the activity of HSP70 and the members of both families act in tandem.[6] Data of Parsell et al have shown that yeast HSP104 in vivo does not protect transfected luciferase from heat inactivation but mediates the enzyme resolubilization from heat shock-induced protein aggregates.[9] In the same paper it was discussed that the major function of HSP104 appears to be the disaggregation of insoluble protein aggregates, while HSP70 mainly prevents the formation of such aggregates. ClpA, a bacterial homologue of HSP104, also acts as a molecular chaperone.[10]

1.1.2. THE HSP90 CLASS OF STRESS PROTEINS

Although the proteins from this family are present in prokaryotes, deletion of the *E. coli* HSP90 gene can be performed without dramatic consequences.[11] On the contrary, HSP90s are essential for eukaryotes.[6] Permanent and abundant expression of the 82-94 kDa HSPs in cells of the higher organisms suggests an important role for these stress proteins. It is known that HSP90 interacts with several protein kinases including pp60src, and steroid hormone receptors, transcription factors, actin, tubulin, and calmodulin (reviewed in refs. 12-14). By binding to these protein substrates, HSP90 carries out regulatory function. In the case of glucocorticoid receptors, transcription factors, and protein kinases, HSP90 is able to suppress their activity; though the presence of this stress protein is necessary for adequate function of both glucocorticoid receptor and pp60src kinase.[15,16] Moreover, HSP90 itself is a substrate for protein kinases[17] and seems to possess autokinase and ATPase activities.[18,19] Whether ATP-binding and ATPase activities are intrinsic to HSP90 of higher eukaryotes is still arguable (Scheibel and Buchner, personal communication). In addition to the direct binding of HSP90 to transcription factors, the HSP90-mediated regulation of the transcriptional activity may be partially based on the ability of

HSP90 to enhance association of histones to DNA, thus strongly condensing the chromatin structure.[20]

In vitro purified HSP90 functions as a molecular chaperone preventing the aggregation of denatured proteins and accelerating the enzyme refolding.[21] Interestingly, besides the above proteins, several others including such members of the chaperone family as immunophilin (HSP56) and HSP70 are also complexed with HSP90.[22] The biological significance of this complexing is not completely understood but possibly means a cooperative action of various chaperones toward the same protein substrates. In higher eukaryotes a member of the HSP90 family, the 94 kDa glucose-regulated protein (GRP94), is present in the endoplasmic reticulum and likely participates in maturation and subunit assembly of secretory proteins.[6,13,14]

Perhaps one of the important functions of HSP90s is their involvement in the regulation of the cytoskeleton dynamics, cell shape and motility. It has been shown that the 90 kDa stress protein binds to actin and tubulin[12] and is enriched in ruffling membranes.[23] In vitro HSP90 increases the low shear viscosity of polymerized actin solution and this activity is inhibited by calmodulin in the presence of Ca^{2+}.[23,24] Thus, HSP90 may be considered a protein which crosslinks actin filaments in a Ca^{2+}-dependent manner. Experiments in a cell-free system reveal that dissociation of HSP90 from F-actin is induced by ATP and therefore HSP90 may regulate actin-myosin interactions via an ATP-dependent mechanism.[25] Likewise it is known that an elevated level of HSP90 expression in CHO cells alters the cell morphology and increases the ability of the cells to migrate.[26]

Yeast HSP90s are required for normal growth at high temperatures but they are not required for tolerance to extreme hyperthermia.[27] At the same time, the role of HSP90 in thermoresistance of mammalian cells was shown to be considerable.[26,28] During heat shock HSP90 is phosphorylated[17] and accumulates in nuclei[29] but the precise molecular mechanism of

HSP90-mediated cytoprotection from thermal exposure remains unknown.

1.1.3. THE HSP70 FAMILY

The HSP70s are ubiquitous chaperoning ATPases which have a highly conserved nucleotide-binding N-terminal domain and a relatively variable peptide-binding C-terminal domain. These proteins are expressed in prokaryotes (a bacterial analog of HSP70 is referred as to DnaK) and are known to be the major molecular chaperones of eukaryotes, being present in all cellular compartments and organelles (see for review refs. 12-14,30). Synthesis of HSP70s sharply increases after heat shock,[2,4,12] but the role of the constitutively-expressed members of this family is also important, since they are actively involved in processes of protein maturation, protein transport, proteolysis, etc. Being the main component of the chaperone machine in eukaryotes, HSP70 associates with nascent polypeptides on ribosomes to control the folding of newly synthesized proteins.[31,32]

The interaction of HSP70 with immature or denatured (unfolded) proteins is carried out via the following K^+, ATP-dependent cyclic mechanism: (1) when chaperoning action is completed, HSP70 binds ATP to release from prefolded protein and ATP hydrolysis occurs; (2) the formed HSP70-ADP complex actively binds to a new substrate (unfolded polypeptide); (3) after prefolding of the bound polypeptide a new molecule of ATP replaces ADP to be hydrolyzed by HSP70 following K^+ dependent dissociation of the chaperone-protein complex (see Fig. 1.2 and refs. 32-36). Herein it is important that, on the one hand, the avidity of HSP70-protein substrate complex is dependent on ATP and ADP,[33,34] and that, on the other hand, interaction of HSP70 with unfolded polypeptides defines the rate of ATP/ADP exchange in the chaperone's nucleotide-binding domain.[35,36] In vivo HSP70 usually participates in the protein folding pathway together with the other two chaperones GrpE and DnaJ (HSP40), which may regulate ATPase activity of HSP70 and its

binding to protein substrates.[32,37] Some chaperoning actions of HSP70 are performed in cooperation with HSP60 or HSP90.[37,38] Among substrates and partners determining multiple functions of HSP70, a variety of various cellular proteins are known, including topoisomerase I, oncosuppressor p53, clathrin, calnexin, and others.[38]

Besides the regulation of polypeptide folding, another important function of the cytosolic HSP70 in unstressed cells is transport and sorting of proteins between compartments (see Fig. 1.3). This includes chaperoning cellular proteins to facilitate their translocation across membranes or through the nuclear membrane pores, and also interaction with clathrin in vesicle recycling. In particular, the constitutively expressed HSP70 carries out disassembly of clathrin cages and release of clathrin from coated vesicles.[13,14,39] The HSP70-mediated "uncoating" of the vesicles, as well as the HSP70-mediated protein redistribution between compartments, requires ATP hydrolysis by the chaperone's ATPase and therefore the transport function of HSP70 is also regulated in an ATP-dependent fashion.[39]

Some part of the cytoplasmic HSP70s is associated with the cytoskeleton. HSP70 was shown to bind to actin stress fibers,[40] microtubular network and intermediate filaments (reviewed in ref. 41). The constitutively expressed HSP70 can affect actin polymerization in *Dictiostellium* by regulating an activity of the actin-binding 32/34 cap protein; this mechanism of the regulation appears to be ATP-dependent.[42] Interaction of HSP70 with intermediate filaments occurs in an ATP-dependent manner as well.[43] The HSP70-cytoskeleton association has not yet been studied in detail but its possible significance for cell resistance to stresses is often discussed.

In addition, HSP70s are components of the cellular proteolytic system. A member of the HSP70 family, the 73 kDa peptide recognizing protein (prp73) is involved in selective proteolyses. Prp73 binds KFERQ or related peptide sequences that target intracellular proteins for lysosomal

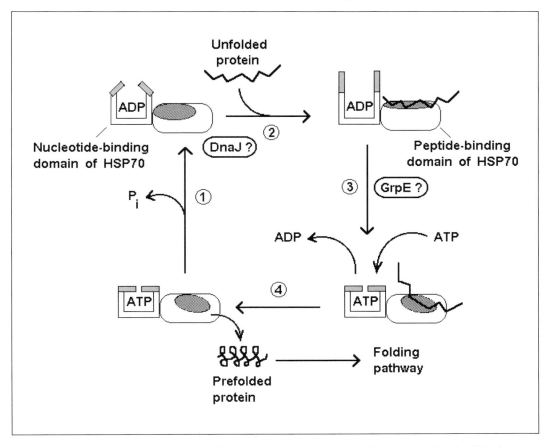

Fig. 1.2. A working circuit of the HSP70 chaperone machine. The model is described in detail in the text.

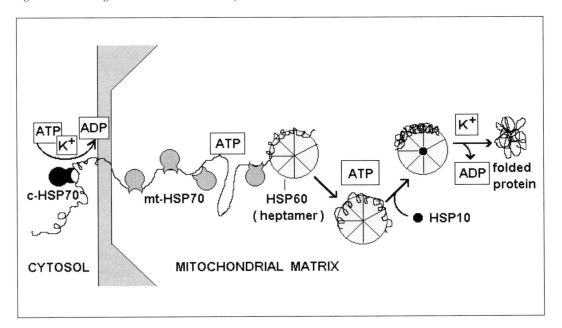

Fig. 1.3. HSP-mediated protein import and folding in mitochondria.

degradation.[13,44] Apparently, prp73 performs unfolding of the bound proteins and their translocation across the lysosomal membrane, thus facilitating selective uptake and cleavage of the KFERQ proteins by lysosomes; this HSP70 function, as well as the functions above, requires ATP.[44] Participation of HSP70s in the ubiquitin proteolysis pathway also seems possible.[45]

In the endoplasmic reticulum the HSP70 family is represented by the 78 kDa protein referred as BiP or glucose-regulated protein 78 (GRP78).[12-14] This chaperone is expressed constitutively and is one of the most abundant proteins in the reticulum lumen; however, its synthesis can be enhanced after heat shock, hypoxia, glucose deprivation, Ca^{2+} overloading and some other stresses disturbing the post-translational modification of secretory proteins.[13,46] The major functions of BiP seem to be an ATP-dependent translocation of proteins across the reticulum membrane and protein refolding and assembly within the reticulum lumen.[38,46,47]

A mitochondrial homologue of HSP70, the 75 kDa chaperoning ATPase (or GRP75), is encoded by a special heat shock gene in the nucleus.[48] This HSP is transported in mitochondria and localized in the mitochondrial matrix where it plays an important role in processes of import, folding and assembly of proteins destined for the mitochondria.[13,49] The function of mitochondrial HSP70 is tightly coupled with the activity of two other members of the chaperone family, the cytoplasmic HSP70 and mitochondrial HSP60 (Fig. 1.3).[49]

Finally, HSP70 appears to be one of the most important regulators of the stress response mechanism in eukaryotes. The constitutively expressed HSP70 inhibits the interaction of the heat shock transcription factor with DNA and, therefore, controls expression of the entire heat shock multigene family (see the next section). Numerous studies with *HSP70* gene overexpression, *HSP70* gene antisense transformation and microinjection of anti-HSP70 antibodies provide strong evidence that HSP70 can define thermoresistance of eukaryotic cells

(summarized in refs. 6, 37, 50, 51). Since the role of HSP70 in chaperoning unfolded proteins in vitro or within a cell at normal conditions has been established, its cytoprotective role in chaperoning cellular proteins damaged by heat shock or other "proteotoxic" stresses seemed to be a logical conclusion.[51-54] Indeed, overexpressed HSP70 in vivo protects a reporter enzyme, luciferase, from thermoinactivation in heat-shocked cells (G.N. Pagoulatos and O. Bensaude, personal communication). Likewise, it is known that in heat-shocked cells HSP70 is accumulated in the nucleus and enriched in nucleoli, probably to repair nucleolar functions afterward.[55] Stege et al[56] have shown that intranuclear accumulation of overexpressed HSP70 protects nuclear proteins against heat-induced aggregation.

1.1.4. THE HSP60 CLASS OF STRESS PROTEINS

A bacterial homologue of HSP60 is known as groEL and it is the main component of the chaperone machine of prokaryotes. GroEL hydrolyzes ATP to regulate protein folding and assembly in cooperation with other chaperones, groES (HSP10) and DnaK (HSP70).[13,32,37] In eukaryotes HSP60 is a nuclear gene product which is constitutively synthesized in the cytoplasm and then translocated into the mitochondria. In the matrix of mitochondria, HSP60 forms a ring-like oligomeric structure from seven subunits; this complex has a weak K^+-dependent ATPase activity and exhibits distinct chaperone properties (Fig. 1.3).[32,49] Expression of this chaperone is enhanced by heating, and besides thermal shock, ischemia is a potent inducer of HSP60 synthesis in cardiomyocytes.[57] Apparently, the function of eukaryotic HSP60 under normal conditions lies mainly in ATP-dependent refolding and in the oligomeric assembly of proteins imported into the mitochondria and also in the folding of proteins synthesized on intramitochondrial ribosomes. The chaperoning actions of HSP60 are carried out in cooperation with mitochondrial HSP70 and HSP10 (Fig. 1.3).[6,32,49,50] In mammals the

60 kDa chaperone is absolutely necessary for normal mitochondrial biogenesis and lack of human HSP60 in fetus was found to be lethal through a failure in heart mitochondria.[58] The cytosolic chaperonin (the T-complex proteins, or TCPs) exists in eukaryotes that possess an HSP60-like activity and similar oligomeric structure and also shows significant homology in amino acid sequence with HSP60.[59]

In vitro HSP60 prevents aggregation of certain enzymes under high temperature and it can restore the activities in the presence of ATP and HSP10 when temperature is reduced.[60] A contribution of eukaryotic HSP60 in the protection of mitochondrial proteins under stressful conditions appears to be possible. At the least, temperature-sensitive mutations in HSP60 of yeast result in the misfolding and aggregation of many oligomeric matrix proteins, including F1-ATPase, at the nonpermissive temperature.[6,61]

1.1.5. THE SMALL HSPs

The small or low molecular weight (15-28 kDa) HSPs are found in mycobacteria and all eukaryotes; these stress proteins are relatively less conserved.[6,14,62] All members of this family contain conserved α-crystallin domains and can be separated into two subfamilies: the 25-28 kDa HSPs (often referred to as HSP27) and so-called crystallins (15-22 kDa). Both HSP27 and crystallins are expressed constitutively in many eukaryotic cells and mammalian tissues but, at the same time, stress-induced activation of the genes is also well-known. Expression of the small HSPs is differentially changed during development and growth cycle (for review, see refs. 62, 63).

In unstressed mammalian cells HSP27 appears to be essential in signal transduction to actin microfilaments. Apparently, the small HSP is known as an inhibitor of actin polymerization which has properties of a barbed-end capping protein and can depolymerize F-actin;[64] moreover, HSP27 enriches regions of active actin polymerization in fibroblasts such as leading edge, lamellipodia and ruffles.[62,65] This functional

activity of HSP27, as well as its oligomerization, is strongly dependent on phosphorylation[62,65] and a variety of stimuli (heat shock, oxidative stress, tumor necrosis factor, serum, mitogens, etc.) activate protein kinase(s) which in turn phosphorylates HSP27.[62,66]

αA- and αB-crystallins are encoded by two distinct genes and widely distributed in tissues of mammals with dramatic prevalence in eye lens.[63] The latter is known to possess a striking sequence homology, as well as similar secondary and tertiary structures, with HSP27; moreover, expression of αB-crystallin increases after stress[63] and this protein interacts with actin and desmin.[67]

In heat-shocked cells the small HSPs form super-aggregated oligomeric complexes ($>10^6$ daltons) that are usually localized inside the nucleus.[62] Cells selectively overexpressing αB-crystallin[68] or HSP27[69] are known to be thermoresistant. The molecular mechanism of HSP27-mediated thermoresistance may involve stabilization of the actin skeleton[70] and/or accelerated recovery from heat-induced nuclear protein aggregation.[71] Interestingly, besides thermoresistance, the cytoprotection against cytochalasin D,[70] tumor necrosis factor (TNFα), hydrogen peroxide,[72] and doxorubicin[73] is also associated with HSP27. Despite the absence of ATPase activity, in a cell free system HSP27 is able to act as typical molecular chaperone, namely by preserving some enzymes from thermal inactivation and aggregation and accelerating their post-denaturation refolding.[62,74]

1.1.6. UBIQUITIN

Besides lysosomal degradation of cellular proteins, an alternative proteolytic pathway via the ubiquitin system occurs in eukaryotes. In this system proteins are digested by a 26S protease complex (proteasome) following specific covalent conjugation with multiubiquitin. The multiubiquitination of protein substrates, as well as their subsequent degradation in proteasomes, requires ATP (for review, see ref. 75). Ubiquitin is a highly conserved low molecular weight (~8 kDa) protein. Its synthesis increases after heat shock, which allows the inclusion of ubiquitin in the HSP family.[6,14,50] In mammalian cells, heat shock sharply stimulates protein degradation that correlates with a rise in multiubiquitin-protein conjugates; furthermore, a temperature-sensitive mutation in one of the enzymes in ubiquitin pathway blocks both the increased proteolysis and normal cell growth following heat shock.[76] The function of the ubiquitin and ubiquitin-conjugating enzymes in stressed cells appears to convert damaged proteins in targets for rapid cleavage by the 26S proteasomes.[6,50]

1.1.7. OTHER STRESS PROTEINS AND CHAPERONES

In addition to the major HSPs mentioned above, environmental stress may induce in some eukaryotic cells other proteins which are sometimes designated as HSPs. So, heat shock, oxidative stress or SH-reagents stimulate in many cell lines the expression of hemeoxygenase, often referred to as HSP32. This stress protein seems to be involved in an antioxidant defense mechanism together with ferritin.[77]

Heating can also elevate the level of a collagen-binding 47 kDa protein (HSP47) in some matrix-producing mammalian cells;[78] apparently, HSP47 participates in procollagen processing and assembly of the collagen triple-helical structure.

Such slow but crucial step in the protein folding pathway as the *cis-trans* isomerization of certain prolyl residues is catalyzed by special enzymes, peptidyl prolyl *cis-trans* isomerases (PPIases). PPIases with MW 56 kDa and 20 kDa are usually termed immunophilins and cyclophilins.[13] At least two immunophilins of *S. cerevisiae* are HSPs whose presence appears to be necessary for a maximal survival of yeast subjected to thermal stress.[79] Cyclophilin 20 was shown to be involved in intramitochondrial protein folding in cooperation with GRP75 and HSP60.[80] The members of the HSP70 and HSP60 chaperone machines, DnaJ and GroES, are

also sometimes designated as HSP40 and HSP10, respectively. In vivo both these proteins act in concert with the major chaperones (HSP70 and HSP60) to regulate their interaction with peptide substrates and ATPase activity.[32,37,38] After heat shock, HSP40 is translocated from the cytoplasm into the nucleus and accumulates in nucleoli, thus colocalizing with HSP70.[81] It seems likely that HSP40 assists HSP70 to recover the nucleolar function (first of all, preribosome formation) in cells which have experienced thermal stress.[81]

1.2. THE REGULATION OF HEAT SHOCK GENE EXPRESSION IN EUKARYOTES

The stress response is a typical example of inducible downregulated gene expression. The major regulatory step of this response in eukaryotes is transcriptional activation of the heat shock genes that is well visualized on polytene chromosomes as a formation of new puffs (see Fig. 1.1 and ref. 1). On the molecular level, the activation is mediated by a specific protein factor known as heat shock transcription factor (HSF). Briefly, in unstressed cells HSF is present in a nonactive form which does have DNA-binding activity. In response to heat shock and some other stressful or physiological exposures, HSF is activated and accumulates in the nucleus where it binds to DNA recognizing a specific target sequence, the so-called heat shock element (HSE), to initiate transcription of the heat shock genes. But how can HSF sense elevation of temperature or other stresses and external signals? Why does the response to environmental stresses occur so quickly? What defines the duration of the heat shock gene expression and afterward turns it off under normal conditions? To clarify these and other relevant questions we refer our readers to the text below where features of HSF and the mechanism of its interaction with DNA are described.

1.2.1. The Family of Heat Shock Transcription Factors (HSFs)

The members of the HSF family are conserved proteins with distinct domain organization (summarized in refs. 82, 83). Only one HSF has been found in yeast and *Drosophila*, while at least three HSFs are known to exist in chicken and tomato, and two, HSF1 and HSF2, are in mice and humans. HSF1 is activated in vivo by heat shock and many other stresses. On the contrary, DNA-binding activity of HSF2 is detected in human K562 cells after hemin treatment, or during cell differentiation in mouse (Fig. 1.4).[82,83]

The stress-induced activation of HSF1 includes oligomerization from a nonactive monomer to trimer, acquisition of specific DNA-binding activity, increase in phosphorylation, and redistribution within the nucleus. As a result, in stressed cells activated HSF1 binds to the promoter of the *HSP70* gene that leads to the transcription response and synthesis of HSPs. In contrast to HSF1, the oligomeric state, properties and localization of HSF2 are not changed in cells under stressful conditions and during heat shock this transcription factor remains in a nonactive form corresponding to a homo- or heterodimer. However, in vivo HSF2 is known to be active as a trimer during early embrionic development, spermatogenesis and hemin treatment (Fig. 1.4).[83] Recent data allow us to consider HSF1 and HSF2 as two different regulatory effectors of heat shock gene transcription: the former is responsible for rapid triggering the transcriptional response to increase HSP level in cells undergoing environmental or pathophysiological stresses; the latter is a developmental activator of nonstress-induced heat shock gene transcription.[82,83]

1.2.2. Structure of Heat Shock Gene Promoters

Activated HSFs associate with DNA, specifically recognizing certain target sequences within the promoter region of heat-inducible genes. These target sequences,

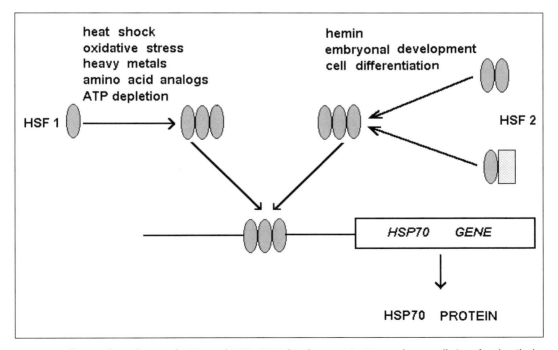

Fig. 1.4. Differential regulation of HSF1 and HSF2 DNA-binding activity in vertebrate cells (see for details the text and ref. 83).

commonly termed heat shock elements (HSE), consist of an array of inverted repeats of the sequence 5'-nGAAn-3'.[84] The nucleotide sequence of the HSE is highly conserved from yeast to humans. Besides the HSE, promoters of heat shock genes contain the TATA box and alternating GA-CT sequences which are recognized by other (unrelated to HSF) transcription factors, namely by TATA-binding protein (TBP) and GAGA factor (Fig. 1.5). In addition, uninduced heat shock promoters usually carry bound RNA polymerase II that was transcriptionally activated but paused after synthesizing about 25 nucleotides.[84] Association of GAGA factor with the target GAGA sequence upstream of the TATA box appears to be critical for the interruption of normal nucleosome packaging in order to "open" heat shock promoters.[85] The binding of TBP to TATA sequence is necessary to regulate the activity of RNA polymerase II (Fig. 1.5).[84] Probably, the binding of both GAGA factor

and TBP to their target sequences, as well as the presence of paused RNA polymerase II in the promoter regions of uninduced heat shock genes, serves as preactivation of these genes to provide their maximally rapid expression in case of stress.

1.2.3. HEAT-INDUCED ACTIVATION OF HSF AND ITS INTERACTION WITH HSE

In yeasts *S. cerevisiae* and *K. lactis* HSFs are present permanently in a trimeric (DNA-binding) form that remains bound to HSE under both heat shock and normal conditions. The transcriptional activity of heat shock genes in these yeasts seems to be stimulated via phosphorylation of HSF.[86] In higher eukaryotes HSF1 is synthesized constitutively and detected in a latent monomeric form in the cytoplasm and the nucleus of unstressed cells. In response to an elevation of temperature or other stresses, HSF is trimerized and accumulated within the nucleus to bind to HSE.[82,83]

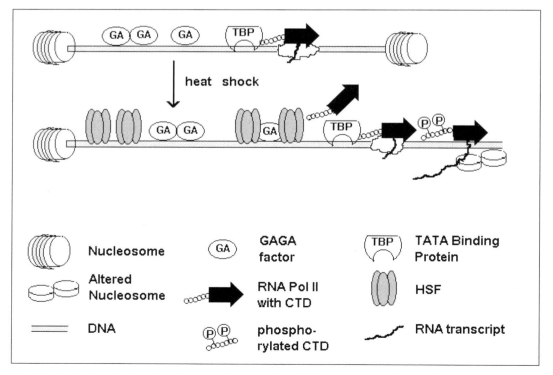

Fig. 1.5. Heat-induced alterations in the architecture of the HSP70 promoter. The mechanism is discussed in the text and ref. 84.

Which features of HSF enable it to react this way? It is generally accepted that at least two regulatory steps take place in the stress-induced activation of HSF1: trimerization and phosphorylation. The DNA-binding activity of HSF can be induced in vitro in unshocked cell extracts by a variety of exposures affecting protein structure and, obviously, the ability of HSF to recognize and to bind HSE is due to certain conformational alterations through the trimerization of HSF monomers (Fig. 1.4).[82-84,87] Larson et al[88] have shown that human HSF per se can directly sense heating: the in vitro purified and deactivated monomeric HSF is able to form DNA-binding trimers in a temperature-dependent fashion with the activity maximum at 43°C (heat shock temperature). Although this unique ability of HSF to adopt trimeric conformation in response to heating seems to be sufficient for acquisition of the DNA-binding activity,[83,87,88] a stress-dependent phosphorylation of serine and threonine residues of HSF may also modulate the transcriptional activation of heat shock genes.[86,89]

The heat shock transcriptional response occurs very quickly because of the rapid HSF activation and the special preactivation of a heat shock promoter (Fig. 1.5).[84] The accelerated binding of HSF to HSE is likely facilitated by changes in the nucleosomal architecture which "open" nucleosome-free promoter regions. Perhaps activated HSF is able to interact directly with the paused RNA polymerase II, and thus stimulate elongation. HSF may also increase the rate of elongation affecting interaction between the paused polymerase and TBP. Another possibility is that HSF is conducive to phosphorylation of the C-terminal domain (CTD) of the paused polymerase, since this phosphorylation may be an important step in stimulating the paused enzyme. Alternatively, HSF may facilitate the entry of new RNA polymerases and then accelerate initiation of the transcription by those additional polymerases (reviewed in ref. 84).

Interestingly, the activation of HSF may not necessarily result in heat shock gene transcription. Apart from HSF, a constitutive HSE-binding factor (CHBF) is known that appears to be involved in the regulation of *HSP70* gene transcription.[90] An inverse correlation between CHBF-DNA binding and *HSP70* gene transcription was found. Under some stressful treatments, CHBF may inhibit the HSF-mediated transcriptional signal despite the stress-induced activation of HSF.[90]

1.2.4. HSPs as Suppressors of Heat Shock Transcriptional Response: The Concepts of "Negative Regulation" and "Proteotoxicity"

Since numerous observations indicate that the elevation of the intracellular HSP level is accompanied by the termination of *HSP* genes' transcription and the deactivation of HSF, it has been speculated that HSPs themselves negatively regulate heat shock gene transcription via an autoregulatory loop.[91-95] Indeed, being molecular chaperones, HSPs might regulate the HSF monomer-trimer transitions, thus autoregulating heat shock transcriptional response. This speculation is supported by experiments involving the artificial manipulation of the cellular HSP levels. Experiments showed that activation of HSF in response to temperature elevation was markedly reduced in cells containing a large amount of HSP70 which had experienced previous heat shock,[94] or the *HSP70* gene overexpression,[95] or the microinjection of exogenous HSP70.[96] The most likely reason that various protein-damaging agents prevent HSF deactivation is because they create new substrates for HSP70 and therefore decrease the cytosolic pool of this chaperone.[94] The data suggested a special role for HSP70 in the mechanism of "negative regulation" as an inhibitor of the HSF-DNA interaction.[91-96] One of the possible scenarios is chaperoning the HSF monomers by excess HSP70 to prevent the trimerization and HSF-HSE binding.[91,92,95] However, stable association between HSP70 and the inactive form of HSF has not yet been demonstrated, whereas complexes of HSP70 with the active HSF trimers have been detected in extracts of heat-shocked cells.[93,94] Hence, it is more likely that the regulatory role of HSP70 is to deactivate activated HSF. This may be realized, for instance, via the inability of the HSP70-trimeric HSF complexes to bind to HSE or via the HSP70-mediated removal of activated HSF from DNA and/or by catalyzing the reverse trimer-monomer HSF transition. Besides HSP70, HSP90 seems also to be involved in the mechanism of "negative regulation," since it can bind to the inactive form of HSF[19] but not to the active form.[94] Mosser et al[95] have proposed a hypothetical scheme for the regulation of HSF activity by both HSP70 and HSP90 simultaneously (see also Fig. 1.6).

The "negative regulation" of heat shock gene transcription implies the possibility of competition between HSF and other protein substrates in interactions with HSP70 (and perhaps HSP90). Since proteins with affected structure are known to be major substrates for HSPs, hyperthermia or any other "proteotoxic" (i.e., protein-damaging) exposure generates many new targets for HSPs within a stressed cell, thus depleting the cytosolic pool of "free" (nonbound to a protein substrate) HSPs. At least for HSP70, uncovered hydrophobic sites[97] and/or unfolded polypeptide loops,[98] which are exposed by thermolabile cellular proteins upon heat shock, have been suggested as targets for the chaperone attack. Such stress-induced sequestration of "free" chaperones may release HSF from the inhibitory influence of HSP70 (and HSP90) and result in the binding of activated HSF to DNA and the transcriptional response (see Fig. 1.6 and refs. 94-96). This is supported

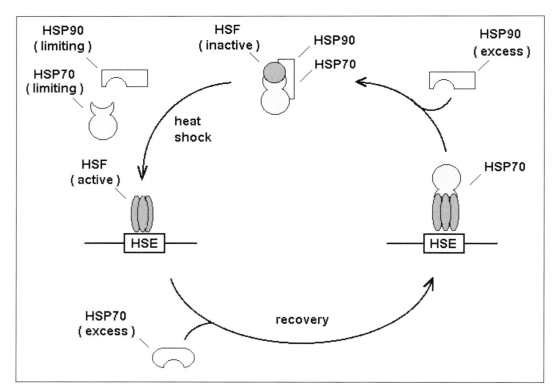

Fig. 1.6. Model for regulation of HSF activity by HSPs. The inactive form of HSF is a monomer that either transiently interacts with or is stably associated with HSP70 (and/or HSP90). The active form is a trimer that is capable of binding to the HSE. HSF trimers associate with HSPs during recovery from heat shock. HSP70 disrupts these trimers and refolds the monomers. Redrawn with permission from Mosser et al, Mol Cell Biol 1993; 13: 5427-5438.

by previous findings that D_2O and glycerol, i.e., agents delaying conformational transitions in proteins, inhibit the heat shock response.[99] Perhaps the most convincing evidence for the key role of HSP70 and damaged proteins in regulating the heat shock transcriptional response has been obtained in experiments with microinjection of exogenous HSP70 and in vitro denatured proteins in *Xenopus laevis* oocytes.[96,100,101] In those works it has been clearly demonstrated that direct introduction of various abnormal (denatured or misfolded) proteins into unstressed oocytes did provoke the heat shock gene expression[100,101] that can be moderated by injection of additional HSP70.[96] Mifflin and Cohen present the interesting fact that it is not the appearance of abnormal proteins per se, but the aggregation of abnormal proteins inside a cell, as well as the size and localization of these aggregates, which is critical in triggering the stress response.[101] According to these observations, it seems logical to accept the process of protein aggregation as an important event likely to affect HSF activity.

The concept of "proteotoxicity" enables one to explain, at least in some cases, the phenomenon of the so-called "cross tolerance," namely, why do such diverse factors as heating, low pH, oxidants, detergents, ethanol, heavy metals, amino acid analogs, etc. induce the same cellular transcriptional (heat shock) response and tolerance to a wide spectrum of harmful exposures. Evidently, the common feature of all the above factors is their "proteotoxicity." All are able to affect protein folding and can therefore induce protein aggregation inside a cell, which in turn promotes transcription of heat shock genes and HSP synthesis. Excess (newly synthesized) HSPs confer tolerance to proteotoxic stresses but, being "negative regulators" of HSF, turn off the transcriptional signal (see Fig. 1.6).

1.2.5. THE TRANSLATIONAL REGULATION OF HSP SYNTHESIS

The transcriptional activity is, undoubtedly, the most important point in regulating HSP expression in higher organisms; however, it should be noted that the mechanisms of the translational control are also known. During thermal stress the translation of preexisting mRNAs is arrested by blocking initiation and elongation[102] or in consequence of polysome disruption.[103] Since heat-stimulated transcription of heat shock genes results in the accumulation of HSP mRNAs in the cytoplasm, polysomes are reformed on the messages and HSPs become the transient, prevalent products of protein synthesis.[103]

Another event favorable for the predominant synthesis of HSP is selective stabilization of HSP mRNA by thermal stress; in particular, the half-life of HSP70 mRNA increases 10-fold in heat-shocked HeLa cells.[104] In addition, long adenine-rich untranslated leader sequences are often present at the 5' end of HSP mRNAs that appear to define the selectivity of the translation process.[105]

Likewise, several proteins seem to be involved in the translational regulation. They are initiation factors elF-2a and elF-4b,[106] the cap-binding protein,[107] and the 25 kDa protein that specifically recognizes the 5' end of HSP70 mRNA.[108] Taken together all the above listed factors, namely the disruption and reformation of polysomes, stabilization of HSP messages and the availability of the special leader sequence in them, and the protein regulators, should determine preferential synthesis of HSPs in heat-shocked cells.

CONCLUDING REMARKS

Biological evolution has created a universal adaptive mechanism that allows any cell to rapidly synthesize special proteins with cytoprotective properties in response to stress. These protective proteins are commonly known as HSPs or molecular chaperones. The protective proteins perform the following functions: (1) control cellular protein folding under normal conditions; (2) prevent the denaturation and aggregation of intracellular proteins during stressful exposures; and (3) accelerate the disaggregation, renaturation and/or degradation of damaged proteins within stressed cells.

Interestingly, being products of distinct genes, HSPs possess some common features. For instance, all chaperones should recognize unfolded, misfolded and aggregated protein. Some chaperones (HSP90, HSP70 and the small HSPs) interact with cytoskeletal proteins and accumulate inside the nucleus upon heat shock. At least three of them (HSP100, HSP70, ubiquitin) are involved in the protein degradation pathway. We would like to emphasize here that all of the major HSPs (HSP100, HSP70, HSP60) require ATP for their chaperoning functions. What happens, however, when ATP is deficient? Can the ATP-dependent HSPs maintain cellular proteins in their normal state under ATP-depleting conditions? These queries need detailed consideration.

There is no doubt that HSP expression is autoregulated. According to recent hypotheses, the aggregation of damaged (or abnormal) proteins within a cell promotes synthesis of HSPs which disaggregate and renaturate injured cellular proteins or facilitate their degradation. After the protein repair and turnover, excess HSPs deactivate HSF and thus interrupt heat shock gene transcription. In the next chapters, we will explore how this autoregulatory mechanism is realized under ischemia-like metabolic stresses, leading to severe depletion of cellular ATP.

RECENT NEWS

One important point in regulation of the heat shock response has become known just recently. When we wrote this manuscript, complexing between HSP70 and non-active HSF was not yet revealed in unstressed cells, while complexes of HSP70 with activated HSF1 were indeed found in extracts from heat shocked cells (see refs 93, 94). This somewhat complicated understanding of HSP70 as a negative regulator in the heat shock transcription response. However, recently Baler et al[109] have identified HSP70/non-activated HSF1 heterodimer in unstressed cells. Likewise, they have shown that both activation (trimerization) of HSF1 and disappearance

of the inactive heterodimers during heat shock are inhibited by overexpressed (or unengaged) HSP70.[109] These data essentially provide evidence of HSP70-mediated negative regulation of HSF1. At the same time, the capacity of HSP70 to bind non-active HSF1, thus preventing its activation, does not exclude the availability of other regulatory functions intrinsic to HSP70. As was discussed above, excess HSP70 may also associate with activated (trimeric) HSF1 to deactivate it (e.g. by blocking HSF-HSE interaction and/or by catalyzing dissociation of trimers to inactive monomers). A role of HSP90 and other HSPs in regulation of HSF1 remains to be defined.

REFERENCES

1. Ritossa F. A new puffing pattern induced by heat shock and DNP in *Drosophila*. Experientia 1962; 18: 571-573.

2. Tissieres A, Mitchell HK, Tracy UM. Protein synthesis in salivary glands of *D. melanogaster*. Relation to chromosome puffs. J Mol Biol 1974; 84: 389-398.

3. McKrnzie SL, Lindquist S, Meselson M. Translation in vitro of *Drosophila* heat-shock messages. J Mol Biol 1977; 117: 279-283.

4. Mirault ME, Goldschmidt-Clermont M, Moran L et al. The effect of heat shock on gene expression in *Drosophila melanogaster*. Cold Spring Harbor Simp Quant Biol 1978; 42: 819-827.

5. Parsell DA, Sanchez Y, Stitzel JD, Lindquist S. HSP104 is a highly conserved protein with two essential nucleotide-binding sites. Nature 1991; 353: 270-273.

6. Parsell DA, Lindquist S. Heat shock proteins and stress tolerance. In: Morimoto RI, Tissieres A, Georgopoulos C, eds. The Biology of Heat Shock Proteins and Molecular Chaperones. Cold Spring Harbor, NY: Cold Spring Harbor Laboratory Press, 1994: 457-494.

7. Sanchez Y, Lindquist SL. HSP104 required for induced thermotolerance. Science 1990; 248: 1112-1115.

8. Sanches Y, Taulien J, Borkovich KA, Lindquist S. HSP104 is required for tolerance to many forms of stress. EMBO J 1992; 11: 2357-2364.

9. Parsell DA, Kowal AS, Singer MA, Lindquist S. Protein disaggregation mediated by heat-shock protein Hsp104. Nature 1994; 372: 475-478.

10. Wickner S, Gottesman S, Skowyra D et al. A molecular chaperone, ClpA, functions like DnaK and DnaJ. Proc Natl Acad Sci USA 1994; 91: 12218-12222.

11. Bradwell JCA, Craig EA. Ancient heat shock gene is dispensable. J Bacteriol 1988; 170: 2977-2983.

12. Lindquist S, Craig EA. The heat-shock proteins. Annu Rev Genet 1988; 22: 631-637.

13. Gething MJ, Sambrook J. Protein folding in the cell. Nature 1992; 355: 33-45.

14. Welch WJ. Mammalian stress response: cell physiology, structure/function of stress proteins, and implications for medicine and disease. Physiol Rev 1992; 72: 1063-1081.

15. Pratt WB. The role of heat shock proteins in regulating the function, folding, and trafficking of the glucocorticoid receptor. J Biol Chem 1993; 268: 21455-21458.

16. Xu Y, Lindquist S. Heat-shock protein hsp90 governs the activity of pp60v-src kinase. Proc Natl Acad Sci USA 1993; 90: 7074-7078.

17. Legagneux V, Morange M, Bensaude O. Heat shock increases turnover of 90-kDa heat shock protein phosphate groups in HeLa cells. FEBS Lett 1991; 291: 359-362.

18. Csermely P, Kahn CR. The 90-kDa heat shock protein (hsp-90) processes an ATP binding site and autophosphorylating activity. J Biol Chem 1991; 266: 4943-4950.

19. Nadeau K, Das A, Walsh ST. Hsp90 chaperonins process ATPase activity and bind heat shock transcriptional factor and peptidyl prolyl isomerases. J Biol Chem 1993; 268: 1479-1487.

20. Csermely P, Kajtar J, Hollosi M et al. The 90 kDa heat shock protein (hsp90) induces the condensation of the chromatin structure. Biochem Biophys Res Commun 1994; 202: 1557-1663.

21. Wiech H, Buchner J, Zimmermann R, Jacob U. Hsp90 chaperones protein folding in vitro. Nature 1992; 358: 169-170.

22. Bohen SP, Yamamoto KR. Modulation of steroid receptor signal transduction by heat shock proteins. In: Morimoto RI, Tissieres A, Georgopoulos C, eds. The Biology of Heat Shock Proteins and Molecular Chaperones. Cold Spring Harbor, NY: Cold Spring Harbor Laboratory Press, 1994: 313-334.

23. Koyasu S, Nishida E. Kadowaki T et al. Two mammalian heat shock proteins, HSP90 and HSP100, are actin-binding proteins. Proc Natl Acad Sci USA 1986; 83: 8054-8058.

24. Nishida E, Koyasu S, Sakai H, Yahara H. Calmodulin-regulated binding of the 90 kDa heat shock protein to actin filaments. J Biol Chem 1986; 261: 16033-16036.

25. Kellermayer MSZ, Csermely P. ATP induces dissociation of the 90 kDa heat shock protein (hsp90) from F-actin: Interference with the binding of heavy meromyosin. Biochem Biophys Res Commun 1995; 211: 166-174.

26. Yahara I, Iida H, Koyasu S. A heat shock-resistant variant of Chinese hamster cell line contitutively expressing heat shock protein of Mr. 90,000 at high level. Cell Struct Funct 1989; 11: 65-73.

27. Borkovich KA, Farrelly FW, Finkelstein DB et al. Hsp82 is an essential protein that is required in higher concentrations for growth of cells at higher temperatures. Mol Cell Biol 1989; 9: 3919-3930.

28. Bansal GS, Norton PM, Latchman DS. The 90-kDa heat shock protein protects mammalian cells from thermal stress but not from viral infection. Exp Cell Res 1991; 195: 303-306.

29. Akner G, Mossberg K, Sundqvist K-G et al. Evidence for reversible, non-microtubule and non-microfilament-dependent nuclear translocation of hsp90 after heat shock in human fibroblasts. Eur J Cell Biol 1992; 58: 356-364.

30. McKay DB, Wilbanks SM, Flaherty KM et al. Stress-70 proteins and their interaction with nucleotides. In: Morimoto RI, Tissieres A, Georgopoulos C, eds. The Biology of Heat Shock Proteins and Molecular Chaperones. Cold Spring Harbor, NY: Cold Spring Harbor Laboratory Press, 1994: 153-177.

31. Beckmann RP, Mizzen LA, Welch WJ. Interaction of HSP 70 with newly synthesized proteins: Implications for protein folding

and assembly. Science 1990; 248: 850-654.

32. Frydman J, Hartl FU. Molecular chaperone functions of hsp70 and hsp60 in proteins folding. In: Morimoto RI, Tissieres A, Georgopoulos C, eds. The Biology of Heat Shock Proteins and Molecular Chaperones. Cold Spring Harbor Laboratory Press, Cold Spring Harbor, NY 1994: 251-283.

33. Palleros D, Welch W, Fink A. Interaction of Hsp70 with unfolded proteins: Effects of temperature and nucleotides on the kinetics of binding. Proc. Natl Acad Sci USA 1991; 88: 5719-5723.

34. Palleros DR, Reid KL, Shi L et al. ATP-induced protein-hsp70 complex dissociation requires K$^+$ and does not involve ATP hydrolysis. Analogy to G proteins. Nature 1993; 365: 664-666.

35. Sadis S, Hightower LE. Unfolded proteins stimulate molecular chaperone Hsc70 ATPase by accelerating ADP/ATP exchange. Biochemistry 1992; 31: 9406-9412.

36. Hightower LE, Sadis SE, Takenaka IM. Interactions of vertebrate hsc70 and hsp70 with unfolded proteins and peptides. In: Morimoto RI, Tissieres A, Georgopoulos C, eds. The Biology of Heat Shock Proteins and Molecular Chaperones. Cold Spring Harbor, NY: Cold Spring Harbor Laboratory Press, 1994: 179-207.

37. Georgopoulos C, Welch WJ. Role of major heat shock proteins as molecular chaperones. Annu Rev Cell Biol 1993; 9: 601-634.

38. Rassow J, Voos W, Pfanner N. Partner proteins determine multiple functions of Hsp70. Trends Cell Biol 1995; 5: 207-212.

39. DeLuca-Flaherty C, McKay DB, Parham P, Hill BL. Uncoating protein (hsc70) binds a conformationally labile domain of clathrin light chain LC to stimulate ATP hydrolysis. Cell 1990; 62: 875-887.

40. La Thangue NB. A major heat-shock protein defined by a monoclonal antibody. EMBO J 1984; 3: 1871-1879.

41. Nover L. In Heat Shock Response. Boca Raton: CRC Press, Inc. 1991: 509.

42. Haus U, Trommler P, Fisher PR et al. The heat shock cognate protein from *Dictyostelium* affects actin polymerization through interaction with the actin-binding protein cap32/34. EMBO J 1993; 12: 3763-3771.

43. Liao J, Lowthert LA, Ghori N, Omary MB. The 70-kDa heat shock proteins associate with grandular intermediate filaments in an ATP-dependent manner. J Biol Chem 1995; 270: 915-922.

44. Dice JF, Agarraberes F, Kirven-Brooks M et al. Heat shock 70-kDa proteins and lysosomal proteolysis. In: Morimoto RI, Tissieres A, Georgopoulos C, eds. The Biology of Heat Shock Proteins and Molecular Chaperones. Cold Spring Harbor, NY: Cold Spring Harbor Laboratory Press, 1994: 137-151.

45. Craig EA, Baxter BK, Becker J et al. Cytosolic hsp70s of *Saccharomyces cerevisiae*: roles in protein synthesis, protein translocation, proteolysis, and regulation. In: Morimoto RI, Tissieres A, Georgopoulos C, eds. The Biology of Heat Shock Proteins and Molecular Chaperones. Cold Spring Harbor, NY: Cold Spring Harbor Laboratory Press, 1994: 31-52.

46. Gething M-J, Blond-Elguindi S, Mori K, Sambrook JF. Structure, function, and regulation of the endoplasmic reticulum chaperone, BiP. In: Morimoto RI, Tissieres A, Georgopoulos C, eds. The Biology of Heat Shock Proteins and Molecular Chaperones. Cold Spring Harbor, NY: Cold Spring Harbor Laboratory Press, 1994: 111-135.

47. Brodsky JL, Schekman R. Heat shock cognate proteins and polypeptide translocation across the endoplasmic reticulum membrane. In: Morimoto RI, Tissieres A, Georgopoulos C, eds. The Biology of Heat Shock Proteins and Molecular Chaperones. Cold Spring Harbor, NY: Cold Spring Harbor Laboratory Press, 1994: 85-109.

48. Bhattacharyya T, Karnezis AN, Muphy SP et al. Cloning and subcellular localization of human mitochondrial hsp70. J Biol Chem 1995; 270: 1705-1710.

49. Langer T, Neupert W. Chaperoning mitochondrial biogenesis. In: Morimoto RI, Tissieres A, Georgopoulos C, eds. The Biology of Heat Shock Proteins and Molecular Chaperones. Cold Spring Harbor, NY: Cold Spring Harbor Laboratory Press, 1994: 53-83.

50. Angelidis CE, Lazaridis I, Pagoulatos GN. Constitutive expression of heat-shock pro-

tein 70 in mammalian cells confers thermoresistance. Eur J Biochem 1991; 199: 35-39.

51. Parsell DA, Lindquist S. The function of heat-shock proteins in stress tolerance: degradation and reactivation of damaged proteins. Annu Rev Genet 1993; 27: 437-496.

52. Pelham HR. Speculations on the functions of the major heat shock and glucose-regulated proteins. Cell 1986; 46: 959-961.

53. Rothman JE. Polypeptide chain binding proteins: Catalysts of protein folding and related processes in cells. Cell 1989; 59: 591-601.

54. Hightower LE. Heat shock, stress proteins, chaperones and proteotoxicity. Cell 1991; 66: 191-197.

55. Laszlo A. The effects of hyperthermia on mammalian cell structure and function. Cell Prolif 1992; 25: 59-87.

56. Stege GJJ, Li L, Kampinga HH et al. Importance of the ATP-binding domain and nucleolar localization domain of HSP72 in the protection of nuclear proteins against heat-induced aggregation. Exp Cell Res 1994; 214: 279-284.

57. Marber MS, Latchman DS, Walker JM, Yellon DM. Cardiac stress protein elevation 24 hours after brief ischemia or heat stress is associated with resistance to myocardial infarction. Circulation 1993; 88:1264-1272.

58. Agsteribbe E, Huckriede A, Veenhuis M et al. A fatal, systemic mitochondrial disease with decreased mitochondrial enzyme activities, abnormal ultrastructure of the mitochondria and deficiency of heat shock protein 60. Biochem Biophys Res Commun 1993; 193: 146-154.

59. Lewis VA, Hynes GM, Zheng D et al. T-complex polypeptide-1 is a subunit of a heteromeric particle in the eukaryotic cytosol. Nature 1992; 358: 249-252.

60. Martin J, Horwich AL, Hartl FU. Prevention of protein denaturation under heat stress by the chaperonin Hsp60. Science 1992; 258: 995-998.

61. Cheng MY, Hartl FU, Martin J et al. Mitochondrial heat-chock protein hsp60 is essential for assembly of proteins imported into yeast mitochondria. Nature 1989; 337: 620-625.

62. Arrigo AP, Landry J. Expression and function of the low-molecular weight heat shock proteins. In: Morimoto RI, Tissieres A and Georgopoulos C, eds. The Biology of Heat Shock Proteins and Molecular Chaperones. Cold Spring Harbor, NY: Cold Spring Harbor Laboratory Press, 1994: 335-373.

63. Sax CM, Piatigorsky J. Expression of the α-crystallin/small heat shock protein/molecular chaperone genes in the lens and other tissues. In: Meister A, ed. Advances in Enzymology and related areas of molecular biology. New York, NY: Cornell University Medical College New York, 1994: 69: 155-201.

64. Miron T, Vancompernolle K, Vanderkerckhove J et al. A 25-kD inhibitor of actin polymerization is a low molecular mass heat shock protein. J Cell Biol 1991; 114: 255-261.

65. Lavoie JN, Hickey E, Weber LA, Landry J. Modulation of actin microfilament dynamics and fluid phase pinocytosis by phosphorylation of heat shock protein 27. J Biol Chem 1993; 268: 24210-24214.

66. Huot J, Lambert H, Lavoie JN et al. Characterization of 45-kDa/54-kDa HSP27 kinase, a stress-sensitive kinase which may activate the phosphorylation-dependent protective function of mammalian 27-kDa heat-shock protein HSP27. Eur J Biochem 1995; 227: 416-427.

67. Bennardini F, Wrzosek A, Chiesi M. αB-Crystallin in cardiac tissue. Association with actin and desmin filaments. Circ Res 1992; 71: 288-294.

68. Aoyama A, Frohli E, Schafer R, Klemenz R. αB-crystallin expression in mouse NIH 3T3 fibroblasts: Glucocorticoid responsiveness and involvement in thermal protection. Mol Cell Biol 1993; 13: 1824-1835.

69. Chretien P, Landry J. Enhanced constitutive expression of the 27-kDa heat shock proteins in heat-resistant variants from Chinese hamster cells. J Cell Physiol 1988; 137: 157-166.

70. Lavoie JN, Gingras-Breton G, Tanguay RM, Landry J. Induction of Chinese hamster HSP27 gene expression in mouse cells confers resistance to heat shock. HSP27 stabilization of the microfilament organization. J Biol Chem 1993; 268: 3420-3429.

71. Kampinga HH, Brunsting JF, Stege GJJ et al. Cells overexpressing Hsp27 show accelerated recovery from heat-induced nuclear protein aggregation. Biochem Biophys Res Commun 1994; 204: 1170-1177.

72. Melhen P, Preville X, Chareyron P et al. Constitutive expression of human hsp27, *Drosophila* hsp27, or human αB-crystallin confers resistance to TNF- and oxidative stress-induced cytotoxicity in stably transfected murine L929 fibroblasts. J Immunol 1995; 154: 363-374.

73. Oesterreich S, Weng C-N, Qiu M et al. The small heat shock protein hsp27 is correlated with growth and drug resistance in human breast cancer cell lines. Cancer Res 1993; 53: 4443-4448.

74. Jacob U, Gaestel M, Engel K, Buchner J. Small heat shock proteins are molecular chaperones. J Biol Chem 1993; 268: 1517-1520.

75. Clechanover A. The ubiquitin-proteasome proteolytic pathway. Cell 1994; 79: 13-21.

76. Parag HA, Raboy B, Kulka RG. Effect of heat shock on protein degradation in mammalian cells: Involvement of the ubiquitin system. EMBO J 1987; 6: 55-61.

77. Vile GF, Basu-Modak S, Waltner C, Tyrrell RM. Hemeoxygenase 1 mediates an adaptive response to oxidative stress in human skin fibroblasts. Proc Natl Acad Sci USA 1994; 91: 2607-2610.

78. Nakai A, Satoh M, Hirayoshi K, Nagata K. Involvement of the stress protein HSP47 in procollagen processing in the endoplasmic reticulum. J Cell Biol 1992; 117: 903-914.

79. Sykes K, Gething M-J, Sambrook J. Proline isomerases functions during heat shock. Proc Natl Acad Sci USA 1993; 90: 5853-5857.

80. Rassow J, Mohrs K, Koidl S, Barthelmess IB et al. Cyclophilin 20 is involved in mitochondrial protein folding in cooperation with molecular chaperones Hsp70 and Hsp60. Mol Cell Biol 1995; 15: 2654-2662.

81. Ohtsuka K, Utsumi KR, Kaneda T, Hattori H. Effect of ATP on the release of hsp 70 and hsp 40 from the nucleus in heat-shocked HeLa cells. Exp Cell Res 1993; 209: 357-366.

82. Wu C, Clos J, Giorgi G, Haroun RI et al. Structure and regulation of heat shock transcription factor. In: Morimoto RI, Tissieres A, and Georgopoulos C, eds. The Biology of Heat Shock Proteins and Molecular Chaperones. Cold Spring Harbor, NY: Cold Spring Harbor Laboratory Press, 1994: 395-416.

83. Morimoto RI, Jurivich DA, Kroeger PE et al. Regulation of heat shock gene transcription by a family of heat shock factors. In: Morimoto RI, Tissieres A, Georgopoulos C, eds. The Biology of Heat Shock Proteins and Molecular Chaperones. Cold Spring Harbor, NY: Cold Spring Harbor Laboratory Press, 1994: 417-455.

84. Fernandes M, O'Brien T, Lis JT. Structure and regulation of heat shock gene promoters. In: Morimoto RI, Tissieres A., Georgopoulos C, eds. The Biology of Heat Shock Proteins and Molecular Chaperones. Cold Spring Harbor, NY: Cold Spring Harbor Laboratory Press, 1994: 375-393.

85. Tsukiyama T, Becker PB, Wu C. ATP-dependent nucleosome disruption at a heat-shock promoter mediated by binding of GAGA transcription factor. Nature 1994; 367: 525-532.

86. Sorger PK. Yeast heat shock factor contains separable transient and sustained response transcriptional activators. Cell 1990; 62: 793-805.

87. Westwood JT, Wu C. Activation of *Drosophila* heat shock factor: Conformational change associated with a monomer-to-trimer transition. Mol Cell Biol 1993; 13: 3481-3486.

88. Larson JS, Schuetz TJ, Kingston RE. In vitro activation of purified human heat shock factor by heat. Biochemistry 1995; 34: 1902-1911.

89. Sarge K, Murphy SP, Morimoto RI. Activation of heat shock transcription by HSF1 involves oligomerization, acquisition of DNA-binding activity, and nuclear localization and can occur in the absence of stress. Mol Cell Biol 1993; 13: 1392-1407.

90. Liu RY, Kim D, Yang S-H, Li GC. Dual control of heat shock response: Involvement of a constitutive heat shock element-binding factor. Proc Natl Acad Sci USA 1993; 90: 3078-3082.

91. Craig EA, Gross CA. Is hsp70 the cellular thermometer? Trends Biochem Sci 1991; 16: 135-140.

92. Hightower LE. Heat shock, stress proteins, chaperones, and proteotoxicity: meeting review. Cell 1991; 66: 191-197.

93. Abravaya K, Myers MP, Murphy SP, Morimoto RI. The human heat shock protein hsp70 interacts with HSF, the transcription factor that regulates heat shock gene expression. Genes Dev 1992; 6: 1153-1164.

94. Baler R, Welch WJ, Voelmy R. Heat shock gene regulation by nascent polypeptides and denatured proteins: hsp70 as a potential autoregulatory factor. J Cell Biol 1992; 117: 1151-1159.

95. Mosser DD, Duchaine J, Massie B. The DNA-binding activity of the human heat shock transcription factor is regulated in vivo by hsp70. Mol Cell Biol 1993; 13: 5427-5438.

96. Mifflin LC, Cohen RE. hsc70 moderates the heat shock (stress) response in *Xenopus laevis* oocytes and binds to denatured protein inducers. J Biol Chem 1994; 269: 15718-15723.

97. Pelham RHB. Speculations on the functions of the major heat shock and glucose-regulated proteins. Cell 1986; 46: 959-961.

98. Rothman JE. Polypeptide chain binding proteins: catalysts of protein folding and related processes in cells. Cell 1989; 59: 591-602.

99. Edington BV, Whelan SA, Hightower LE. Inhibition of heat shock (stress) protein induction by deuterium oxide and glycerol: additional support for the abnormal protein hypothesis of induction. J Cell Physiol 1989; 139: 219-228.

100. Anathan J, Goldberg AL, Voellmy R. Abnormal proteins serve as eukaryotic stress signals and trigger the acivation of heat shock genes. Science 1986; 232: 252-254.

101. Mifflin LC, Cohen RE. Characterization of denatured protein inducers of the heat shock (stress) response in *Xenopus laevis* oocytes. J Biol Chem 1994; 269: 15710-15717,

102. Ballinger DG, Pardue ML. The control of protein synthesis during heat shock in *Drosophila* cells involves altered polypeptide elongation rates. Cell 1983; 33: 103-114.

103. Lindquist S. Varying patterns of protein synthesis in *Drosophila* during heat shock: implications for regulation. Dev Biol 1980; 77: 463-469.

104. Theodorakis NG, Morimoto RI. Post-transcriptional regulation of HSP 70 expression in human cells: Effects of heat shock, inhibition of protein synthesis, and adenovirus infection on translation and mRNA stability. Mol Cell Biol 1987; 7: 4357-4368.

105. Klemenz R, Hultmark D, Gehring W. Selective translation of heat shock mRNA in *Drosophila melanogaster* depends on sequence information in the leader. EMBO J 1985; 4: 2053-2060.

106. Duncan RF, Hershey JWB. Heat shock-induced translational alterations in HeLa cells. Initiation factor modifications and the inhibition of translation. J Biol Chem 1984; 259: 11882-11889.

107. Maroto FG, Sierra JM. Translational control in heat-shocked *Drosophila* embrios. J Biol Chem 1988; 263: 15720-15725.

108. Yost HJ, Petersen RB, Lindquist S. Post-transcriptional regulation of heat shock protein synthesis in *Drosophila*. In: Morimoto RI, Tissieres A, Georgopoulos C, eds. Stress Proteins in Biology and Medicine. Cold Spring Harbor, NY: Cold Spring Harbor Laboratory Press, 1990: 379-409.

109. Baler R, Zou J, Voellmy R. Evidence for a role of Hsp70 in the regulation of the heat shock response in mammalian cells. Cell Stress Chaperones 1996; 1:33-39.

CHAPTER 2

ATP HOMEOSTASIS, IONIC BALANCE AND CELL VIABILITY

As discussed in the introduction, the main problems that will be addressed in this book are: (1) why does ATP decrease within a cell lead to HSP induction and (2) how can HSPs rescue ATP-depleted mammalian cells from death. To answer these questions it is necessary to first examine the mechanisms of ATP homeostasis and the effect of ATP depletion on some important cellular functions and viability.

2.1. ATP HOMEOSTASIS IN MAMMALIAN CELLS

ATP is the main source of energy in a cell and has an extremely fast turnover time (e.g., 1-2 minutes in Ehrlich carcinoma cells[1]). Glycolytic and oxidative phosphorylation are major processes of ATP generation. In most of the mammalian cells, respiration is the main pathway for ATP synthesis due to its high efficiency, i.e., 25-36 moles of ATP formed per mole of glucose oxidized (see ref. 2 for review). However, some normal (erythrocytes, retinal cells) and many rapidly growing tumor cells can obtain their energy requirements exclusively or mainly through glycolysis, even though this process has drastically lower energy efficiency (2 moles of ATP per mole of glucose).

Usually the rate of ATP generation is tightly coupled with its consumption rate, and when the energy requirement of a cell is elevated (e.g., in the stimulated muscle cells), the ATP generation rate is also increased and the cellular ATP level does not fall (see ref. 2 for review). In some circumstances, however, the generation of ATP cannot provide cellular energy requirements. The most typical situations that occur in vivo are hypoxia or anoxia (i.e., deprivation of oxygen), or ischemia (i.e., deprivation of both oxygen and nutrients due to cessation of blood flow). As a model of in vitro anoxia, several inhibitors of oxidative phosphorylation with different sites of action are usually used together (Table 2.1) to simulate ischemic conditions, nutrient-free medium and/or inhibitors of glycolysis (deoxyglucose, iodacetic acid, iodacetamide) together with mitochondrial inhibitors. As we shall discuss further, these in vitro models are widely applied in the study of both HSP induction and HSP protective functions.

Table 2.1. Inhibitors of oxidative phosphorylation used to simulate anoxia/ischemia in vitro

Inhibitors	Concentrations Used[1]	Site of Action
I. Respiratory chain inhibitors:		
1. rotenone	1-20 µM	Complex I
2. antimycin	1-20 µM	Complex III
3. cyanide	1-5 mM	Complex IV
II. Uncouplers:		
1. 2,4-dinitrophenol	0.2-2 mM	Inner membrane
2. CCCP	1-10 µM	Inner membrane
III. Inhibitor of H+-ATPase:		
oligomycin	1-10 µM	H+-ATPase

[1]Higher concentrations within these ranges are used in serum-containing media.

2.1.1. CREATINE-PHOSPHATE AND GLYCOLYSIS

To maintain the ATP level under anoxic or ischemic conditions, cells can utilize several means. One of them is a creatine-phosphate (CrP) buffering system which generates ATP through creatine-kinase reaction:

$$CrP + ADP \leftrightarrow Creatine + ATP$$

This system operates mainly in muscle cells during short periods of intensive work. Since the content of CrP in muscles exceeds ATP content only about three- to four-fold, and it is lower in other tissues,[3] maintenance of ATP level by creatine-phosphate under energy deficiency can proceed only for several minutes.

A more important system of ATP stabilization under hypoxia or ischemia is the glycolytic system. In the absence of oxidative phosphorylation, some cells can maintain their ATP level exclusively through glycolytic phosphorylation. Certainly, this system can operate only when glycolytic substrates (either exogenous or endogenous) are available. As an example, Fig. 2.1 illustrates the effect of inhibitors of oxidative phosphorylation on the ATP level in normal cells (thymocytes) and tumor cells (Ehrlich ascites carcinoma, EAC). Inhibitors have no effect on ATP levels in EAC

cells in the presence of glucose, since glycolysis in these cells is very high and can fully compensate for blocked oxidative phosphorylation (Table 2.2). However, even in the presence of glucose, thymocytes lose ATP under the same conditions due to their limited glycolytic capability (Table 2.2). At the same time, in the absence of glucose both EAC cells and thymocytes can only maintain their ATP levels through oxidative phosphorylation since any inhibitor results in rapid ATP depletion (Fig. 2.1).[4,5]

In contrast to EAC cells and thymocytes which apparently have no endogenous glycolytic substrates, other cells which have endogenous glycolysis can play a significant role in maintaining ATP. For example, the hepatocytes of fed rats (i.e., with high glycogen content) had a markedly slower rate of ATP depletion than the hepatocytes of starved rats (with low glycogen content). In addition, in these cells fructose rather than glucose is the best exogenous substrate for glycolysis because it prevents ATP decrease in anoxic conditions.[6] In rat cardiomyocytes under severe hypoxia ($PO_2 < 0.1$ torr) CrP and endogenous glycolysis can provide essential ATP generation only for the first 10 minutes, but thereafter the ATP level decreases to 20% of the initial level after 30 minutes

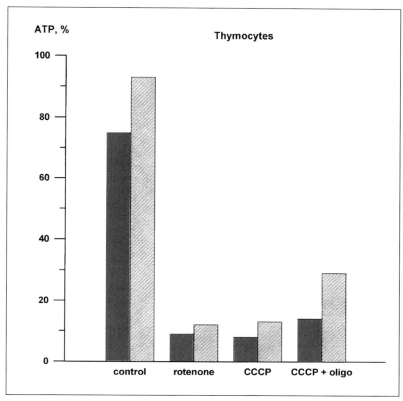

A

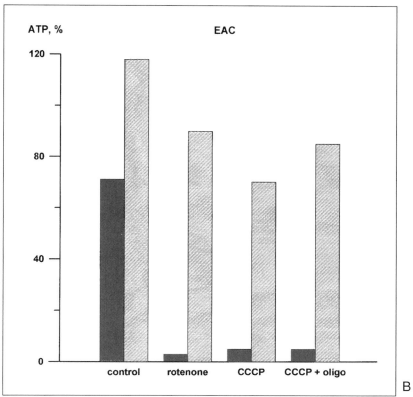

B

Fig. 2.1. Effect of mitochondrial inhibitors and glucose on the ATP level in thymocytes (A) and EAC cells (B). Dark columns, without glucose; bright columns, with 10 mM glucose; rotenone, 2 μM; CCCP, 2 μM; oligomycin, 2 μg/ml. Cells were incubated in Hanks' balanced salt solution for 30 minutes and their ATP level was determined by luciferine-luciferase assay as described elsewhere.[4,5]

Table 2.2. Effect of inhibitors of oxidative phosphorylation on respiration and glycolysis in thymocytes and EAC cells

Conditions	Glucose	Glycolysis[1] Thymocytes	Glycolysis[1] EAC	Respiration[2] Thymocytes	Respiration[2] EAC
Control	–	0.0	0.0	0.11	1.40
	+	0.08	6.0	0.11	0.77
Rotenone	–	0.0	0.0	0.0	0.0
	+	0.18	11.0	0.0	0.0
CCCP	–	0.0	0.0	0.22	3.85
	+	0.18	14.0	0.25	3.16
CCCP +	–	0.0	0.0	0.22	4.48
oligomycin	+	0.18	9.5	0.22	2.10

[1] Glycolysis was determined by lactate formation and
[2] respiration by O_2 consumption (nmole/min per 10^6 cells).
Cells were incubated in Hanks' balanced salt solution either without or with 10 mM glucose; rotenone concentration was 2 μM; CCCP, 2 μM; oligomycin, 2 μg/ml.[5]

of hypoxia.[7] Although endogenous glycolysis in these cells was activated about 10-fold under these conditions, the contribution of anaerobic ATP production was nearly 10% of demand.[7] A similar situation occurs in the brain which is also strictly dependent on aerobic ATP production (see, for example, ref. 8).

2.1.2. SUPPRESSION OF BIOSYNTHESIS

Another possibility to maintain the ATP level under unfavorable conditions may be shutting off unnecessary cellular reactions which utilize ATP, for instance some biosynthetic reactions. One example of such adaptation is the inactivation of lipid biosynthesis in hepatocytes by AMP-activated protein kinase. When ATP falls and ADP rises, it is always accompanied by a marked elevation in the AMP level which is the result of the presence of adenylate kinase. Adenylate kinase accomplishes the reaction $2ADP \leftrightarrow AMP + ATP$ close to equilibrium by conversion of ADP excess to ATP and AMP.[3] A rise in the AMP level activates a protein kinase which, in turn, inhibits the activities of its two targets, the kinase of acetyl-coenzyme A carboxylase and 3-hydroxy-3-methylglutaryl-

coenzyme A reductase, the key regulatory enzymes of lipid biosynthesis.[9,10] However, the importance of this inhibition to ATP stabilization has not been assessed, but it is probably not significant (see below).

Several studies have been done on mammalian cells to evaluate the relative contribution of different cellular processes to ATP consumption under steady-state conditions. The experiments showed that, in quiescent pig spleen lymphocytes, protein synthesis consumes about 28% of the ATP generated; Ca^{2+}-ATPase, 27%; Na^+/K^+-ATPase, 19%; RNA synthesis, 15%; and proteolysis, 8%.[11] In K562 human erythroleukemia cells, protein synthesis consumes 35%; Ca^{2+}-ATPase, 28%; Na^+/K^+-ATPase, 19%; RNA synthesis, 8.5%; and DNA synthesis, 8%.[12] These data indicate that about half of the cellular ATP is consumed for maintaining ion homeostasis. Therefore, suppression of all of the biosynthetic reactions cannot slow down ATP depletion more than two-fold, and this share of ATP consumption, in contrast to many biosynthetic reactions, may be necessary for cell viability both in the active and resting state (see section 2.3 for details). Interestingly, in a recent study on concanavalin A-stimulated thymocytes

Buttgereit and Brand demonstrated a clear hierarchy of the response of different energy-consuming reactions to changes in energy supply. Pathways of macromolecular biosynthesis (protein synthesis and RNA/DNA synthesis) were most sensitive to energy supply, followed by sodium cycling and then calcium cycling across the plasma membrane.[13] Apparently, cells suppress unnecessary biosynthetic reactions when ATP generation is reduced.

2.1.3. METABOLIC ARREST

In his excellent review,[14] Hochachka discussed possible defense strategies of organisms against hypoxia. He concludes that the most favorable strategy is metabolic arrest. Namely, ATP turnover rates (energy demand) decrease during the oxygen-limiting period. This strategy is widely used by many invertebrates and by lower vertebrates, and data show that such mechanisms also operate in mammals. For example, metabolic arrest for hypoxic resistance has been demonstrated in mammalian kidney. The main ATP-consuming process in this organ is membrane-coupled ion translocation, and the medullary thick ascending limb of Henle's loop is the most hypoxia-sensitive segment of the nephron. However, the hypoxia sensitivity can be greatly reduced by perfusion with ouabain, a specific inhibitor of Na^+/K^+-ATPase or by reducing the ion-pumping work by prevention of glomerular filtration (see ref. 14 for review). Apparently, ouabain-induced ionic imbalance (i.e., efflux of K^+ and influx of Na^+) is nontoxic by itself (see also section 2.3 below).

For various cells and organs, main ATP-consuming processes vary greatly. As an example, in our studies on tumor cells (EL-4 thymoma) we failed to observe any protective effect of ouabain on rotenone-induced ATP depletion[15] and cell survival (Gabai et al, unpublished data). In myocardium, the main ATP-utilizing function is mechanical work. It has been calculated that a 50% reduction in coronary blood flow completely depletes CrP and ATP in less than 1 minute unless myocardial ATP

consumption is reduced during ischemia. Really, ATP utilization in the ischemic heart is markedly slowed down compared to the nonischemic heart because of its reversible left ventricular dysfunction (so called "myocardial hibernation", see ref. 16 for review). This adaptive mechanism may greatly affect heart sensitivity to ischemia, and prolonged steady, as large as 50%, reduction in coronary blood flow over 1 hour does not lead to infarction.[17] As Arai et al showed, the mechanism of reduction of myocardial contractility under mild ischemia is not obviously associated with diminution of CrP, ATP, or lactic acidosis.[18] Webster and Bishopric[19] observed in cardiomyocytes adapted to chronic hypoxia (4-8 mm Hg) that a decrease in beating frequency coincided with a reduction in the cAMP level. They also showed that the chronic administration of cAMP elevating drugs was toxic specifically to hypoxic cardiomyocytes; this has apparently resulted from an increase in the contraction rate and acceleration of ATP loss in the dying cells.[18] Therefore, reduction of mechanical work both in the heart or in cultured cardiomyocytes may have protective effect against mild ischemia. However, when ischemia is severe, marked ATP depletion is accompanied by irreversible contractile dysfunction (myofibril contracture) with subsequent cell death[19] (see also section 2.3 below).

2.1.4. MITOCHONDRIAL ATPASE

One of the cellular ATPases, the mitochondrial H^+-ATPase, is the main enzyme of ATP generation during oxidative phosphorylation that carries out the reverse reaction (ATP $\rightarrow$ ADP + P_i), thus consuming ATP in order to maintain mitochondrial membrane potential when cellular respiration is blocked. As we have showed in collaboration with Chernyak's lab, this ATPase alone in its fully active state can theoretically deplete all of the cellular ATP in EAC or thymocytes within 1-2 minutes.[4,5] However, there are several mechanisms of H^+-ATPase inactivation that suppress the ATP-hydrolysing activity of ATPase. One of them has an inhibitory effect on ADP during ATP hydrolysis, and another one is due to the action of a special inhibitory protein, IF_1 (see ref. 20 for review).

We have studied the role of IF_1 in the inhibition of ATP hydrolysis in thymocytes and EAC cells under in vitro "ischemia" (as shown in Table 2.2) using a special assay allowing determination of IF_1 activity within cells (refs. 4, 5). Both types of cells possess almost the same inhibitory action of IF_1 (70%) in de-energized mitochondria (i.e., mitochondria lacking membrane potential due to the cessation of respiration or uncoupling).[4] However, total prevention of mitochondrial ATP hydrolysis by oligomycin in EAC cells without oxidative and glycolytic phosphorylation did not slow down ATP depletion, whereas in thymocytes this exposure decelerated ATP depletion about two-fold.[4,5] In accordance with the data, oligomycin slowed down cell death of uncoupler- or rotenone-treated thymocytes but not EAC cells.[5]

In other normal cells (rat cardiomyocytes), treated with 2,4-dinitrophenol, rotenone or cyanide, inhibition of ATP hydrolysis by oligomycin also significantly slowed down ATP depletion.[7] As Jennings et al[21] established, approximately 35% of the ATP utilization observed during the first 90 minutes of total ischemia in the canine heart is due to mitochondrial ATPase activity. Recently Vuorinen and co-workers[22] found that in the ischemic rat myocardium oligomycin inhibited ATP decrease; moreover, they observed an inhibition of mitochondrial ATPase during ischemic preconditioning (3 minutes of ischemia with 9 minutes of reperfusion) and suggested that this effect may be responsible for the protective action of the preconditioning on ischemic myocardium (see chapters 3 and 5 for detailed discussion of the preconditioning effect). When hearts were arrested with hyperkalemia, the protective effect of preconditioning on ATP content during total ischemia in dogs was not eliminated, indicating that a decrease in energy demand in preconditioned myocardium was not due to its reduced con-

tractility.[23] No inhibition of mitochondrial ATPase during preconditioning (5 minutes ischemia/5 minutes reperfusion) was found by Kobara et al[24] on the rat heart or Vanderheide et al[25] on the canine heart, although prolonged ischemia (15-30 minutes) did cause its marked suppression in both studies. On submitochondrial particles (SMP) Rouslin and co-workers[26] detected an increase in IF_1 binding in SMP isolated from ischemic rabbit but not rat hearts. They concluded that the data corroborated the fundamental differences in mitochondrial ATPase regulation between slow heart-rate species (rabbits, dogs, humans) and fast heart-rate ones (rats, mice) (see refs. 20, 24 and references therein). It is quite possible that the relative contribution of mitochondrial ATPase in ATP depletion under myocardial ischemia depends on several factors, one of them being the intensity of contractile work (i.e., activity of myosin ATPase).

Thus, there are several ways for cells to maintain their ATP level under energetically unfavorable conditions, including: (1) creatine phosphate buffering system; (2) activating of alternative pathways of ATP generation (glycolysis); (3) inhibition of ATP consumption by some unnecessary biosynthetic enzymes and (4) suppression of metabolic activity. However, when cellular protective systems for maintaining the ATP level under anoxia or ischemia are exhausted, ATP depletion occurs with dramatic consequences for cells.

2.2. CELL VIABILITY AND ATP DEPLETION

It seems trivial that energy-deprived cells eventually die, but surprisingly, the vulnerability of different mammalian cells to ischemia varies greatly—from only several minutes (in neurons)[27] to many hours (e.g., some tumor cells or quiescent fibroblasts).[28,29] In canine myocardium, low flow ischemia (10-12% of control flow) results in significant cell death within 40-60 minutes.[30] Some differences in the sensitivity of cells to ischemia may be due to the above described cellular protective systems of ATP stabilization, but apparently the main distinctions are associated not with the rate of ATP depletion in these cells but to cellular sensitivity to ATP deprivation (see Fig. 2.2 below).

There are two main modes of cell death, namely reproductive death and interphase death; the latter includes necrosis and apoptosis. Reproductive death can be demonstrated only in proliferating cells and is usually assessed by suppression of colony-forming ability; this form of cell death is associated with irreparable damage to genomic DNA[31,32] or mitotic apparatus.[33,34] Rotin and co-workers[35] studied the effect of ATP depletion (anoxia in glucose-free medium) on colony-forming ability of Chinese hamster ovary cells. No effect on cell survival was found within the first 3 hours of incubation under these conditions, but after 6 hours, when the ATP level decreased to 2.5% of the initial, cell survival decreased to 5%. In EAC cells, anoxia in a glucose-free medium resulted in marked DNA degradation of intact (double-stranded) DNA (to 20% of initial) after 3 hours.[36] This DNA damage can be responsible for reproductive death of ATP-depleted cells.

2.2.1. NECROSIS

Interphase death can occur both in resting cells (cardiomyocytes, neurons, hepatocytes) and proliferating cells. The most typical consequence of prolonged ischemia is necrotic cell death, which is characterized by cell swelling, blebbing of the plasma membrane, nuclear chromatin clumping and shrinkage, swelling of organelles and eventually breakage of the plasma membrane.[37,38] Naturally, a cell with a disrupted plasma membrane is not able to maintain ionic balance and some of its cytoplasmic constituents (e.g., enzymes) are lost. Therefore, such cells can be considered necrotic because the most reliable sign of cell necrosis is the plasma membrane breakage.

At the present time, there are several assays for quantification of necrotic death both in vivo and in vitro (Table 2.3). Most

of these assays are based on either loss of cytoplasmic enzymes (e.g., release of lactate dehydrogenase or creatine kinase) and special labels (e.g., ^{51}Cr) or penetration of dyes such as trypan blue through the disrupted plasma membrane. As a rule, these assays give very similar results in the same systems, although in ischemic cardiomyocytes creatine kinase release occurs during the reversible phase of injury[19] and is greater than trypan blue staining.[39] However, for quantification of necrotic zone in ischemic tissues (e.g., myocardium), the above methods cannot be used since it is difficult to determine plasma membrane damage in situ. In this case, assays based on the reduction of tetrazolium salts by mitochondria can be employed (Table 2.3). Viable cells convert these salts to the colored product (formazan) while nonviable cells cannot. The limitation of the latter assay, however, is that in some cases (e.g., in tumor cells) mitochondrial poisons totally prevent formazan formation without affecting cell viability. In these circumstances, other methods should be used.

All these assays can usually discriminate between viable and nonviable cells. However, thus far there are no reliable tests to evaluate viable, but irreversibly injured, cells which ultimately die. For instance, we observed that the plasma membrane blebbing is irreversible in EL-4 thymoma, but reversible in EAC carcinoma[40-42] (see chapter 3 for consideration of the blebbing mechanism). According to Trump and Berezesky,[38] the irreversible stage ("point of no return") is characterized by markedly swollen mitochondria which contain dense inclusions in their matrix (see also section 2.4 below). However, this mitochondrial damage within a cell can only be observed by electron microscopy and cannot be used as a simple quantitative method.

Kristensen[28] has evaluated the dependence of necrotic death (assayed by lactate dehydrogenase release) in quiescent human embryonal lung fibroblasts on their ATP level and energy charge, namely (ATP + 1/2 ADP)/(ATP+ADP+AMP). When these cells were treated with oligomycin in a

Fig. 2.2. Effect of rotenone on ATP depletion (A) and viability (B) of different cells. Cells were incubated in a glucose-free medium with 2 µM rotenone. The ATP level was determined by a luciferine-luciferase assay[4,5] and viability by Trypan blue exclusion test.

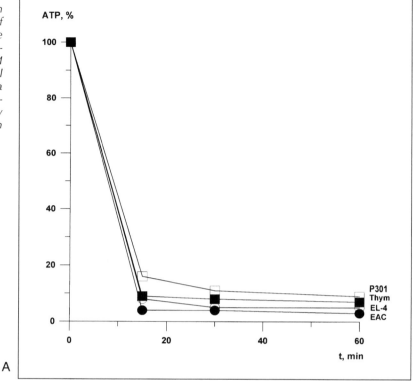

A

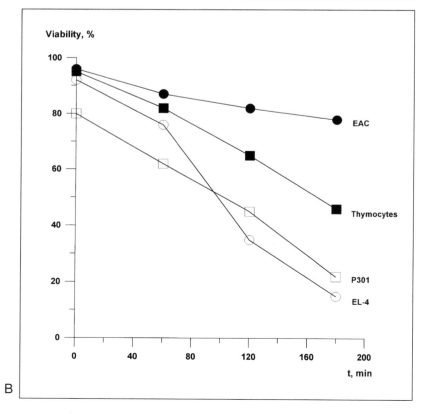

B

Table 2.3. Methods for evaluation of necrotic cell death

1. Loss of cytoplasmic enzymes:
   1.1. Lactate dehydrogenase
   1.2. Creatine kinase

2. Loss of mitochondrial activity:
   Loss of reduction of tetrazolium salts (e.g., triphenyltetrazolium chloride)

3. Loss of intracellular labels:
   3.1. ^{51}Cr
   3.2. Fluorescent dyes (e.g., fluorescein diacetate)

4. Staining by postmortal dyes:
   4.1. Trypan blue, nigrosin, eosin, etc.
   4.2. DNA-binding fluorochromes (ethidium bromide, propidium iodide).

glucose-free medium, a rapid depletion of ATP occurred during the first hours which was followed by a slow protracted decrease over several subsequent hours. Cell death began after only 12 hours of incubation when ATP was 1-2% of the initial level and about 60% of the cells were dead after 15 hours. The energy charge decreased to about 0.35 after 3 hours and remained at that level for 15 hours. There was no change when cell death began. Almost the same results were obtained when 2,4-dinitrophenol was used instead of oligomycin. In the presence of antimycin, ATP depletion was faster and energy charge decreased to one-tenth after 3 hours, but the time course of cell death was very similar to that after oligomycin or 2,4-dinitrophenol treatment. If cells were incubated with the above inhibitors in a medium containing glucose, they did not experience significant ATP decrease or cell death. Interestingly, the addition of glucose to the inhibitor-treated cells before their death resulted in both complete ATP recovery and rescue from necrosis. The author concluded that no simple correlation between ATP or energy charge and cell survival exists.[28]

We have investigated the dependence of cell necrosis (assayed by trypan blue exclusion) on ATP depletion in different cell types including: thymocytes, lymphoid tumors (EL-4 thymoma, P_3O_1 myeloma) and EAC carcinoma (see refs. 5, 29, 40, 43). As one can see from Figure 2.2, the ATP depletion rate after rotenone treatment was fastest in EAC cells, but the rate of death in these cells was slowest; in contrast, myeloma cells had the slowest rate of ATP depletion, but their death was much faster than that of EAC. No clear difference in cell sensitivity to ATP depletion was found between normal and tumor cells. Therefore, although there are no doubts that ATP depletion can result in cell necrosis, no clear correlation exists between the ATP level and cell death.

2.2.2. APOPTOSIS

Apart from necrosis, apoptosis is often referred to as physiological or programmed cell death, since this type of cell elimination occurs in normal embryogenesis, tissue homeostasis, immune response, etc. (see refs. 44, 45 for recent reviews). In many cases (especially in lymphoid cells), apoptosis can be easily discriminated from necrosis by some of its cardinal features such as cell shrinkage, chromatin condensation, fragmentation of the nucleus and DNA (usually into oligonucleosomes, giving a characteristic "ladder" pattern on agarose gels) and the formation of apoptotic bodies.[37,38,46] Of great physiological importance is that apoptosis in vivo is not accompanied by plasma membrane breakage and results in no inflammatory reaction, since apoptotic bodies are rapidly phagocytosed by macrophages or adjacent cells.[37,46]

Several methods are employed to quantify apoptosis including: light, fluorescent and electron microscopy, flow cytometry, DNA electrophoresis, etc. (Table 2.4).[47,48] At times, however, apoptosis cannot be easily discriminated from necrosis (e.g., ref. 49), especially in nonlymphoid cells. We believe that delayed plasma membrane permeability may be a good criterion for apoptosis in such cases.

Now it is clear that apoptotic death both in vitro and in vivo can be induced not only by physiological factors, but also by many injurious exposures such as radiation, oxidative stress, hyperthermia, anticancer drugs and other cytotoxines.[46] Recently, a number of studies showed that short-term or partial ischemia in various organs and cell cultures can also induce apoptotic death. In 1990, using light and electron microscopy Gobe et al initially observed that renal atrophy in rats after partial ischemia was associated with both necrosis and apoptosis.[50] In 1992, Schumer et al demonstrated apoptotic body formation and oligonucleosomal DNA fragmentation at 12-24 hours of reperfusion after 5-45 minutes of complete renal ischemia.[51] In 1993, Okamoto and coworkers reported internucleosomal DNA cleavage in pyramidal neurons of the hippocampal CA1 area at 48-54 hours after 5 minutes of brain ischemia.[52] During the last 2 years, several

papers have appeared which establish, by various assays, that apoptotic death occurs after ischemia-reperfusion in myocardium, kidney, and brain (see chapters 6, 8). In our experiments on EAC cells, we have also observed apoptotic death after transient ATP depletion.[42] In these cells, a short-term energy deprivation (30-60 minutes) resulted in apoptosis while more prolonged ATP decrease caused necrosis.[53,54] Regarding all the data, Richter and co-workers have recently proposed that the cellular ATP level is an important determinant of cell death, either by apoptosis or necrosis. When ATP falls below a certain ATP level, apoptosis ensues provided that enough ATP is still available for some energy-requiring apoptotic processes; however, under severe ATP depletion, controlled cell death ceases and necrosis begins (see also chapters 6, 8 for further discussion).[55]

From the above data, we can conclude that ATP depletion results in reproductive, apoptotic and necrotic death. Apparently, depending on the duration of energy deprivation, the resistance to ATP depletion varies greatly between various cells and organs.

2.3. SIGNIFICANCE OF IONIC IMBALANCE FOR THE DEATH OF ATP-DEPLETED CELLS

As we described above, nearly one-half of cellular ATP is consumed by ion-trans-

---

Table 2.4. Methods of evaluation of apoptotic cell death

1. Light microscopy:
   hematoxylin-eosin staining

2. Fluorescent microscopy:
   DNA-binding fluorochromes (e.g., propidium iodide, diamidino-2-phenylindole, Hoechst 33342, etc.)

3. DNA fragmentation and strand breaks:
   3.1. Measurement of soluble DNA (e.g., by diphenylamine reagent, DNA-binding fluorochromes)
   3.2. Agarose gel electrophoresis of DNA
   3.3. Terminal deoxynucleotide transferase (TdT)-mediated dUTP-biotin nick end labeling (TUNEL) method for assessing DNA strand breaks

4. Flow cytometry:
   4.1. Forward and side light scattering (cell size and granularity)
   4.2. DNA content (propidium iodide, Hoechst 33342)
   4.3. DNA strand breaks (TUNEL method)

porting ATPases to maintain ion gradients across the plasmalemma and membranes of the organelles (see section 2.1). Therefore, it is not surprising that the cessation of ATP generation may drastically affect ion distribution between a cell and the surrounding medium as well as between the cytoplasm and the cellular compartments. Indeed, an increase in intracellular $[Ca^{2+}]$, $[Na^+]$, $[H^+]$ and decrease in $[K^+]$ are often observed in ATP-depleted cells (see also refs. 38, 56 for recent reviews). The question arises as to whether this ion redistribution can be fatal for a cell?

2.3.1. K^+/Na^+ IMBALANCE

Loss of K^+ and accumulation of Na^+ during ATP deprivation are believed to be the result of the inhibition of plasma membrane Na^+/K^+-ATPase[38] whose K_m for ATP is rather high (0.2-0.3 mM) (see ref. 57). Therefore, a five- to ten-fold decrease in the ATP level can greatly affect Na^+/K^+-ATPase activity, thus affecting both K^+ and Na^+ redistribution. For example, when rat hepatocytes were subjected to 1 hour of anoxia, their ATP level fell by 66% and $[Na^+]_i$ increased about two-fold (from 15.9 mM to 32.2 mM).[57] Anoxic treatment of EAC cells in a glucose-free medium caused ATP to fall to 2% after 3 hours; this was accompanied by a three-fold decrease in $[K^+]_i$.[58] The same ATP depletion in quiescent fibroblasts evoked only two-fold decrease in $[K^+]_i$ after 8-12 hours.[28] In cultured rat ventricular myocytes poisoned by iodacetic acid, ATP depletion to 6% after 3 hours brought about a four-fold increase in $[Na^+]_i$ and a seven-fold decrease in $[K^+]_i$.[59] The similar elevation for $[Na^+]_i$ was made in isolated rabbit hearts after 1 hour of ischemia.[60] Therefore, all the data show that K^+/Na^+ imbalance occurs during ATP deprivation.

In addition to inhibition of Na^+/K^+-ATPase, K^+ loss due to ATP decrease may also be attributed to the opening of ATP-sensitive K^+-channels which were found in a number of cells (e.g., myocardial).[56] These channels are closed when the ATP level is high but open when ATP decreases and

ADP increases. It is suggested that the K^+-channel functioning is associated with a cardioprotective mechanism of ischemic preconditioning, apparently through a shortening of the action potential duration and preservation of myocardial ATP level.[56,61,62] Regarding Na^+ imbalance, Anderson and co-workers have recently shown that an increase in $[Na^+]_i$ during myocardial ischemia may be partly mediated by a Na^+/H^+-exchanger.[60]

However, when Na^+/K^+-ATPase was greatly suppressed by its specific inhibitor, ouabain, no effect on the viability of various cells was observed for many hours despite the severe K^+/Na^+ imbalance.[63,64] In our studies, we also failed to observe any cytotoxic effect of prolonged (3-5 hours) incubation with ouabain (1 mM) on EL-4 or EAC tumor cells (Gabai et al, unpublished data). Therefore, elevation of $[Na^+]_i$ and loss of K^+ observed in energy-deprived cells cannot be markedly cytotoxic by itself. Moreover, ouabain has a protective effect during renal ischemia owing to ATP preservation (see section 2.1), as do openers of K^+-channels (nicorandil, pinacidil) during myocardial ischemia.[56]

At the same time, in some circumstances ouabain is able to have a rather rapid cytotoxic effect. Talbot and co-workers[65] found that transformation of human HOS cells with retroviral oncogenes (v-ras, v-mos, v-src) resulted in the rapid ouabain-induced death of transformed cells (assayed by ^{51}Cr release) under alkaline (pH 8.0), but not under neutral, pH. However, the mechanism of this unusual effect remains unknown.

2.3.2. pH CHANGES

A change in concentration of H^+ was also observed during ATP depletion. In anoxic hepatocytes, pH_i decreased from 7.41 to 7.04 during the first 10 minutes;[57] in cyanide-intoxicated hepatocytes, pH_i fell immediately from 7.3 to approximately 7.0,[66] and a similar intracellular acidification was observed in rat neonatal cardiomyocytes during "chemical ischemia" (cyanide + 2-deoxyglucose).[67] Several mechanisms are responsible for such acidification during ischemia. In isolated cells, this may be caused by H^+ generation owing to ATP hydrolysis.[2,68] In ischemic tissues, pH decrease is additionally associated with lactate accumulation due to stimulation of glycolysis.

Although some time ago many researchers considered ischemic acidosis a detrimental event, nowadays this seems unlikely. For instance, in perfused rat hearts low-flow (0.5 ml/min) ischemia decreased pH_i to 6.87 (as measured by ^{31}P-NMR spectroscopy) and resulted in rigor contracture (loss of mechanical activity) within 5 minutes.[69] When glucose was supplied throughout ischemia, no contracture occurred after 30 minutes, although pH_i decreased to 6.60.[70] Likewise, a reduction of lactate accumulation during global ischemia in rat hearts by prior perfusion with glucagon (a hormone which decreases glycogen content) did not attenuate lethal ischemic injury (creatine kinase release),[71] whereas the glycogen-containing hearts were more resistant to brief ischemia.[69]

There are several studies showing that mild cytosolic acidification provides powerful protection during many cellular stresses, including ATP depletion. The first indication for its protective role was obtained by Pentilla and Trump as early as 1974. They observed that necrotic death of EAC and rat renal cortex cells under anoxia was significantly delayed at extracellular pH 5.9.[58] In recent years, the protective role of low pH has been intensively studied in Lemasters' lab in both in vitro and in vivo models of ischemia. An interesting phenomenon called "pH-paradox" has been found in this lab. When ATP-deprived hepatocytes or cardiomyocytes were incubated at low pH (6.0-6.5), no significant necrosis happened, but placing these cells in a physiological pH (7.3-7.4) promptly evoked cell death.[67,72] This phenomenon was also demonstrated in vivo in perfused rat liver during anoxia/reoxygenation. The release of lactate dehydrogenase as an indicator of necrosis (see Table 2.3 above) was greatly accelerated when the pH of the perfused medium was increased from

6.1 to 7.3.[73] Thereafter, these authors clearly demonstrated that it is the intracellular pH which is responsible for the phenomena, and that it is possible to delay cell death after returning to normal pH by inhibitors of the Na^+/H^+ exchanger (i.e., by prevention of intracellular alkalinization).[67,68] Likewise, these drugs (amiloride and its analogs) in addition to mild acidosis exert a cardioprotective effect during ischemia/reperfusion.[19,56] Conversely, the prevention of intracellular acidification in ATP-depleted hepatocytes by monensin (Na^+/H^+ exchanger) greatly accelerated cell death.[68]

In our studies on tumor cells, we also observed the protective action of acidification on the integrity of the plasma membrane in energy-deprived EAC and EL-4 tumor cells (ref. 36; Gabai, unpublished data). At extracellular pH 6.4, ATP depletion in anoxic and rotenone-treated EL-4 thymoma was slightly slowed down.[15] However, the deceleration of the ATP level decrease was not as significant as the delay of cell death. Therefore, it is obvious that the protective effect of acidification is not associated with ATP loss.

At the same time, ischemia-induced DNA degradation was not suppressed at an acidic pH in either EAC or EL-4 cells (ref. 36; Proskuryakov, Gabai, unpublished data). The increased DNA degradation at low pH_i may be associated with activation of DNase II, which has an acidic pH optimum (5.5); this nuclease may be involved in apoptotic cell death.[74,75] Likewise, reproductive cell death (colony-forming ability) of CHO and human bladder cancer cells under hypoxia in a glucose-free medium was accelerated at pH 6.5-6.0.[35] Furthermore, the drugs which equilibrate pH_i with external pH (nigericin, protonophore CCCP) markedly suppressed colony formation at low, but not neutral, values of external pH.[76,77]

Therefore, an increase in intracellular [H^+] occurring in ATP-depleted cells has, obviously, a favorable rather than harmful effect on their plasma membrane integrity, thus delaying necrotic death. At the same time, no protective effect of acidification on DNA integrity was found; conversely, a marked acidification may provoke apoptotic or reproductive cell death.

2.3.3. Ca^{2+} Imbalance

Apart from K^+, Na^+ and H^+ ion perturbations during energy deprivation, great attention has been paid to alterations of Ca^{2+} homeostasis, but up to now the results obtained are rather controversial and need more careful consideration. Ca^{2+} is intimately involved in signal transduction as a second messenger; therefore several mechanisms exist for fine regulation of its concentration within a cell (see ref. 78 for recent review).

Normally, the concentration of free Ca^{2+} within a cell is about 100-200 nM, i.e., about four orders of magnitude lower than within the extracellular medium (1-2 mM). Such a low $[Ca^{2+}]_i$ is usually maintained by Ca^{2+}-transporting systems of the plasma membrane and the endoplasmic reticulum (ER); in some pathological conditions, an excess of cytoplasmic Ca^{2+} can also be removed by mitochondria but they have much lower affinity for Ca^{2+} than the above two systems.[79,80] Ca^{2+} molecules enter into cells through the special gated or ungated channels of the plasmalemma while its efflux is performed by Ca^{2+}-ATPase or, in some cells, also by the Na^+/Ca^{2+} exchange system. The major intracellular source of Ca^{2+} is the ER whose uptake system consists of the Ca^{2+}-ATPase, whereas Ca^{2+} release from the ER occurs mainly through inositol-3-phosphate-gated or Ca^{2+}-gated channels.[78,80]

Theoretically, Ca^{2+} deregulation following injury may lead to Ca^{2+} overload with fatal consequences for a cell, since several destructive enzymes (phospholipases, proteases, and nucleases) can be activated by high Ca^{2+} (see ref. 81). Such Ca^{2+} deregulation may be the result of any combination of the following events: (1) influx from the extracellular space; (2) efflux from the ER and (3) efflux from mitochondria (refs. 27, 37). Indeed, many researchers from a variety of labs found an increase in $[Ca^{2+}]_i$

following ATP depletion. In rat hepatocytes, anoxia increased $[Ca^{2+}]_i$ in two distinct phases: (1) the first rise occurred within 15 minutes (from 127 nM to 390 nM) and (2) the second peak reached a maximum (1450 nM) after 1 hour; the latter $[Ca^{2+}]_i$ rise, in contrast to the former, was not observed in the Ca^{2+}-free medium thus indicating extracellular Ca^{2+} influx.[57] Gasbarrini and co-workers suggested that the initial rise in $[Ca^{2+}]_i$ is evoked by its loss by deenergized mitochondria, as was demonstrated by other researchers[82,83] (see also next section). Elevation of $[Ca^{2+}]_i$ in hypoxic hepatocytes through Ca^{2+} influx was also found by Brecht et al.[84] However, the effect of extracellular Ca^{2+} depletion on the viability of ATP-depleted hepatocytes was quite different from that observed by Gasbarrini et al who demonstrated the suppression of lactate dehydrogenase release in a Ca^{2+}-free medium.[57,85] In contrast, Brecht et al found that cell death (assessed by propidium iodide staining) was slightly accelerated. Moreover, in the latter study Ca^{2+}-depletion per se resulted in cell death even under normoxic conditions.[84] The reason for such contradictory results is not clear.

At the same time, Lemasters and co-workers found no $[Ca^{2+}]_i$ increase before hepatocyte blebbing and necrosis under "chemical hypoxia" (KCN + iodacetate).[86,87] More importantly, when $[Ca^{2+}]_i$ in hepatocytes rises to more than 2500 nM (i.e., greater than 10-fold) during incubation with exogenous ATP, which increases $[Ca^{2+}]_i$ through its efflux from the ER,[88,89] no significant cell death occurred within several hours.[90] In rabbit proximal tubule cells, uncoupler 1799 caused severe ATP depletion but did not increase $[Ca^{2+}]_i$ before loss of viability as assayed by trypan blue staining. Preventing this increase with low Ca^{2+} (100 nM) medium did not protect 1799-treated cells. In addition, a reduced pH suppressed cell death without ameliorating the increase in $[Ca^{2+}]_i$ (see ref. 91). In the neuronal cell line PC12, cyanide-induced ATP depletion increased $[Ca^{2+}]_i$ 2.5-fold within 10 minutes, but after 30 minutes $[Ca^{2+}]_i$ was slightly below that observed in untreated cells although the ATP level fell by 92%.[92] In another neuronal line, SK-N-SH neuroblastoma, ATP depletion (cyanide + 2-deoxyglucose) also did not appreciably increase $[Ca^{2+}]_i$ until the time of cell death, when the loss of plasma membrane integrity allows unimpeded influx of extracellular Ca^{2+} (ref. 93). Thus, all the above data indicate that necrotic death of various cells under ATP depletion can be independent of $[Ca^{2+}]_i$ elevation.

It may seem strange that in some cells ATP depletion has not affected Ca^{2+} homeostasis for a rather long time. However, the K_m for ATP of plasmalemmal Ca^{2+}-ATPases is very low (1-3 µM for a high affinity site and 150-180 µM for a low affinity site).[94] Therefore, even a 10-fold depletion of intracellular ATP (from 3-5 mM to 0.3-0.5 mM) may not significantly inhibit Ca^{2+}-transporting ATPases.

The significance of $[Ca^{2+}]_i$ deregulation in cell injuries has been intensively studied in the heart and on isolated cardiomyocytes. It is clear that cardiomyocytes in situ accumulate an abnormally large amount of Ca^{2+} during the reperfusion phase following prolonged ischemia.[95,96] However, the reason for this is not fully understood. Wang et al,[97] using patch clamp techniques, observed on adult rat ventricular myocytes an approximately 10-fold increase in open probability of specific Ca^{2+} leak channel when these cells were poisoned with iodoacetic acid and 2-deoxyglucose. When the same cells were subjected to prolonged anoxia or metabolic inhibitors (cyanide + 2-deoxyglucose), no significant increase in $[Ca^{2+}]_i$ was found during the first 30-60 minutes of incubation despite marked ATP depletion.[98-101] Thereafter, $[Ca^{2+}]_i$ began to increase and eventually the cells were killed.[98,101] Prior to $[Ca^{2+}]_i$ elevation, the normal elongated cells spontaneously shortened to about 60% of their original length. The important finding of the study of Allshire and collegues[98] is that the ability of cells to survive and restore Ca^{2+} homeostasis upon reoxygenation is exceeded once $[Ca^{2+}]_i$

reaches 5 µM; the recovery in isolated cardiomyocytes is only possible when $[Ca^{2+}]_i$ is less than 1.5 µM and Ca^{2+}-ATPase of the ER is not suppressed by caffeine. From these data, the researchers concluded that Ca^{2+}-independent shortening of anoxic cardiomyocytes reflects the onset of rigor, which triggers the loss of $[Ca^{2+}]_i$ homeostasis. In later studies that monitored ATP concentrations, the same group found that single cardiomyocytes injected with luciferase experienced rigor associated with sudden cytosolic ATP depletion, apparently due to the activation of myosin ATPase.[99,100] However, it is not clear from the research whether the Ca^{2+} influx is necessary for irreversible cardiomyocyte damage and death.

Pierce and Czubryt[56] suggested that Ca^{2+} influx during ischemia-reperfusion is mediated by Na^+/Ca^{2+}-exchanger subsequent to Na^+ accumulation through Na^+/H^+-exchanger, which could explain the cardioprotective effect of Na^+/H^+-blockers (see above). Bond and coworkers studied changes in $[Ca^{2+}]_i$ during "pH-paradox" and found that reperfusion of ATP-depleted cardiomyocytes with Ca^{2+}-free solution did prevent an increase in $[Ca^{2+}]_i$ but it did not postpone cell necrosis.[67]

At the same time, a Ca^{2+}-overload of cardiomyocytes is undoubtedly highly toxic. When isolated hearts are perfused for a brief period (2-10 minutes) with Ca^{2+}-free solution and then reperfused with normal Ca^{2+} containing medium, this leads to a pronounced Ca^{2+} accumulation and severe cardiomyocyte damage, the phenomenon referred to as "Ca^{2+}-paradox".[102-104] Apparently, the depletion of extracellular Ca^{2+} increases Ca^{2+} permeability of the sarcolemma and thus evokes a massive Ca^{2+} influx after Ca^{2+} repletion.[102] In turn, Ca^{2+}-overload may irreversibly damage mitochondria (see the next section) and the myofilament system (e.g., through activation of a Ca^{2+}-dependent neutral protease, calpain). This protease was shown to be involved in the degradation of calspectrin (nonerythroid spectrin or fodrin) during reperfusion after brief ischemia of rat heart.[105] However,

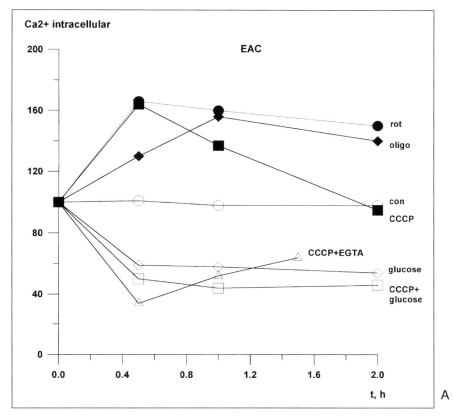

A

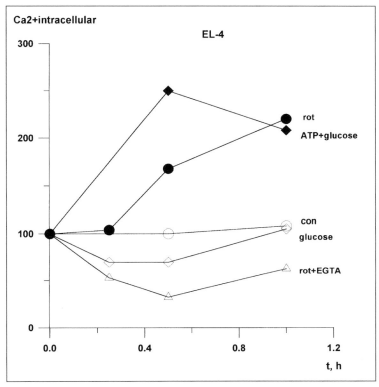

B

Fig. 2.3. Effect of mitochondrial inhibitors on alteration of $[Ca^{2+}]_i$ in EAC (A) and EL-4 (B) cells. $[Ca^{2+}]_i$ was determined using Quin-2AM fluorescent probe as described elsewhere.[40,41] Cells were incubated with or without glucose (10 mM); rotenone, 2 µM; CCCP, 2 µM; oligomycin (oligo), 2 µg/ml; ATP, 1 mM; EGTA, 1 mM in a Ca^{2+} free medium.

when Atsma and coworkers examined the role of calpain in necrotic death of cultured neonatal rat cardiomyocytes during metabolic inhibition (cyanide + 2-deoxyglucose), they found that despite calpain activation before cell death, calpain inhibitors did not attenuate it.[101] From all the data, we believe that $[Ca^{2+}]_i$ elevation observed in energy-deprived cardiomyocytes may likely aggravate cell death, but is not necessary for triggering irreversible cell injuries.

In our studies on EL-4 thymoma and EAC cells, we observed that any inhibitor of oxidative phosphorylation (rotenone, uncoupler CCCP, or oligomycin) resulted in an approximate two-fold rise in $[Ca^{2+}]_i$, which was prevented by glucose or the Ca^{2+}-chelator EGTA (Fig. 2.3). Although glucose completely suppressed cell necrosis, preventing ATP depletion, no protective effect of EGTA on the viability of ATP-depleted cells was found in EL-4 thymoma[40,41] or EAC cells (Fig. 2.4). This implies that $[Ca^{2+}]_i$ elevation is not necessary for tumor cell death. However, a Ca^{2+}-ionophore (A23187) was toxic both for EL-4[40] and EAC cells, and its cytotoxic effect was precipitated by energy deprivation (Fig. 2.4). The similar aggravation of A23187-induced necrosis by anoxia (in glucose-free medium) was also observed previously in EAC cells by Laiho et al.[106] In contrast to rotenone and other mitochondrial inhibitors, EGTA completely prevents the toxic effect of A23187 (Fig. 2.4) indicating that it is exogenous Ca^{2+} that is responsible for ionophore-induced cell killing.[40,107] Interestingly, glucose also protects tumor cells from toxicity of A23187 (Fig. 2.4). This suggests that glycolytic ATP generation can maintain Ca^{2+}-homeostasis even in A23187-treated cells, whereas mitochondria are damaged by high $[Ca^{2+}]_i$ (see the next section). Indeed, A23187 treatment in a glucose-free medium sharply decreases the ATP level but glucose prevents this effect.[40] Toxicity of exogenous ATP, which increases $[Ca^{2+}]_i$ in EL-4 cells about three-fold (Fig. 2.3), is also prevented by the presence of glucose and precipitated in ATP-depleted cells (Fig. 2.4).

From our data, we believe that there are at least two distinct mechanisms of tumor cell necrosis: (1) Ca^{2+}-independent and associated with pronounced ATP depletion, and (2) Ca^{2+}-dependent (e.g., in the A23187 and ATP-treated cells).[40] Later, a similar conclusion was drawn by Kamendulis and Corcoran[108] after their study of hepatocyte necrosis.

It is obvious that the effect of energy deprivation on $[Ca^{2+}]_i$ will depend on many factors such as Ca^{2+}-permeability of the plasma membrane, release of Ca^{2+} from the ER and mitochondria, activity of Ca^{2+}/Na^+-antiporter and Ca^{2+}-ATPase etc. At the present time, however, the contribution of these factors in Ca^{2+} imbalance during ATP depletion is largely unknown. It seems likely that the cells with a high density of Ca^{2+}-channels in the plasma membrane (e.g., neurons) will have a much greater vulnerability to ATP deprivation because of the massive efflux of extracellular Ca^{2+} (see ref. 27 and chapter 6).

Data on the role of ionic imbalance in injury and death of ATP-depleted cells allows the following conclusions: (1) ATP depletion is usually accompanied by an efflux of intracellular K^+, influx of Na^+ and accumulation of H^+; (2) Na^+/K^+ imbalance per se has no effect on cell viability whereas intracellular acidification exerts a powerful protective effect against necrosis in ATP-depleted cells; (3) In most cases of energy deprivation, Ca^{2+} elevation by itself is neither necessary nor sufficient for necrotic death, but it may accelerate this process.

2.4. MITOCHONDRIAL DYSFUNCTION AND DAMAGE DURING ISCHEMIA

As we have mentioned in section 2.1, some cells such as mature mammalian erythrocytes can live for a long time without any mitochondria. On the contrary, cerebral and myocardial cells are strictly dependent on aerobic (mitochondrial) ATP generation. As a result, irreversible damage of their mitochondria can lead to a loss of viability even if the oxygen supply after ischemia is resumed. There is no doubt

Fig. 2.4. Effect of energy inhibitors and Ca^{2+} overload on viability of EAC (A) and EL-4 (B) cells. A23187, 10 µM; for other concentrations see Fig. 2.3. Cell viability was determined by the Trypan blue exclusion test.

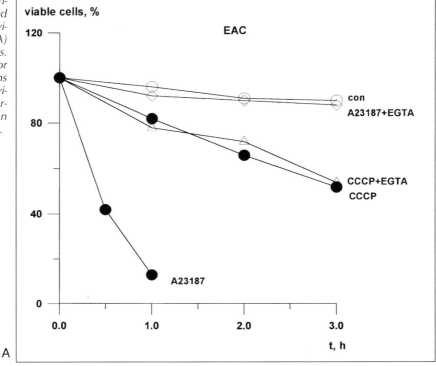

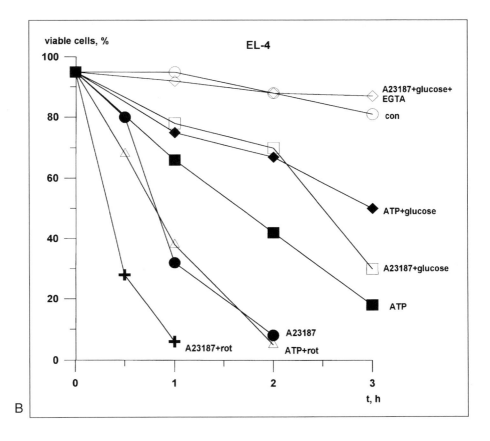

that in post-ischemic necrotic tissues mitochondria are harshly damaged, but the problem is whether this damage is the reason for necrosis or only its consequence. A long time ago, it was suggested that the "point of no return" during necrotic death is associated with irreversible mitochondrial impairment (in particular, high amplitude swelling) (see ref. 38 for review). Below we will discuss some recent data on mitochondrial dysfunction during ischemia and ischemia-reperfusion as well as the significance of mitochondrial failure for necrotic cell death.

2.4.1. MITOCHONDRIAL DYSFUNCTION IN ISCHEMIC TISSUES

There are several parameters which characterize the functions of isolated mitochondria: respiratory rate, respiratory control ratio (RCR), P/O ratio, Ca^{2+}-accumulation capacity, ATPase activity, etc. The respiratory rate indicates the activity of the respiratory chain. RCR shows the extent of coupling between respiration and phosphorylation (in uncoupled mitochondria RCR is decreased). The ADP/O (or P/O) ratio indicates the efficiency of ATP formation per oxygen consumed (maximal efficiency may be 6M ATP per 1M O_2, i.e., ADP/O = 3). Calcium accumulation capacity characterizes mitochondrial Ca^{2+} transport, and in injured mitochondria this capacity is usually decreased.

A number of studies on mitochondria from different tissues show their dysfunction during anoxia/ischemia. For instance, a 30 minute exposure of rat hepatocytes to anoxia resulted in a nearly two-fold decrease in the respiratory rate after reoxygenation, and this respiratory inhibition was associated with the loss of mitochondrial Ca^{2+} and repression of mitochondrial H^+-ATPase (ref. 82, see also section 2.1). In isolated rat liver mitochondria, anoxia also brought about release of mitochondrial Ca^{2+}, phospholipid hydrolysis, and uncoupling of mitochondria.[109] As we established on EL-4 and EAC tumor cells, the uncoupler-induced loss of mitochondrial Ca^{2+} markedly suppresses NADH-dependent

respiration of these cells.[110] A more than two-fold decrease in the respiratory rate and the ADP/O ratio was also observed in rat kidney cortex mitochondria.[111] The most severe mitochondrial defects after ischemia were found by Sciamanna and coworkers in rat forebrain mitochondria. After 15-minute ischemia, the respiratory rate decreased by 70%, and respiratory control was almost totally lost (i.e., RCR was close to 1).[112] Since the addition of Ca^{2+} chelators to mitochondria fully restored mitochondrial functions, the authors concluded that it is the ischemia-induced disturbance of cellular Ca^{2+}-homeostasis that leads to mitochondrial failure.[112,113]

Mitochondria isolated from ischemic myocardium are also heavily impaired. They have decreased respiratory activity (especially in the complex I of respiratory chain), low RCR and Ca^{2+} accumulation capacity, and a diminished activity of adenine nucleotide translocase (see ref. 19 for review). Among these impairments, the inability of mitochondria to retain Ca^{2+} may be especially fatal for cells, as we discussed in the previous section. However, until recently the mechanism by which the disruption of mitochondrial Ca^{2+} homeostasis can bring about cell necrosis has not been clear. The discovery of mitochondrial permeability transition (MPT) may shed some light.

2.4.2. MITOCHONDRIAL PERMEABILITY TRANSITION

When mitochondria are exposed to a number of stressful conditions (e.g., oxidative stress), they lose intramitochondrial constituents with MW <1500 through nonspecific pores in the inner mitochondrial membrane; the opening of this pore is highly sensitive to the inhibition of an immunosuppressive drug, cyclosporine A (CsA) (see refs. 114, 115 for review). Importantly, after some time the pore opening becomes irreversible and is accompanied by high amplitude swelling of mitochondria; such swelled mitochondria are one of the characteristics of necrotic cell death (see section 2.2). Therefore, it was suggested that prevention of MPT in some way (e.g., CsA) may protect various cells from death. Indeed, in 1993 Pastorino and other co-workers from Farber's lab found an inhibitory effect of CsA on anoxic death of cultured hepatocytes; moreover, they observed that this inhibition was not associated with the prevention of ATP depletion.[116] In subsequent studies, these researchers observed that intracellular Ca^{2+} chelation by BAPTA-AM or removal of extracellular Ca^{2+} also protected hepatocytes from anoxia or rotenone, and these treatments prevented MPT.[117] From the data, the authors concluded that it is mitochondrial pore opening rather than ATP depletion under anoxia that results in cell death. Furthermore, they did not observe the toxic effect of oligomycin which decreases ATP but preserves membrane potential.[115] In our studies on EAC cells, we found a similar toxic effect of uncoupler CCCP and oligomycin on their viability.[5]

In Lemaster's studies on hepatocytes, MPT was also observed under oxidative stress or "chemical hypoxia" (cyanide + iodoacetate).[118,119] The authors demonstrated, however, that deenergization of mitochondria with uncoupler CCCP did not result in MPT but hepatocytes rapidly died due to ATP loss.[119] In our studies on EAC cells, we also did not detect a protective effect of CsA on rotenone- or CCCP-induced necrotic death (Gabai et al, unpublished). Therefore, the significance of MPT in cell death on different cells or under different stresses requires further evaluation. At the present time, it is unclear how MPT may result in plasma membrane disruption.

There are several studies on ischemic cardiomyocytes or hearts where the protective effect of CsA on cell survival and recovery of myocardial functional properties has been observed. Initially, Nazareth et al showed that CsA retarded cardiomyocyte injury induced by substrate-free anoxia.[120] This prolongation of cardiomyocyte viability was later shown to be associated with the preservation of mitochondrial inner membrane potential (as judged by changes

in rhodamine 123 fluorescence).[121] In 1993, Griffiths and Halestrap showed that 0.2 µM of CsA in isolated perfused rat hearts subjected to 30 minute ischemia and 15 minute reperfusion restored the ATP/ADP ratio to the pre-ischemic value. More importantly, CsA-treated hearts demonstrate a significantly improved functional recovery (left ventricular developed pressure) during reperfusion.[122] In a recent study by the authors, however, no MPT was observed during ischemia, but this occurred during reperfusion, although CsA did enhance recovery of heart function.[123]

In this connection, it is important to emphasize that not only ischemia, but subsequent reperfusion as well, can severely damage myocardial cells through oxidative stress (refs. 124, 125 for recent review). Reperfusion-induced oxidative stress is associated with the generation of reactive oxygen species by an overreduced mitochondrial respiratory chain.[119,125-127] Since several studies clearly demonstrate the protective effect of CsA against oxidative stress-induced MPT and cell death,[119,128-130] the question arises whether MPT is a prerequisite for ischemic or reperfusion injuries. In any case, the data on mitochondrial pore openings may be important for an understanding of the mechanisms of cell death from ischemia or ischemia-reperfusion.

Recent studies by Kroemer's lab unexpectedly show that MPT may also be involved in triggering apoptosis (see section 2.2 and ref. 131 for review). In particular, Marchetti et al found that apoptosis of U937 tumor cells may be induced by energy deficiency and mitochondria undergoing permeability transition release of some soluble factor that is capable of inducing chromatin condensation, a hallmark of apoptosis, in isolated nuclei in vitro.[132] Furthermore, the induction of apoptosis by uncoupler is prevented by inhibitors of MPT (bongkrekic acid) or bcl-2, an anti-apoptotic protein localized in mitochondria.[133] We believe that the study of the role of mitochondria in post-ischemic apoptosis in various tissues is an exciting area of future research (see also chapter 6).

Summarizing, it seems likely that mitochondrial damage may play an important role in the death of the cells which are strictly dependent on aerobic energy production. The preservation of the mitochondrial function during the post-ischemic period is obviously necessary for the viability and functional recovery of such cells. However, mitochondrial dysfunction and injury alone are not prerequisite for cell death; cells are capable of counterbalancing the disturbed mitochondrial energy production via glycolysis.

CONCLUDING REMARKS

In this chapter, we have discussed several mechanisms by which cells can withstand ischemic stress. They include the creatine phosphate buffering system, activation of endogenous glycolysis, suppression of metabolic activity and the shutting off of unnecessary biosynthetic pathways (section 2.1). When the mechanisms of ischemic tolerance are exhausted, ATP depletion occurs and can lead to cell death. Depending on the duration and extent of ATP deprivation, cells either die by reproductive death (associated with DNA or mitotic apparatus damage) or by interphase death (apoptotic or necrotic). Resistance to ischemia is greatly varied among different cells and tissues: from several minutes to many hours. This resistance is often not associated with the preservation of energy stores. Furthermore, there is no direct correlation between cell death and the magnitude of ATP deprivation. Therefore, it is very important to find the parameter(s) which can result in the great variability in cell sensitivity to ischemia. Energy imbalance within a cell may change many of the biochemical reactions and impair various cellular structures. Most likely, some of the changes are reversible and can be effectively repaired after the restoration of cellular ATP; at the same time, some injuries are irreversible and cells with such injuries eventually die.

The earliest consequence of cellular energy deprivation is ionic imbalance, namely, loss of K^+ and the accumulation of Na^+ and H^+; frequently, Ca^{2+} also increases (Fig. 2.5). Although such disturbances of ion homeostasis are not lethal per se, and in some cases even have a protective effect (e.g., K^+ loss and H^+ accumulation during myocardial ischemia), they can also aggravate cell damage. Calcium overload may be especially toxic for ATP-depleted cells because it can activate several destructive enzymes and evoke mitochondrial damage by opening nonspecific pores in the inner mitochondrial membrane. Intracellular acidification may produce DNA breaks via activation of DNase II, thus stimulating reproductive or apoptotic death. Sodium accumulation which is accompanied by water uptake and cell swelling may accelerate necrotic death of the cells whose cortical actin skeleton is disrupted (see the next chapter). In addition, K^+ loss may impair the function of the major cellular chaperone, HSP70, by preventing its release from substrates, since this chaperone requires K^+ for activity (see chapters 1 and 3).

Therefore, such marked distinctions in cell sensitivity to ATP depletion may be partly due to the differences in their ion homeostasis. However, as we will discuss in the next chapter, the main reason for early cell death under energy deprivation is the disruption of the cellular cytoskeleton and protein aggregation, and it is distinction in these parameters that is crucial for cell tolerance to ischemic stress.

Occurring within ATP-depleted cells, ionic imbalance may be important for the induction of HSP synthesis. On the other hand, the protective functions of HSPs under energy deprivation may somehow be associated with the attenuation of ionic imbalance or its consequences and the preservation of intracellular structures. The possibilities will be discussed in more detail in chapters 4 and 7, respectively.

RECENT NEWS

An interesting paper considering molecular mechanisms of cell tolerance to hypoxia has been recently published by Hochachka and co-workers.[134] They suggest

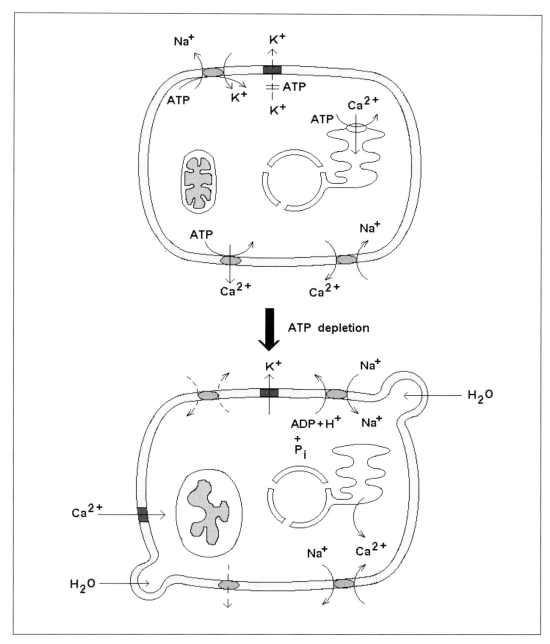

Fig. 2.5. Ionic imbalance in ATP-depleted cells. Following ATP depletion, the operation of ion-transporting ATPases ceases (dashed lines) which leads to the accumulation of Na^+ and Ca^{2+}; H^+ is increased mainly through ATP hydrolysis; loss of K^+ occurs via the ATP-regulated channel. Such ionic imbalance is accompanied by cell swelling (see also text for further explanation).

that during very early (defense) phase of hypoxia, global decline in protein synthesis and membrane permeability decreases ATP demand about 10-fold. During the second (rescue) phase, a heme protein based, signal transduction pathway is activated which turns on the genes for sustained survival at low ATP turnover rate (e.g., glycolytic enzymes) and turns off genes for less required enzymes (e.g., Krebs cycle and gluconeogenesis). The authors consider chaperones as constituents of hypoxia-tolerant cells.

Several papers demonstrated that, besides necrosis, apoptosis can substantially contribute to myocardial cell death after infarction in rats and humans (refs. 135-137). Considering the role of mitochondria in apoptosis, Skulachev[138] hypothesizes that mitochondrial permeability transition (see section 2.4) is a protective device to eliminate superoxide-producing mitochondria; when gross mitochondrial damage occures, mitochondria release apoptosis-inducing protein(s) to eliminate superoxide-producing cells and rescue normal cells.

REFERENCES

1. Skog S, Tribukait B, Sundius G. Energy metabolism and ATP turnover time during the cell cycle of Ehrlich ascites tumour cells. Exp Cell Res 1982; 141: 23-29.
2. Hochachka PW, Somero GN. Biochemical adaptation. 1984. Princeton Univ Press, NJ.
3. Leninger AL. Principles of Biochemistry. 1982. Worth Publishers Inc.
4. Chernyak BV, Dedov VN, Gabai VL. Mitochondrial ATP hydrolysis and ATP depletion in thymocytes and Ehrlich ascites carcinoma cells. FEBS Lett 1994; 337: 56-59.
5. Dedov VN, Gabai VL, Chernyak BV. Action of mitochondrial ATPase inhibitor protein (IF1) in intact rat thymocytes and Ehrlich ascites carcinoma cells. Biochemistry (Moscow) 1995; 60: 865-870.
6. Anundi I, de Groot H. Hypoxic liver cell death: critical PO_2 and dependence of viability on glycolysis. Am J Physiol 1989; 257: G58-G64.
7. Noll T, Koop A, Piper HM. Mitochondrial ATP-synthase activity in cardiomyocytes after aerobic-anaerobic metabolic transition. Am J Physiol 1992; 262: C1297-C1303.
8. Ueda H, Hashimoto T, Furuya E et al. Changes in aerobic and anaerobic ATP-synthesizing activities in hypoxic mouse brain. J Biochem 1988; 104: 81-86.
9. Moore F, Weekes J, Hardie DG. Evidence that AMP triggers phosphorylation as well as direct allosteric activation of rat liver AMP-activated protein kinase. Eur J Biochem 1991; 199: 691-697.
10. Corton JM, Gillespie JG, Hardie DG. Role of the AMP-activated protein kinase in the cellular stress response. Current Biology 1994; 4: 315-324.
11. Buttgereit F, Muller M, Papoport SM. Quantification of ATP-producing and consuming processes in quiescent pig spleen lymphocytes. Biochem Int 1991; 24: 59-67.
12. Siems WG, Schmidt H, Gruner S et al. Balancing of energy-consuming processes of K 562 cells. Cell Biochem Funct 1992; 20: 61-66.
13. Buttgereit F, Brand MD. A hierarchy of ATP-consuming processes in mammalian cells. Biochem J 1995; 312: 163-167.
14. Hochachka PW. Defense strategies against hypoxia and hypothermia. Science 1986; 231: 234-241.
15. Gabai VL, Mosin AF. Changes in energetics of ascites tumour cells upon pH decrease. Biochemistry (Moscow) 1991; 56: 1652-1660.
16. Bristow JD, Arai AE, Anselone CG et al. Response to myocardial ischemia as a regulated process. Circulation 1991; 84: 2580-2587.
17. Neill WA, Ingwall JS, Andrews E et al. Stabilization of a derangement in adenosine triphosphate metabolism during sustained, partial ischemia in the dog heart. J Am Coll Cardiol 1986; 8: 894-900.
18. Arai AE, Gauer SE, Anselone CG et al. Metabolic adaptation to a gradual reduction in myocardial blood flow. Circulation 1995; 92: 244-252.
19. Piper HM. Metabolic processes leading to myocardial cell death. Meth Achiev Exp Pathol 1988; 13: 144-180.
20. Harris DA, Das AM. Control of mitochondrial ATP synthesis in the heart. Biochem J 1991; 280: 561-573.
21. Jennings RB, Reimer KA, Steenbergen C. Effect of inhibition of the mitochondrial ATPase on net myocardial ATP in total ischemia. J Mol Cell Cardiol 1991; 23: 1383-1395.
22. Vuorinen K, Ylitalo K, Peuhkurinen K et al. Mechanisms of ischemic preconditioning in rat myocardium. Circulation 1995; 91: 2810-2818.

23. Jennings RB, Murry CE, Reimer KA. Energy metabolism in preconditioned and control myocardium: effect of total ischemia. J Mol Cell Cardiol 1991; 23: 1449-1458.

24. Kobara M, Tatsumi T, Matoba S et al. Effect of ischemic preconditioning on mitochondrial oxidative phosphorylation and high energy phosphates in rat hearts. J Mol Cell Cardiol 1996; 28: 417-428.

25. Vanderheide RS, Hill ML, Reimer KA et al. Effect of reversible ischemia on the activity of the mitochondrial ATPase: relationship to ischemic preconditioning. J Mol Cell Cardiol 1996; 28: 103-112.

26. Rouslin W, Brode CW, Guerrieri F et al. ATPase activity, IF_1 content, and proton conductivity of ESMP from control and ischemic slow and fast heart-rate hearts. J Bioenerg Biomem 1995; 27: 459-466.

27. Siesjo BK. Calcium and cell death. Magnesium 1989; 8: 223-237.

28. Kristensen SR. A critical appraisal of the association between energy charge and cell damage. Biochim Biophys Acta 1989; 1012: 272-278.

29. Gabai VL, Kabakov AE. Tumor cell resistance to energy deprivation and hyperthermia can be determined by the actin skeleton stability. Cancer Lett 1993; 70: 25-31.

30. Jennings RB, Reimer KA. Lethal myocardial injury. Am J Pathol 1981; 102: 241-255.

31. Radford IR. The level of induced DNA double-strand breakage correlates with cell killing after X-radiation. Int J Rad Biol 48: 45-54.

32. Eastman A, Barry M. The origins of DNA breaks: a consequence of DNA damage, DNA repair, or apoptosis. Cancer Invest 1992; 10: 229-240.

33. Vidair CA, Dewey W. Two distinct modes of hyperthermic cell death. Radiat Res 1988; 116: 157-171.

34. Vidair CA, Doxsey SJ, Dewey WC. Thermotolerant cells possess an enhanced capacity to repair heat-induced alterations to centrosome structure and function. J Cell Physiol 1995; 163: 194-203.

35. Rotin D, Robinson B, Tannock IF. Influence of hypoxia and an acidic environment on the metabolism and viability of cultured cells: potential implications for cell death in tumors. Cancer Res 1986; 46:2821-2826.

36. Proskuryakov SYA, Gabai VL, Mosin AF et al. DNA degradation, and changes in permeability and morphology of cells of the Ehrlich ascite tumor under anaerobic incubation without glucose. Cytology 1989; 31: 690-695.

37. Smith CA, McCarthy NJ, Williams GW. Cell recognition of apoptotic cells. In: Horton MA, ed. Blood cell biochemistry, volume 5: Macrophages and related cells. New York: Plenum Press, 1993; 393-421.

38. Trump BF, Berezesky IK. Cellular and molecular basis of toxic cell injury. In: Acosta D Jr, ed. Cardiovascular toxicology. New York: Raven Press, Ltd., 1992: 75-113.

39. Huser M, Stegemann E, Kammermeir H. Is enzyme release a sign of irreversible injury of cardiomyocytes? Life Sciences 1996; 58: 545-550.

40. Gabai VL, Makarova YuM, Gulyaev VA et al. Dependence of damage and death of ascites tumor cells under energy deprivation on cellular content of ATP and free calcium. Cytology 1990; 32: 712-719.

41. Gabai VL, Kabakov AE, Mosin AF. Association of blebbing with assembly of cytoskeletal proteins in ATP-depleted EL-4 ascites tumour cells. Tissue Cell 1992; 24: 171-177.

42. Gabai VL, Zamulaeva IV, Mosin AF et al. Resistance of Ehrlich tumor cells to apoptosis can be due to accumulation of heat shock proteins. FEBS Lett 1995; 375: 21-26.

43. Kabakov AE, Gabai VL. Heat shock-induced accumulation of 70-kDa stress protein (HSP70) can protect ATP depleted tumor cells from necrosis. Exp Cell Res 1995; 217: 15-21.

44. Steller H. Mechanisms and genes of cellular suicide. Science 1995; 267: 1445-1449.

45. Thompson CB. Apoptosis in the pathogenesis and treatment of disease. Science 1995; 267: 1456-1462.

46. Kerr JFR. Neglected opportunities in apoptosis research. Trend Cell Biol 1995; 5: 55-57.

47. Darzynkiewicz Z, Bruno S, Del Bino G et al. Features of apoptotic cells measured by flow cytometry. Cytometry 1992; 13: 795-808.

48. Dive C, Gregory CD, Phipps DJ et al. Analysis and discrimination of necrosis and apoptosis (programmed cell death) by multiparameter

flow cytometry. Biochim Biophys Acta 1992: 275-285.

49. Columbano A. Cell death: current difficulties in discrimination apoptosis from necrosis in the contex of pathological processes in vivo. J Cell Biochem 1995; 58: 181-190.

50. Gobe GC, Axelsen RA, Searle JW. Cellular events in experimental unilateral ischemic renal atrophy and regeneration after collateral nephrectomy. Lab Invest 1990; 63: 770-779.

51. Shumer M, Colombel MC, Sawczuk IS et al. Morphological, biochemical, and molecular evidence of apoptosis during the reperfusion phase after brief periods of renal ischemia. Am J Pathol 1992; 140: 831-838.

52. Okamoto M, Matsumoto M, Ohtsuki T et al. Internucleosomal DNA cleavage involved in ischemia-induced neuronal death. Biochem Biophys Res Comm 1993; 196: 1356-1362.

53. Gabai VL, Kabakov AE. Rise in heat-shock protein level confers tolerance to energy deprivation. FEBS Lett 1993; 327: 247-250.

54. Gabai VL, Kabakov AE, Makarova YuM et al. DNA fragmentation in Ehrlich ascites carcinoma under exposures causing cytoskeletal protein aggregation. Biochemistry (Moscow) 1994; 59: 399-404.

55. Richter C, Schweizer M, Cossarizza A et al. Control of apoptosis by the cellular ATP level. FEBS Lett 1996; 378: 107-110.

56. Pierce GN, Czubryt MP. The contribution of ionic imbalance to ischemia/reperfusion-induced injury. J Mol Cell Cardiol 1995; 27: 53-63.

57. Gasbarrini A, Borle AB, Farghali et al. Effect of anoxia on intracellular ATP, Na^+, Ca^{2+}, Mg^{2+}, and cytotoxicity in rat hepatocytes. J Biol Chem 1992; 267: 6654-6663.

58. Penttila A, Trump B. Extracellular acidosis protect Ehrlich ascites tumor cells and rat renal cortex against anoxia injury. Science 1974; 185: 277-278.

59. Jones RL, Miller JC, Hagler HK et al. Association between inhibition of arachidonic acid release and prevention of calcium loading during ATP depletion in cultured rat cardiac myocytes. Am J Pathology 1989; 135: 541-556.

60. Anderson SE, Dickinson CZ, Liu H et al. Effects of Na-K-2Cl cotransport inhibition on myocardial Na and Ca during ischemia and reperfusion. Am J Physiol (Cell Physiol) 1996; 39: C608-C618.

61. Terzic A, Jahangir A, Kurachi Y. Cardiac ATP-sensitive K+ channels: Regulation by intracellular nucleotides and K+ channel-opening drugs. Am J Physiol (Cell Physiol) 1995; 38: C525-C545.

62. Strasser R, Vogt A, Schaper W. Ischemic preconditioning. Experimental results and clinical studies. Zeit Kardiol 1996; 85: 79-89.

63. Jurkowitz-Alexander MS, Altschuld RA, Hohl CM et al. Cell swelling, blebbing, and death are dependent on ATP depletion and independent of calcium during chemical hypoxia in a glial cell line (ROC-1). J Neurochem 1992; 59: 344-352.

64. Nakagawa Y, Rivera V, Larner AC. A role for the Na/K-ATPase in the control of human c-fos and c-jun transcription. J Biol Chem 1992; 267: 8785-8788.

65. Talbot N, Tagliaferri P, Yanagihara K et al. A pH-dependent differential cytotoxicity of ouabain for human cells transformed by certain oncogenes. Oncogene 1988; 3: 23-26.

66. Masaki N, Thomas AP, Hoek JB et al. Intracellular acidosis protects cultured hepatocytes from the toxic consequences of a loss of mitochondrial energization. Arch Biochem Biophys 1989; 272: 152-161.

67. Bond JM, Chacon E, Herman B et al. Intracellular pH and Ca^{2+} homeostasis in the pH paradox of reperfusion injury to neonatal rat cardiac myocytes. Am J Physiol 1993; 265: C129-C137.

68. Gores GJ, Nieminen A-L, Wray BE et al. Intracellular pH during "chemical hypoxia" in cultured rat hepatocytes. J Clin Invest 1989; 83: 386-396.

69. Cross HR, Opie LH, Radda GK et al. Is a high glycogen content beneficial or detrimental to the ischemic rat heart? A controversy resolved. Circ Res 1996; 78: 482-491.

70. Cross HR, Clarke K, Opie LH et al. Is lactate-induced myocardial ischemic injury mediated by decreased pH or increased intracellular lactate? J Mol Cell Cardiol 1995; 27: 1369-1381.

71. Vanderheide RS, Delyani JA, Jennings RB et al. Reducing lactate accumulation does not attenuate lethal ischemic injury in iso-

lated perfused rat hearts. Am J Physiol (Heart Circ Physiol) 1996; 39: H38-H44.

72. Gores GJ, Nieminen A-L, Fleishman KE et al. Extracellular acidosis delays onset of cell death in ATP-depleted hepatocytes. Am J Physiol 1988; 255: C315-C322.

73. Currin RT, Gores GJ, Thurman RG et al. Protection by acidotic pH against anoxic cell killing in perfused rat liver: evidence for a pH paradox. FASEB J 1991; 5: 207-210.

74. Barry MA, Eastman A. Endonuclease activation during apoptosis: the role of cytosolic Ca^{2+} and pH. Biochem Biophys Res Comm 1992; 186: 782-789.

75. Barry MA, Eastman A. Identification of deoxyribonuclease II as an endonuclease involved in apoptosis. Arch Biochem Biophys 1993; 300: 440-450.

76. Newell LJ, Tannock IF. Reduction of intracellular pH as a possible mechanism for killing cells in acidic regions of solid tumors: effects of carbonylcyanide-3-chlorophenylhydrazone. Cancer Res 1989; 49: 4477-4482.

77. Tannock IF, Rotin D. Acid pH in tumors and its potential for therapeutic exploitation. Cancer Res 1989; 49: 4373-4384.

78. Clapham DE. Calcium signaling. Cell 1995; 80: 259-268.

79. Carafoli E. Intracellular calcium homeostasis. Ann Rev Biochem 1987; 56: 395-433.

80. Meldolesi J, Madeddu L, Pozzan T. Intracellular Ca^{2+} storage organelles in nonmuscle cell: heterogeneity and functional assignment. Biochim Biophys Acta 1990; 1055: 130-140.

81. Orrenius S, Mc Conkey DJ, Bellomo G et al. Role of Ca^{2+} in toxic cell killing. Trends Pharm Sci 1989; 10: 281-285.

82. Aw TY, Andersson BS, Jones DP. Suppression of mitochondrial respiratory function after short-term anoxia. Am J Physiol 1987; 252: C362-C368.

83. Andersson BS, Aw TY, Jones DP. Mitochondrial transmembrane potential and pH gradient during anoxia. Am J Physiol 1987; 252: C349-C355.

84. Brecht M, Brecht C, De Groot H. Late steady increase in cytosolic Ca+ preceding hypoxic injury in hepatocytes. Biochem J 1992; 283: 399-402.

85. Gasbarrini A, Borle AB, Farghali H et al. Fasting enhances the effects of anoxia on ATP, Ca^{2+} and cell injury in isolated rat hepatocytes. Biochim Biophys Acta 1993; 1178: 9-19.

86. Lemasters JJ, DiGuiseppi J, Nieminen A-L et al. Blebbing, free Ca^{2+} and mitochondrial membrane potential preceding cell death in hepatocytes. Nature (London) 1987; 325: 78-81.

87. Nieminen A-L, Gores GJ, Wray BE et al. Calcium dependence of bleb formation and cell death in hepatocytes. Cell Calcium 1989; 9: 237-246.

88. Artalejo AR, Garcia-Sancho J. Mobilization of intracellular calcium by extracellular ATP and by calcium ionophores in the Ehrlich ascites tumor cell. Biochem Biophys Acta 1988; 941: 48-54.

89. Boynton AL, Cooney RV, Hill TD et al. Extracellular ATP mobilizes intracellular Ca^{2+} in T51B rat liver epithelial cells: a study involving single cell measurements. Exp Cell Res 1989; 181: 245-255.

90. Nagelkerke JF, Dogterom P, DE Bont HJGM et al. Prolonged high intracellular free calcium concentrations induced by ATP are not immediately cytotoxic in isolated rat hepatocytes. Biochem J 1989; 262: 347-353.

91. Weinberg JM, Davis JA, Roeser NF et al. Role of increased cytosolic free calcium in the pathogenesis of rabbit proximal tubule cell and protection by glycine or acidosis. J Clin Invest 1991; 87: 581-590.

92. Carrol JM, Toral-Barza L, Gibson G. Cytosolic free calcium and gene expression during chemical hypoxia. J Neurochem 1992; 59: 1836-1843.

93. Johnson ME, Gores GJ, Uhl B et al. Cytosolic free calcium and cell death during metabolic inhibition in a neuronal cell line. J Neurosci 1994; 14: 4040-4049.

94. Carafoli E. Calcium pump of the plasma membrane. Physiol Rev 1991; 71: 129-153.

95. Eis JS, Hons BSc, Nayler WG. Calcium gain during post-ischemic reperfusion. The effect of 2,4-dinitrophenol. Am J Pathol 1988; 131: 137-145.

96. Buja LM, Hagler HK, Willerson JT. Altered calcium homeostasis in the pathogen-

esis of myocardial ischemic and hypoxic injury. Cell Calcium 1988; 9: 205-217.

97. Wang SY, Claque JR, Langer GA. Increase in calcium leak channel activity by metabolic inhibition or hydrogen peroxide in rat ventricular myocytes and its inhibition by polycation. J Mol Cell Cardiol 1995; 27: 211-222.

98. Allshire A, Piper HM, Cuthbertson KSR et al. Cytosolic free Ca^{2+} in single rat heart cells during anoxia and reoxygenation. Biochem J 1987; 244: 381-385.

99. Bowers KS, Allshire AP, Cobbold PH. Bioluminescent measurement in single cardiomyocytes of sudden cytosolic ATP depletion coincident with rigor. J Mol Cell Cardiol 1992; 24: 213-218.

100. Bowers KS, Allshire AP, Cobbold PH. Continuous measurement of cytoplasmic ATP in single cardiomyocytes during simulation of the "oxygen paradox". Cardiovasc Res 1993; 27: 1836-1839.

101. Astma DE, Bastiaanse EML, Jerzewski A et al. Role of calcium-activated neutral protease (calpain) in cell death in cultured neonatal rat cardiomyocytes during metabolic inhibition. Circ Res 1995; 76: 1071-1078.

102. Duncan CJ. Biochemical events associated with rapid cellular damage during the oxygen- and calcium-paradoxes of the mammalian heart. Experientia 1990; 46: 41-48.

103. Siegmunt B, Schluter KD, Piper HM. Calcium and the oxygen paradox. Cardiovasc Res 1993; 27: 1778-1783

104. Daniels S, Duncan CJ. Dual activation of the damage system that causes the release of cytosolic proteins in the perfused rat heart. Cell Physiol Biochem 1995; 5: 330-343.

105. Yoshida K, Inui M, Harada K et al. Reperfusion of rat heart after brief ischemia induces proteolysis of calspectin (nonerythroid spectrin or fodrin) by calpain. Circ Res 1995; 77: 603-610.

106. Laiho KU, Berezesky IK, Trump BF. The role of calcium in cell injury. Surv Synth Path Res 1983; 2: 170-183.

107. Mosin AF, Gabai VL, Makarova YuM et al. Damage and interphase death of the Ehrlich ascite carcinoma tumor cells, being at different growth phases, due to energy depri-

vation and heat shock. Cytology 1994; 36: 384-391.

108. Kamendulis LM, Corcoran GB. Independence and additivity of cultured hepatocyte killing by Ca^{2+} overload and ATP depletion. Toxicol Lett 1992; 63: 277-287

109. Nishida T, Inoue T, Kamiike W et al. Involvement of Ca^{2+} release and activation of phospholipase A_2 in mitochondrial dysfunction during anoxia. J Biochem 1989; 106: 533-538.

110. Gabai BL. Inhibition of uncoupled respiration in tumor cells. A possible role of mitochondrial Ca^{2+} efflux. FEBS 1993; 329: 67-71.

111. Henke W, Nickel E. The contribution of adenine nucleotide loss to ischemia-induced impairment of rat kidney cortex mitochondria. Biochim Biophys Acta 1992; 1098: 233-239.

112. Sciamanna MA, Zinkel J, Fabi AY et al. Ischemic injury to rat forebrain mitochondria and cellular calcium homeostasis. Biochim Biophys Acta 1992; 1134:223-232.

113. Sciamanna MA, Lee CP. Ischemia/reperfusion-induced injury of forebrain mitochondria and protection by ascorbate. Arch Biochem Biophys 1993; 305: 215-224.

114. Gunter KK, Gunter TE. Transport of calcium by mitochondria. J Bioenerg Biomem 1994; 26: 471-485.

115. Bernardi P, Broekemeier KM, Pfeiffer DR. Recent progress on regulation of the mitochondrial permeability transition pore: a cyclosporin-sensitive pore in the inner mitochondrial membrane. J Bioenerg Biomem 1994; 26: 509-517.

116. Pastorino JG, Snyder JW, Serroni A ey al. Cyclosporin and carnitine prevent the anoxic death of cultured hepatocytes by inhibiting the mitochondrial permeability transition. J Biol Chem 1993; 268: 13791-13798.

117. Pastorino JG, Snyder JW, Hoek JB et al. Ca^{2+} depletion prevents anoxic death of hepatocytes by inhibiting mitochondrial permeability transition. Am J Physiol (Cell Physiol) 1995; 37: C676-C685.

118. Zahrebelski G, Nieminen A-L, Alghoul K et al. Progression of subcellular changes during chemical hypoxia to cultured rat hepatocytes: a laser scanning confocal mi-

croscopic study. Hepatology 1995; 21: 1361-1372.

119. Nieminen A-L, Saylor AK, Tesfai SA et al. Contribution of the mitochondrial permeability transition to lethal injury after exposure of hepatocytes to t-butylhydroperoxide. Biochem J 1995; 307: 99-106.

120. Nazareth W, Yafei N, Crompton M. Inhibition of anoxia-induced injury in heart myocytes by cyclosporin A. J Mol Cell Cardiol 1991; 23: 1351-1354.

121. Crompton M, McGuinness O, Nazareth W. The involvement of cyclosporin A binding proteins in regulating and uncoupling mitochondrial energy transduction. Biochim et Biophys Acta 1992; 1101: 214-217.

122. Griffiths EJ, Halestrap AP. Protection by cyclosporin A of ischemia/reperfusion-induced damage in isolated rat hearts. J Mol Cell Cardiol 1993; 25: 1461-1469.

123. Griffiths EJ, Halestrap AP. Mitochondrial non-specific pores remain closed during cardiac ischaemia, but open upon reperfusion. Biochem J 1995; 307: 93-96.

124. Vandenhoek TL, Shao ZH, Li CQ et al. Reperfusion injury in cardiac myocytes after simulated ischemia. Am J Physiol (Heart Circ Physiol) 1996; 39: H1334-H1341.

125. Steare SE, Yellon DM. The potential for endogenous myocardial antioxidants to protect the mycocardium against ischaemia-reperfusion injury: refreshing the parts exogenous antioxdants cannot reach? J Mol Cell Cardiol 1995; 27: 65-74.

126. Caraceni P, Ryu HS, Vanthiel DH et al. Source of oxygen free radicals produced by rat hepatocytes during postanoxic reoxygenation. Biochim Biophys Acta 1995; 1268: 249-254.

127. Smith DR, Stone D, Darleyusmar VM. Stimulation of mitochondrial oxygen consumption in cardiomyocytes after hypoxia-reoxygenation. Free Radical Res 1996; 24: 159-166.

128. Kass GEN, Juedes MJ, Orrenius S. Cyclosporin A protects hepatocytes against prooxidant-induced cell killing. A study on the role of mitochondrial Ca2+ cycling in

cytotoxicity. Biochem Pharmacol 1992; 44: 1995-2003.

129. Takeyama N. Matsuo N, Tanaka T. Oxidative damage to mitochondria is mediated by the Ca2+-dependent innermembrane permeability transition. Biochem J 1993; 294: 719-725.

130. Gogvadze V, Richter C. Cyclosporine A protects mitochondria in an in vitro model of hypoxia/reperfusion injury. FEBS 1993; 333: 334-338.

131. Kroemer G, Petit P, Zamzani N et al. The biochemistry of programmed cell death. FASEB J 1995; 9: 1277-1287.

132. Marchetti P, Susin SA, Decaudin D et al. Apoptosis-associated derangement of mitochondrial function in cells lacking mitochondrial DNA. Cancer Res 1996; 2033-2038.

133. Zamzani N, Susin SA, Marchetti P et al. Mitochondrial control of nuclear apoptosis. J Exp Med 1996; 183: 1533-1544.

134. Hochahka PW, Buck LT, Doll CJ, Land SC. Unifying theory of hypoxia tolerance: Molecular/metabolic defense and rescue mechanisms for surviving oxygen lack. Proc Natl Acad Sci USA 1996; 93: 9493-9498.

135. Kajstura J, Cheng W, Reiss K et al. Apoptotic and necrotic myocyte cell death are independent contributing variables of infarct size in rats. Lab Invest 1996; 74: 86-107.

136. Bardales RH, Hailey LS, Xie SS et al. In situ apoptosis assay for the detection of early acute myocardial infarction. Am J Path 1996; 149: 821-829.

137. Olivetti G, Quaini F, Sala R et al. Acute myocardial infarction in humans is associated with activation of programmed myocyte cell death in the surviving portion of the heart. J Mol Cell Cardiol 1996; 28: 2005-2016.

138. Skulachev VP. Why are mitochondria involved in apoptosis? Permeability transition pores and apoptosis as selective mechanisms to eliminate superoxide-producing mitochondria and cell. FEBS Lett 1996; 397: 7-10.

CHAPTER 3

"PROTEOTOXICITY" OF ATP DEPLETION: DISRUPTION OF THE CYTOSKELETON, PROTEIN AGGREGATION AND INVOLVEMENT OF MOLECULAR CHAPERONES

The majority of cell-stressing exposures affect proteins. Typical "proteotoxic" exposures such as heating, ultra-violet irradiation, decrease in pH and treatment with oxidants or heavy metals are able to damage various proteins both in vitro (in a solution) and in vivo (in a cell). Lack of ATP in artificially prepared protein solutions does not seem to be very critical for the stability of the soluted proteins; in contrast, the depletion of intracellular ATP destroys the cytoskeletal framework and evokes aggregation (or insolubility) of many cellular proteins including HSPs. Although this proteotoxic component is only one of the many harmful effects of ATP depletion on mammalian cells, we consider it the most crucial event coupling the mechanisms of cell injury and cell adaptation under metabolic (or ischemic) stress. That is why this phenomenon is considered substantially here.

3.1. THE CYTOSKELETON UNDER ATP DEPRIVATION

The cell architecture of higher eukaryotes is mainly defined by the endoskeleton comprising three different constituents: (1) actin microfilaments; (2) microtubules and (3) intermediate filaments. Being a very dynamic system, the cytoskeleton has many points of regulation and its state depends on a variety of factors including the level of nucleotide tri- and diphosphates in the cytosol. Apparently, among cellular structures, the cytoskeleton is one of the most sensitive to a drop in the ATP level. In this section, we differentially analyze perturbations of the cytoskeletal framework which can occur in ATP-depleted mammalian cells and may be significant for cell viability or functioning.

3.1.1. ACTIN, ACTIN-ASSOCIATED PROTEINS AND MICROFILAMENTS

Actin, a major cytoskeletal component, exists as a monomeric form (the 43 kDa globular protein designated G-actin) and as double-helical filamentous structures (termed F-actin). In myocytes, F-actin forms so-called thin filaments which when interacting with thick myosin filaments immediately perform a contractile function (for review see ref. 1). Nonmuscle F-actin is a major insoluble element of 8 nm microfilaments, which are mainly present in a cell as a fine cortical network under the plasma membrane and also in long, thick bundles (stress fibers). In contrast, G-actin is a major soluble protein of the cytosol. The actin polymerization and depolymerization (i.e., G$\leftrightarrow$F-actin transitions) play an important role in cell motility and are regulated by special actin-binding proteins (reviewed in refs. 2,3).

Likewise, actin dynamics appear to be dependent on the ATP/ADP ratio. Actin can bind both ATP and ADP, which is necessary for rapid polymerization; however, F-actin consisting of ATP-actin monomers is more stable than that of ADP-actin.[4-6] Two actin-binding proteins, profilin and thymosin β_4, cooperatively regulate the actin dynamics which are dependent on the ATP/ADP ratio.[7-9] This mechanism suggests that actin polymerization increases when the ATP/ADP proportion decreases within living cells.[8] On the contrary, myosin, another actin-binding protein, was shown in vitro to destabilize F-actin in the absence of ATP.[10] Recent data reveal that sequestering actin and severing the activities of gelsolin are also dependent on the ATP concentration.[11] The real behavior of actin and microfilaments in ATP-depleted mammalian cells is reviewed below.

It has been known for a long time that the depletion of cellular ATP leads to a dramatic rearrangement of the actin skeleton. As far back as 1980, Bershadsky and co-workers reported that inhibitors of energy metabolism (2,4-dinitrophenol, azide) cause a gradual disorganization of actin microfilament bundles (stress fibers) in mouse embryo fibroblasts.[12] Later rapid and reversible disintegration of stress fibers has been demonstrated in various cultured adherent cells subjected to ATP depletion.[13-21] Hinshaw and co-workers have determined the average threshold level of ATP, near 40% of the control level, at which microfilament disruption is initiated in bovine pulmonary artery endothelial cells (Fig. 3.1).[20]

However, the mechanism of microfilament disorganization resulting from ATP depletion varies in the different cases described. For instance, some researchers asserted that the actin skeleton disassembly in ATP-depleted cells is due to the fragmentation of F-actin without its depolymerization (i.e., without an increase in G-actin following the destruction of microfilaments)[12-14,19] and that it might be carried out by actin-severing protein(s),[14] possibly gelsolin.[11,19] At the same time, Wang showed that the ATP decrease in mouse 3T3 fibroblasts alters the distribution of α-actinin and vinculin within adhesion plaque and causes reversible formation of actin-myosin rigor complexes that might disturb the normal microfilament architecture.[15] Moreover, actin aggregation is induced by cellular energy deprivation. Thus, a decrease in the cytosolic G-actin pool and a formation of abnormal side-to-side actin aggregates were observed in the P388D$_1$ macrophage-like cells losing ATP.[18] Bundled aggregates and clumps of F-actin have been found in energy-depleted artery endothelial cells.[17,19] It was also documented that as the cytoskeletal framework of ATP-depleted kidney epithelial cells disintegrates, polymerized actin aggregates in the cytoplasm.[21] Finally, the decrease in cellular ATP induces polymerization of actin, since increases in the F-actin pools were found in both endothelial[20] and epithelial cells[23] exposed to ATP depletion. It seems likely that the stress-induced actin polymerization is mediated, on the one hand, by profilin and thymosin β_4 which can stimulate the addition of monomers to actin barbed-end under the decrease in

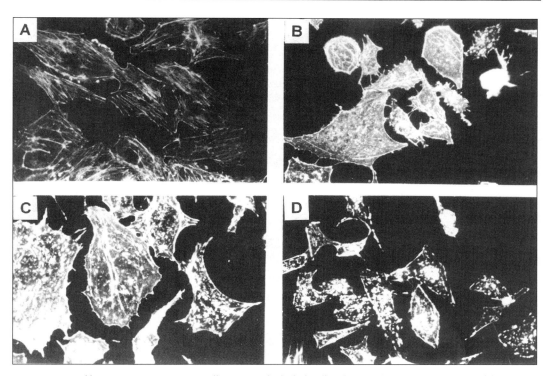

Fig. 3.1. Microfilament organization in adherent endothelial cells 2 hours after exposure to variable degrees of ATP depletion. (A) Control unstressed cells. Staining with rhodamine phalloidin shows typical patterns of intact actin filament bundles (stress fibers). (B) Cells exposed to 650 nM oligomycin (an inhibitor of $H^+ATPase$) plus 5% of control glucose concentration (275 μM) for 2 hours. Note the thinning of stress fibers and bright peripheral staining. (C) Cells exposed to 650 nM oligomycin plus 1% of control glucose concentration (55 μM) for 2 hours. Note the apparent fragmentation of some stress fibers within cells. (D) Cells exposed to 650 nM oligomycin without added glucose for 2 hours. Note the fragmentation/disruption of nearly all stress fibers and their replacement by a chaotic distribution of microfilament fragments and aggregates. (x400 magnification). Reproduced with permission from Hinshaw et al Am J Physiol 1993; 264: C1171-C1179.

ATP/ADP ratio;[8] and, on the other hand, by an appearance of new actin barbed-ends following the fragmentation of microfilaments in ATP-depleted cells.

Some cancer cells (e.g., cells of ascites tumors) do not form tight cell-cell and cell-substrate contacts, and hence they have a spherical shape. Cytoskeletal proteins of such cells are weakly structured: there are no stress fibers across the cytoplasm and F-actin constitutes only a cortical microfilament network, whereas their G/F-actin ratio is higher than that in spread cells. Nevertheless, according to our own observations, actin and certain actin-associated proteins of murine ascites tumor cells are very sensitive to the drop in the cellular ATP level. To estimate the stress-induced

redistribution of cytoskeletal proteins within nonadherent cells we used a simple but effective method, namely the extraction of a cell suspension with a nonionic detergent, Triton X-100 (Fig. 3.2).[24-30] In our studies, we showed that ATP depletion sharply decreases the Triton-extractable actin in EL-4 thymoma,[24-26,28] Ehrlich carcinoma[25-30] and P_3O_1 myeloma cells (see Table 3.1).[29] DNase I-binding assays demonstrated that the insolubility is accompanied by a strong depletion of the cytosolic G-actin pool in the energy-deprived tumor cells (Gabai et al, unpublished data). Since cytochalasin B, an inhibitor of actin polymerization, does not suppress the stress-induced actin insolubility,[25] the phenomenon is due to a misassembling (perhaps

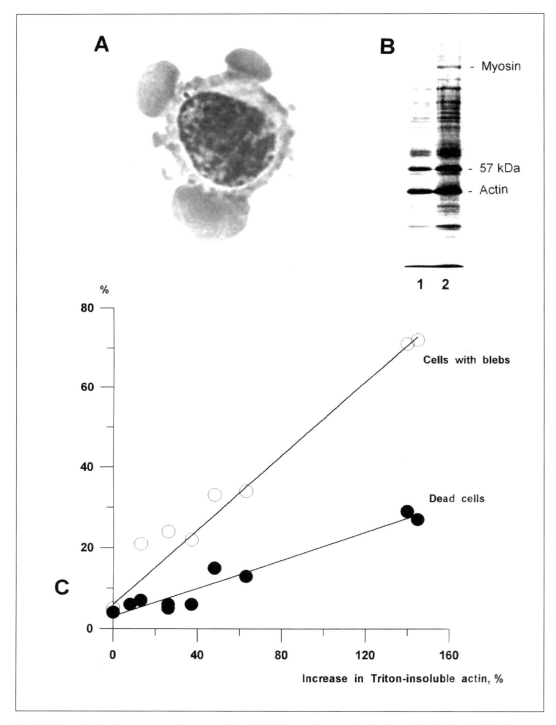

Fig. 3.2. Correlation between actin aggregation, blebbing and necrotic death in ATP-depleted Ehrlich ascites tumor cells. (A) Blebs formed on the cell surface 1 hour after exposure to 2 μM rotenone in the absence of glucose. (B) SDS gel electrophoresis of Triton X-100-insoluble fractions elicited from equal number (10^6) of unstressed (1) and ATP-depleted (2) cells. The stress-induced insolubility of actin and myosin is clearly seen here. (C) Graphs demonstrating direct correlations of actin aggregation with the intensity of blebbing and death-rate in the rotenone-treated cells (see for details refs. 25, 26).

Table 3.1. Protein aggregation and death in ATP-depleted ascites tumor cells

Cell Line	Aggregated (Triton-Insoluble) Protein (% of Control) After 1 Hour of ATP Depletion[1]					Dead Cells (%)[2]	
	Total	Actin	Myosin	p57[3]	HSP70	1 Hour	2 Hours
EL-4	253*	230*	274*	253*	180*	22	61
EAC	148*	163*	400*	163*	212*	13	17
P_3O_1	189*	216*	n.d.[4]	n.d.	200*	22	59

[1] EL-4 thymoma, Ehrlich ascites carcinoma (EAC) and P3O1 myeloma cells were treated with 2 µM rotenone (a respiratory inhibitor) in a glucose-free medium. One hour after the beginning of the ATP-depleting treatment, the cells were fractionated with Triton X-100 and relative increases in the insoluble proteins were detected by immunoblotting (see refs. 25-29 for details).

[2] The number of dead cells was determined by trypan blue exclusive test one hour and two hours after the ATP-depleting treatment.

[3] The 57 kDa Triton-insoluble protein (p57) appears to be a major component of intermediate filament, vimentin (see Fig. 3.2).

[4] n.d., not determined

*The data presented are means of three to six independent measurements; significant difference from control value (P <0.05).

side-to-side aggregation) of preexisting soluble monomers and short fragments rather than as additional actin polymerization. However, both the former[18,21] and latter mechanisms[20,23] have been described above for ATP-depleted spread cells. Besides, actin, vinculin[24] and myosin[26,28] also lost Triton solubility in ATP-depleted ascites tumor cells (see Fig. 3.2), suggesting a relocalization of vinculin and a formation of actin-myosin rigor complexes. The findings support Wang's observations made previously on spread cells.[15] We should note that, for at least Ehrlich carcinoma cells, the described effects were reversible and the resolubilization of actin and myosin up to their basal levels did occur following cellular ATP replenishment. In regard to a possible role of myosin in the destruction of F-actin structures under ATP deficiency,[10,15] we also discussed it as a factor disorganizing the cortical actin skeleton in ATP-deprived ascites cells.[26] At the same time, we cannot exclude the possibility that the stress-induced association of myosin with actin microfilaments stabilizes them to maintain the integrity of the actin framework during ATP deple-

tion.[28] This controversial question needs further examination.

Taking into account that the in vivo level of cellular ATP usually falls during an ischemic insult (see chapter 2), injury of the cytoskeleton in ischemic tissues should also be described here. A variety of cytoskeletal lesions including the disturbance of actin structures in ischemic and anoxic cardiomyocytes are excellently reviewed by Ganote and Armstrong.[1] The disturbance is manifested in rigor actin-myosin contracture and detachment of actin from the fascia adherens junction within the intercalated discs of the anoxic heart (Fig. 3.3).[1,31] Moreover, anoxia, ischemia and metabolic inhibition affect the distribution of vinculin[32-34] and α-actinin[35] in cardiomyocytes, which corresponds with the data obtained for nonmuscle ATP-depleted cells.[15,24] In vitro ischemic-like metabolic stress causes disorganization of the myofibrils in cultured rat ventricular myocytes and of the expression of contractile protein genes necessary for post-stress recovery of the contractile function.[36]

Experimental ischemic injury of the kidney leads to rapid F-actin loss in proxi-

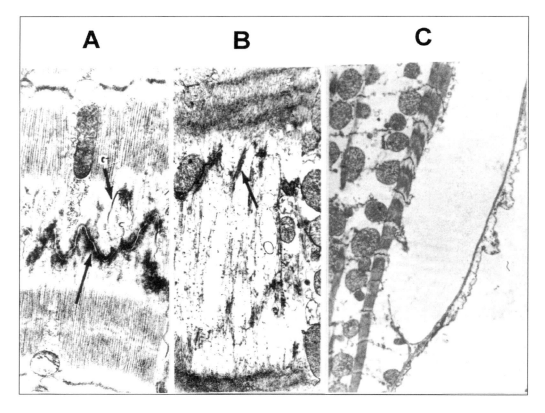

Fig. 3.3. Breakdown of the plasma membrane-actin filament contacts in anoxic cardiomyocytes. Intercalated disc region of a control myocyte (A) as compared to the intercalated disc region in a rat heart damaged by anoxic perfusion (B). The fascia adherens junction (arrow) and a gap junction (G) of the control myocyte are indicated. The intercalated disc of the anoxic heart shows severe damage, with apparent detachments of actin from the fascia adherens junction (arrow). Reprinted from Ganote et al, Cardiovasc Res 1993; 27: 1387-1403 with kind permission of Elsevier Science - NL, SaraBurgerhartstraat 25, 1055 RV Amsterdam, The Netherlands. (C) Electron micrograph of a myocyte from an anoxic perfused heart. There is a large subsarcolemmal bleb which has formed with lifting and detachment of the plasma membrane from the sarcomeric attachments at lateral costamere junctions. Reproduced with permission from Ganote et al, Am J Pathol 1987; 129: 327-344.

mal tubule cells that is actually connected with ATP depletion.[37,38] Analogous disruption of proximal tubule microfilaments has been observed under tissue ATP depletion caused by maleic acid.[39]

In summarizing all the above data, we conclude that four parallel and likely interrelated processes affect the actin skeleton in ATP-depleted mammalian cells: (1) redistribution of certain actin-associated proteins, particularly myosin, vinculin and α-actinin; (2) fragmentation of F-actin and disintegration of microfilaments; (3) aggregation and bundling of actin; and (4) polymerization of actin. Evidently, the two

latter events directly lead to a depletion of the cytosolic G-actin pool.

3.1.2. MICROTUBULES AND INTERMEDIATE FILAMENTS

Microtubular and intermediate filamentous structures form 20 and 12 nm diameter fibers, respectively. In addition to a framework function, they play an important role in cell polarization, organelle traffic, anchorage of integral membrane proteins, etc. The microtubular network consists mainly of polymerized tubulin; in vivo both assembly and disassembly of microtubules require ATP.[14] The data describing the effects of ATP depletion on microtubules are rather contradictory. For instance, Maro and Bornens[40] reported that blocking the mitochondrial ATP generation by an uncoupler of oxidative phosphorylation results in the disruption of microtubules in HeLa cells, but later no effect of ATP depletion on microtubule integrity was observed in fibroblasts[14] and the kidney epithelial cell line.[21] Meanwhile, a reduced intensity of antitubulin immunofluorescence was revealed by flow cytometric analysis in ischemic and metabolically inhibited cardiomyocytes[33] which implies a redistribution or degradation of the antigen. Similar to this, decreases in the immunoreactivity of tubulin and of a microtubule-associated protein (MAP) are characteristic signs of ischemic injury in neurons (reviewed in chapter 6).

Major components of intermediate filaments are tissue-specific proteins such as vimentin, desmin, keratins, etc. In contrast to microtubules, the filamentous structures assemble and disassemble in an ATP-independent manner. Nevertheless, they seem to be sensitive to the depletion of cellular ATP: collapse and aggregation of intermediate filaments occur under uncoupling of oxidative phosphorylation in HeLa cells[40] and under inhibition of respiration by rotenone in murine ascites tumor cells (see Fig. 3.2, Table 3.1 and refs. 26, 27). Detachments of desmin filaments from both the sarcolemma and the outer nuclear membrane were found in anoxic myocytes.[1,35] Interestingly, the intermediate filament network serves also as the attachment for mitochondria. Disruption of the organelle-cytoskeleton contacts is one of the consequences of anoxic myocardial injury.[1,35]

3.1.3. THE CYTOSKELETON, THE PLASMA MEMBRANE AND CELL DEATH DUE TO ATP LOSS

Assembled cytoskeletal structures are involved in practically all vital processes and functions of mammalian cells. It is well-known that specific drugs selectively damaging the cytoskeleton (cytochalasins, phalloidin, colcemid, etc.) are highly toxic. At the same time, the cytoskeleton seems

to be one of the primary targets for the injurious actions of heating,[25,41,42] oxidative stress[18,43] and energy depletion (see two previous subsections). It is evident that irreversible damage of the cytoskeleton as a result of stress can kill a cell. However, the relationship between cytoskeleton dysfunction and cell death upon ATP deprivation requires further consideration.

Despite the numerous well-documented facts on cytoskeleton destruction in ATP-depleted cells, molecular mechanisms of this phenomenon are still poorly understood. Apparently, many different intracellular events including ionic imbalance, changes in the ATP/ADP/AMP ratio, shifts in protein kinase/phosphatase machinery, etc. participate in the stress-induced rearrangement of cytoskeletal structures. As was described above, the disintegration of the cytoskeleton in energy-deprived cells is reversible and the normal cytoskeletal framework is again restored after ATP replenishment.

Nevertheless, we believe that even a transient disturbance of the cytoskeleton-plasma membrane interactions can promote necrotic death in metabolically stressed mammalian cells.[44] Similar speculations have been also put forth by Steenbergen et al[45] and Gannote and Armstrong.[1] These ideas are based on close correlations between the cytoskeleton perturbations, damage of the plasma membrane and necrosis in ATP-depleted (ischemic or anoxic) cells and tissues. Actually, bleb formation on the cell surface (blebbing) is an early sign of ATP depletion in many cell types including ascites tumor cells (Fig. 3.2)[24-26] and cardiomyocytes (Fig. 3.3).[1,35] Such blebbing appears to be a direct consequence of the cortical cytoskeleton detachment from the plasma membrane. Most of the integral plasma membrane proteins sever their cytoskeletal linkages in ATP-depleted cultured epithelial cells.[46] Loss of both actin and desmin filament contacts with sarcolemmal blebs occurs in anoxic cardiomyocytes (Fig. 3.3).[1,35] A relationship of blebbing with changes in actin and its accessory proteins (vinculin, α-actinin, ankyrin, fodrin) involved in microfilament-plasma membrane anchoring has been discussed many times for various mammalian cells undergoing ATP depletion or ischemia.[1,24-26,32,35,44-46] Direct correlations between actin aggregation (insolubility), cell surface blebbing and necrotic death have been established for ATP-depleted ascites tumor cells (see Fig. 3.2 and Table 3.1).[24-26] Hence, actin skeleton stability has been suggested as determinative of cell resistance to energy deprivation.[25,44]

It is easy to accept the following sequence of events: decrease in the ATP/ADP ratio · disruption of microfilament clamps and cortical cytoskeleton collapse → destabilization and blebbing of the plasmalemma → perforation of blebs or their detachment from cell body, the last being fatal for the stressed cell (Fig. 3.4). As a matter of fact, molecular processes involved in the mechanisms of necrotic death under energy depletion are rather intricate. For example, it is not known precisely how the detachment of cortical cytoskeleton from the plasma membrane occurs. Formation of rigor actin-myosin complexes (or myofibril contracture) evoked by ATP drop[15,26,31] may be conducive to the detachment of actin filaments from their membrane anchorages. Likewise, stress-activated proteases and lipases may play a role; these enzymes can break protein-protein and protein-lipid coupling between the cytoskeleton and plasmalemma constituents. Calpain, a calcium-dependent neutral protease, which selectively digests proteins of the cytoskeleton-plasmalemma contacts (vinculin, talin, ankyrin, fodrin) is actually activated in the metabolically-inhibited cardiomyocytes.[47] It is in agreement with the described disappearance of vinculin from costamers of cardiomyocytes undergoing oxygen starvation.[1,32-34] An increase in phospholipase activity has also been detected in anoxic cardiomyocytes[31] and ATP-depleted astroglial cells.[48]

Blebbing in ATP-deprived cells as well as in the cytoskeleton disassembly appears to be reversible if the plasma membrane breakdown does not occur in the blebs.[25] Perforation of the plasmalemma is un-

doubtedly lethal for a cell, since it results in efflux of cytosolic proteins, i.e., necrotic death. The molecular mechanism of the formation and expansion of blebs and their perforation or detachment upon ATP depletion remains unknown. It has been proposed that elevated free Ca^{2+} concentrations in the cytosol may initiate bleb formation and fatally activate submembrane proteases and/or phospholipases which destroy the plasmalemma.[49,50] However, no significant increase in Ca^{2+} has been found in ATP-depleted ascites tumor cells[24,26] or within blebs of metabolically inhibited hepatocytes.[51] Moreover, death of the ATP-depleted cells does not correlate with the increase in calpain activity[47] and is not prevented by inhibitors of phospholipases.[48] Therefore, the proposed involvement of the calcium-activated degradative enzymes in cell death under metabolic stress is not yet evident (see chapter 2 for review).

We hypothesize that increased ionic permeability of the bleb envelope disturbs ionic flows and provokes water influx into blebs, thus creating a local osmotic stress; such progressing intrableb tension in the absence of cytoskeletal clamps can eventually break the cell membrane (Fig. 3.4).[44,45] The role of water entry as an injurious factor under ATP depletion was confirmed by the fact that swelling, blebbing and death of ATP-deprived ROC-1 hybrid glial cells are suppressed by polyethylene glycol, a water-sequestering agent.[52] What causes change in ionic flows across the plasmalemma of ATP-depleted cells? On the one hand, a severe decrease in the ATP/ADP ratio in the submembrane layer of the cytoplasm alone may be conducive to a dysfunction of the ATP-dependent ion channels through deficiency of substrate (ATP) and/ or their inhibition by increased ADP. On the other hand, intact cytoskeletal structures seem to be necessary for anchorage and normal functioning of membrane ion pumps. Hence, the disorganization of the cortical cytoskeleton resulting from metabolic stress may be critical for cellular ionic balance. Indeed, actin can regulate the hydrolytic activity of Na^+,K^+-ATPase.[53]

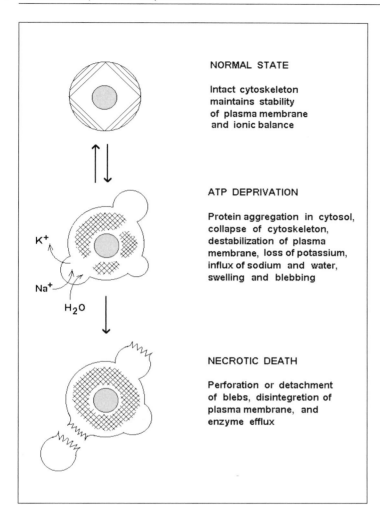

NORMAL STATE

Intact cytoskeleton
maintains stability
of plasma membrane
and ionic balance

ATP DEPRIVATION

Protein aggregation in cytosol,
collapse of cytoskeleton,
destabilization of plasma
membrane, loss of potassium,
influx of sodium and water,
swelling and blebbing

K^+

Na^+

H_2O

NECROTIC DEATH

Perforation or detachment
of blebs, disintegretion of
plasma membrane, and
enzyme efflux

Fig. 3.4. The probable sequence of events leading to rapid cell death through energy deprivation (for details see the text and ref. 44).

Additionally, the depletion of cellular ATP causes cell surface redistribution of Na^+,K^+-ATPase and its dissociation from the membrane complex with the actin framework.[23] The Na^+/K^+ imbalance for lack of the actin skeleton-mediated control of ion transport has been suggested as one of the consequences of ATP depletion in kidney ischemia.[38]

To summarize, one can easily imagine that in cells devoid of ATP abnormal K^+ efflux and Na^+ influx take place, the latter stimulating water entry and elevating interior osmotic tension. In chapter 2, we wrote that the transient dysfunction of Na^+,K^+-ATPase per se is not lethal for a cell; this conclusion was based on the absence of cytotoxic effects under the enzyme inhibition by ouabain. In contrast to ATP

depletion, the ouabain treatment does not affect the actin skeleton although it does enhance the cytotoxic effect of cytochalasin B, a microfilament-disrupting drug (Mosina, Gabai and Kabakov, unpublished observations). According to our theory, neither sodium imbalance nor cytoskeleton collapse alone have the same effect on cells as they do when they occur in combination. Together they cause local osmotic stress, cell swelling and bleb expansion, and ultimately necrotic death (Fig. 3.4).[44] Thus, in ATP-deprived cells, destruction of the actin framework, dysfunction of ion pumps and blebbing of the plasma membrane appear to be tightly interrelated processes which play a major role in the mechanism of necrotic death of ATP-depleted mammalian cells.

Irreversible damage to microtubules and/or intermediate filaments should kill a cell as well but this is considered reproductive or apoptotic death (see chapters 2 and 8). At present, nothing is known about the injuries to the non-actin endoskeleton which might be lethal for metabolically-stressed cells.

3.2. PROTEIN AGGREGATION IN ENERGY-DEPRIVED CELLS

In vivo protein aggregation usually implies an abnormal intermolecular binding which is due to hydrophobic interactions between immature unfolded or misfolded proteins; this process is often accompanied by the insolubility of the proteins and the loss of their functional activity.[54-59] However, we believe that the aggregation is more than simply an undesirable hydrophobic protein-protein liaison. In our opinion, the insoluble aggregates are also formed as a result of the misassembly of supramolecular protein structures like filamentous bundles or multioligomeric complexes. Additionally, the particular case of protein aggregation is a provoked association of cytosolic proteins with such low-soluble intracellular structures as the cytoskeleton, chromatin and nuclear matrix. Abnormal clustering of integral membrane proteins can also be considered aggregation. The major ways protein aggregation occurs in ATP-depleted cells are represented in Figure 3.5. Besides hydrophobic bonds, other types of intermolecular interactions appear to participate in the formation of protein aggregates.

What is the evidence that cellular proteins aggregate under ATP deficiency? Since aggregated proteins are less extractable from cells by a nonionic detergent, Triton X-100, the aggregation in ATP-depleted cells was established on an increase in total Triton X-100-insoluble cellular protein as well as on the insolubility of certain proteins.[24-30,59-61] Moreover, the proteins aggregated for lack of ATP have been directly visualized in preparations of the stressed cells by using various microscopic techniques.[17-19,21,62,63] Some protein aggregates or misassembled oligomeric

complexes do not sediment in Triton X-100 but they can be isolated from ATP-depleted cells by special physicochemical procedures (fractionation, gel chromatography, centrifugation in gradient of density, etc.). Below we explore the nature and biological significance of protein aggregation occurring in ATP-depleted mammalian cells.

3.2.1. WHY IS DEPLETION OF CELLULAR ATP PROTEOTOXIC?

Apart from the cytoskeleton disruption, the proteotoxic effect of ATP depletion is shown in the aggregation (or insolubility) of cellular proteins,[24-30,59-63] misassembling of cytoskeletal structures[17-19,21] and the inactivation of transfected reporter enzymes.[36,59] Likewise, the necessity of myosin light chain gene expression for the post-stress recovery of contractile function in metabolically-inhibited myocytes[36] suggests the substitution of contractile protein(s) damaged as a result of energy deprivation. Some intracellular proteins becoming insoluble during ATP depletion undergo rapid degradation when the ATP level is restored (see chapter 8), which implies their irreversible damage. Finally, both the insolubility of HSP70 in mammalian cells lacking ATP[26,28,59,60] and ubiquitination of cellular proteins in ischemic tissues[64-67] suggest the in vivo proteotoxicity of ATP depletion.

For all that, the question arises: what accounts for the proteotoxic effects of ATP-depleting metabolic stresses? Clearly one consequence of prolonged ATP deficiency as ionic imbalance, especially acidification of the cytoplasm, is the proteotoxic factor, since it can affect some proteins whose native conformations are sensitive to changes in the ionic environment. In the case of ischemia/reperfusion, the postanoxic reoxygenation is accompanied by a burst of free radicals (oxidative stress) that damage many cellular proteins despite ATP replenishment (see chapters 2, 4 and 5). However, a powerful aggregation of cellular proteins already takes place in the early period of ATP deprivation before significant shifts in the cytosolic pH and without an increase in the free radical level.[24-29]

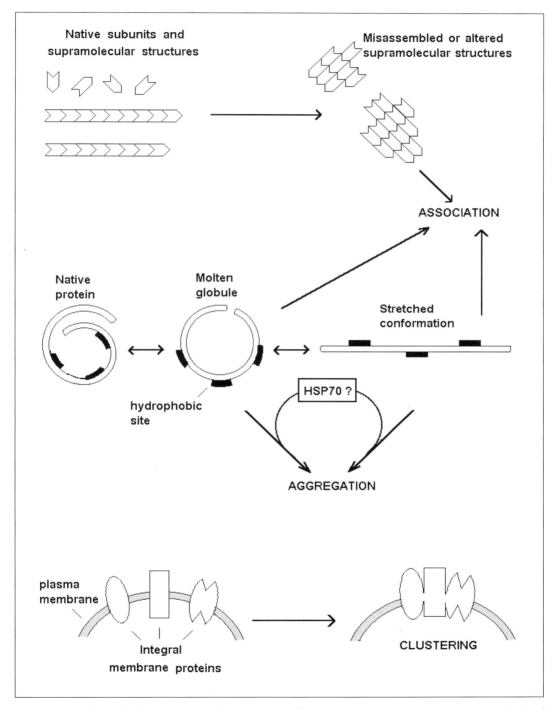

Fig. 3.5. Hypothetical scheme showing the major ways for protein aggregation to occur in ATP-depleted mammalian cells (see the text for details).

This early aggregation occurs immediately as the result of a lack of ATP. This is confirmed by an impetuous dissolution of the stress-insolubilized protein matter when cell-extracting Triton buffer contains ATP.[26,28,60] In vivo restoration of the ATP level in Ehrlich tumor cells, HeLa cells and LZ1 murine cells is also accompanied by a rapid disaggregation of the insoluble protein excess (Kabakov, Gabai, Nguyen and Bensaude, unpublished).

To solve this intriguing problem, we postulate the following: the high energy phosphate homeostasis per se is obligatory for the prevention of protein aggregation in a cell. Firstly, ATP, ADP and AMP are substrates and allosteric effectors of many enzymes and nucleotide-binding proteins including kinases and phosphatases, constituents of the chromatin, cyto- and nucleoskeleton together with regulators of assembly and functions of these cytostructures. Severe shifts in the ATP/ADP/AMP ratio occurring in an energy-deprived cell can destabilize conformations of some ATP-binding proteins and alter topology or surface charges of the subunits and supramolecular constructions, thus being conducive to aggregation (see Fig. 3.5).

Secondly, the normal ATP level is known to be necessary for natural turnover of all cellular proteins from their synthesis, folding and monomer assembly to their terminal degradation by lysosomes and proteasomes. It seems quite likely that the expected block of these ATP-dependent processes in energy-depleted cells leads to a misassembly of the oligomeric complexes and an accumulation of aberrant proteins. Such failure in protein turnover can actually provide material for the aggregation upon metabolic stress. The possible dysfunction of chaperones in an ATP-depleted cell represents a great interest in the context of the protein aggregation, since it is the ATP-dependent chaperone machinery that maintains polypeptides in a soluble state and regulates their folding, transport across cell membranes, and assembly in oligomeric complexes. Molecular chaperones appear to be the main device allowing a diversity of proteins to exist in a cell at extremely high concentrations without aggregation. At least three major chaperones, HSP60, HSP70 and HSP100, should hydrolyze ATP to fold polypeptide chains and to disintegrate the aggregated protein matter (see chapter 1). Hence, the expected breakdown in chaperone machinery due to ATP deficiency may also be considered one of the proteotoxic factors of ATP-depleting metabolic stress. Furthermore, it has been proposed that in the absence of ATP, HSP70 itself can promote the aggregation of certain proteins (see the next subsections and ref. 59).

3.2.2. WHICH OF THE CELLULAR PROTEINS AGGREGATES FOR LACK OF ATP AND WHY?

There are certain proteins which constitute a special class of cellular "ATP sensors." The cytosolic proteins fail to be Triton-extractable in response to the depletion of cellular ATP. This suggests their aggregation or multioligomerization, or association with insoluble cytostructures. Below we characterize the major members of this class of proteins.

First of all, some ATPases and nucleotide-binding proteins including myosin, actin, HSP70, etc. are such ATP sensors, since they are rapidly made insoluble in case of ATP deprivation.[24-30,59,60] Among the major proteins of mammalian cells, myosin is one of the most sensitive indicators of cellular ATP depletion. This is demonstrated in myofibril contracture in metabolically inhibited myocytes[1,31] and in the rapid sequestration of the cytosolic myosin pool in nonmuscle cells losing ATP (Fig. 3.2).[15,26,28] In ascites tumor cells, a decrease in the myosin solubility is detected in the early stage of energy deprivation when the ATP level drops to 70%, while further ATP depletion leads to total insolubility of myosin.[28] These events may be considered as a specific aggregation of contractile proteins due to the stress-provoked formation of rigor actin-myosin complexes and the assembly of abnormal filamentous-like myosin structures.

Actin is also an apparent ATP sensor which becomes insoluble within ATP-deprived cells (see Fig. 3.2 and refs. 17-19, 21, 24-30, 44). In the actin aggregation, some actin-associated proteins, in particular myosin[26,28] and vinculin can be involved.[24] The mechanism of actin aggregation under ATP depletion remains unclear. Actin is an ATP- and ADP-binding protein. Therefore, the stress-induced drop in the ATP/ADP ratio should increase the content of ADP-actin complexes. Moreover, ATP depletion stimulates actin phosphorylation at tyrosine-53, which is localized to a site critical for actin polymerization.[68] Perhaps taken together, the tyrosine phosphorylation and the ATP-ADP exchange alter the topology and the charge of actin monomer globules, and in turn promote the side-to-side misassembling and bundling of G- and F-actin in ATP-depleted cells (Fig. 3.5). Such factors as ionic imbalance, dysfunction of myosin ATPase, detachment of microfilaments from the plasma membrane and F-actin fragmentation may be additional stimuli to actin aggregation in cells lacking ATP.

Notably, in response to the depletion of cellular ATP, the intermediate filaments collapse, forming tight clumps around the nucleus.[40] This perinuclear clumping may also be considered a particular case of cytoskeletal protein aggregation provoked by ATP depletion. A major component of the intermediate filament network, the 57 kDa protein (probably vimentin), is markedly insoluble in energy-deprived ascites tumor cells (see Fig. 3.2 and Table 3.1). However, the mechanism of ATP loss-induced aggregation of intermediate filaments remains to be established.

Another group of ATP sensors appears to consist of certain cytosolic enzymes which lose the detergent solubility and aggregate as a result of ATP depletion. Bensaude and colleagues, who have developed an excellent method of exploring protein stability in living mammalian cells,[54-59] gained general recognition for finding that cellular ATP decrease results in enzyme aggregation. This method is based on differential analyses of the solubility and activity of intracellular reporter enzymes during cellular stress and recovery. In particular, the interferon-induced 68 kDa dsRNA-dependent protein kinase (p68) and the two foreign transfected enzymes, firefly luciferase and $E.$ $coli$ β-galactosidase, become insoluble in ATP-depleted mammalian cells (Fig. 3.6).[59] Why do the reporter enzymes, being soluble in unstressed cells, aggregate under ATP deprivation? One may hypothesize that both ATP-consuming enzymes, p68 and firefly luciferase, missing their substrate (ATP), become less stable and therefore aggregate in the first place during ATP depletion. However, β-galactosidase is not an ATP-binding protein at all, though it clearly loses its solubility in ATP-depleted cells (see Fig. 3.6), while two ATP-consuming enzymes, the 42 kDa MAP kinase and the 100 kDa (2'-5') A_n synthetase, remain soluble despite the cellular ATP loss.[59] Taking into consideration that the same three enzymes, namely, p68, luciferase and β-galactosidase, were made insoluble during heat shock,[54-58] Bensaude has suggested that primarily thermolabile cytosolic proteins aggregate upon ATP depletion.[59]

Indeed, molecules of thermolabile proteins have as a rule, nontight folding without cross-links between domains, which defines high conformational mobility and elevated risk of unfolding. Thus, the accessibility of hydrophobic sites may determine aggregation. Heating causes the rapid denaturation of a thermolabile protein and may occur in a two-step mechanism. Firstly, a globular protein is transformed from the native conformation (with masked hydrophobic sites) into a so-called "molten globule" which is still compact but hydrophobic sites are exposed. The molten globule then becomes completely unfolded up to the "stretched" conformation, the former and the latter abnormal structures being able to form aggregates (Fig. 3.5).

How can this scheme be realized in vivo at nonstressing temperature? In regard to thermolabile proteins, conformation

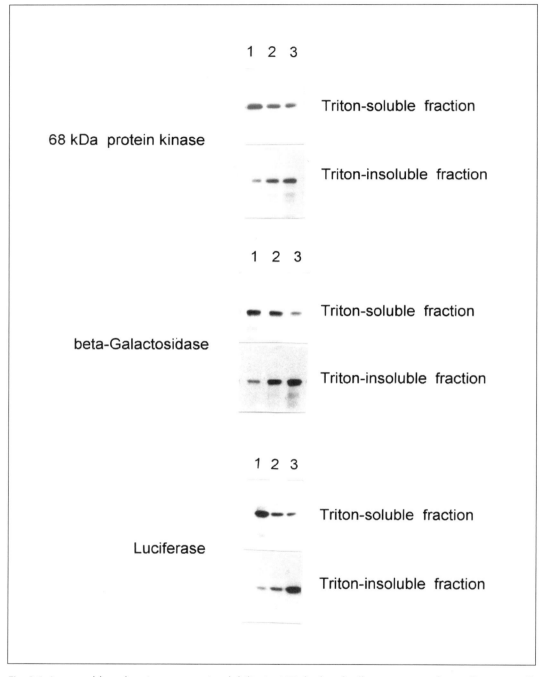

Fig. 3.6. *Immunoblots showing enzyme insolubility in ATP-depleted adherent mammalian cells. HeLa cells expressing the 68 kDa dsRNA-dependent protein kinase and murine NIH 3T3 fibroblasts overexpressing E. coli β-galactosidase and P. pyralis luciferase[59] were incubated in normal growth medium (1) or in a glucose-free medium with 20 μM CCCP (an uncoupler) for a half hour (2) and 1 hour (3). Equal numbers of the unstressed (1) and ATP-depleted (2,3) cells were extracted with 1% Triton X-100 and the enzymes were detected in aliquots of Triton-soluble and insoluble fractions by immunoblotting as described.[59] The stress-induced disappearance of the antigen bands from the soluble fractions and their accumulation in the insoluble pellets are clearly seen here. Obtained by Kabakov, Nguyen and Bensaude (not previously published).*

transitions to the molten globule or even to the stretched state can happen stochastically, without any heating. Nevertheless, such spontaneous unfolding does not lead to aggregation in unstressed cells. An apparent absence of the spontaneous aggregation is due to a very low concentration of the randomly unfolded proteins and relatively high level of molecular chaperones. The latter ones recognize both molten globule and stretched conformations, and bind and refold such unfolded proteins in an K^+, ATP-dependent cyclic mechanism (see subsection 3.3.1. below and chapter 1). Since HSP70 is recognized as a major eukaryotic chaperone and ATP is essential for the release of HSP70 from its protein substrate,[69-71] Bensaude has proposed that in vivo randomly unfolded proteins are immediately trapped by HSP70. However, in case of ATP deficiency, their refolding is sharply delayed and the chaperone-protein complexes do not dissociate.[59] In fact, this may shift conformational equilibrium toward the unfolded state, thus provoking protein aggregation (see Fig. 3.5 and subsection 3.3.1.). Therefore, in ATP-depleted cells, HSP70 may be directly involved in protein aggregation and, unexpectedly, it may play a role as a proteotoxic factor.

Bensaude's reasoning seems correct and is supported by the finding that HSP70 insolubility is induced in various mammalian cells by ATP deprivation.[26,28,59,60] But is HSP70 the sole "trapper" of spontaneously unfolded proteins during energy starvation? We do not think so and assume the same is true for the cell constituents with low solubility; namely, the cytoskeleton, chromatin and nuclear matrix are altered as a result of ATP depletion. Being in the native state a natural anchorage for immobilization of many soluble proteins and multienzyme complexes, these supramolecular cytostructures can be damaged during stress, which increases the risk of undesirable interprotein associations. The stress-induced rearrangements of the cyto- and nucleoskeletons may lead to changes in their configuration and charge, which renders them more adhesive for cytosolic proteins and especially for unfolded or misfolded proteins. In the cytoplasm of ATP-depleted cells, the disorganized cytoskeleton and its abundant debris may catch those soluble proteins which turn out in nonstable conformations, while the affected nuclear matrix and chromatin may expose the adhesive sites promoting aggregation in the nucleus (see Fig. 3.5). For example, the sequestration of vinculin from the soluble pool in ATP-depleted EL-4 thymoma cells[24] may occur because of the vinculin-actin co-aggregation in the cytoplasm; the insolubility of U1 small nuclear ribonucleoprotein (U1RNP) in the same cells devoid of ATP (Kabakov and Gabai, unpublished) may be due to the U1RNP-nuclear matrix association enhanced as a result of the stress. Our suggestion that energy deprivation "chains" some cytosolic proteins to preexisting insoluble cytostructures is strongly supported by recent work of Barbato et al.[72] They documented the association of two cytosolic proteins, glyceraldehyde-3-phosphate dehydrogenase (GAPDH) and αB-crystallin, with myofibrils in ischemic rat hearts. Most likely, acidosis accompanying ischemic insult is the real cause of the αB-crystallin-myofibril association (see also subsection 3.3.2. below), whereas the binding of GAPDH to myofibrils was directly due to ischemic ATP depletion.[72]

An important question is how intracellular proteins localized to different compartments respond to stress. To compare the stability of cytoplasmic and nuclear proteins in living cells, Michels and Bensaude along with their colleagues constructed plasmid vectors for overexpression of reporter enzymes with programmed subcellular localization. The transfected firefly luciferase carrying nuclear localization sequence (NLS) was targeted to the nucleus and exhibited lower thermostability in vivo than the enzyme with cytoplasmic localization.[73] In collaboration with Bensaude and Nguyen, one of us (AEK) employed an analogous model for the study of protein aggregation in the cytoplasm and the nucleus of ATP-deprived mammalian cells.

The transiently overexpressed NLS luciferase and NLS β-galactosidase aggregated during ATP depletion more rapidly than the same enzymes localized in the cytoplasm. Intriguingly, at least two molecular chaperones, HSP70 and HSP27, are also made insoluble within nuclei as a result of ATP loss (see subsections 3.3.1. and 3.3.2. below). Moreover, HSP70 is able to bind to NLS.[74] Summarizing, we suggest that both the insoluble cytostructures and molecular chaperones are involved in intracellular protein aggregation during ATP depletion.

In addition, aged and defective (incomplete, unfolded, misfolded, oxidized, etc. proteins aggregate in ATP-depleted cells. Being the natural waste of macromolecular metabolism or products of gene mutations, such aberrant proteins permanently appear in any cell and are usually quickly digested in lysosomes or via the ubiquitin-proteasome system. Since both translocation of proteins into lysosomes and ubiquitin-mediated proteolysis require ATP, its depletion should in fact block the degradation of abnormal proteins and thus prolong their sojourn in a cell.[75] In addition, ATP deficiency may inactivate the 26S protease complexes which degrade ubiquitinated proteins.[76] Although a share of this pool in total cellular protein seems insignificant, the preexisting defective protein molecules may play a role in priming the aggregation cascade in ATP-deprived cells. Apparently, cell recovery following the ATP level restoration includes degradation of the preexisting abnormal proteins and the proteins irreversibly damaged because of ATP deprivation.

Protein aggregation due to cellular ATP deprivation is certainly not limited by the insolubility of the above listed proteins. We believe that, besides various cytoplasmic and nuclear proteins, some proteins situated in mitochondria and the endoplasmic reticulum are able to aggregate as the result of a local decrease in ATP within these compartments. Moreover, the stress-induced aggregation of cellular proteins occurs not only in a water phase but also in the lipid phase of cell membranes

(see Fig. 3.5). The latter assertion is based on documented clustering of sarcolemmal transmembrane proteins in ischemic (or ATP-depleted) rat cardiomyocytes.[62,63] This phenomenon was mainly explored using freeze fracture electron microscopy and hence its molecular nature is poorly understood. The integral proteins aggregating within the sarcolemma under ischemia have not yet been identified; nevertheless, their clustering occurs without the loosening of their cytoskeletal linkages.[63] Causes of such protein aggregation and its role in the mechanism of ischemic injury remain enigmatic.

3.2.3. PROTEIN AGGREGATION AND THE VIABILITY OF ATP-DEPLETED CELLS

In characterizing protein aggregation in injured cells,[26] we have introduced the novel term "aggregability" which reflects the missing Triton X-100 solubility of cellular proteins in response to either stress. In our opinion, it is the basic parameter determining the stress-sensitivity of eukaryotic cells. In ATP-depleted murine ascites tumor cells, the extent of total protein aggregation (excess in Triton-insoluble cellular protein) is correlated with the plasma membrane damage (blebbing) and necrotic death (see Fig. 3.2 and Table 3.1).[24-30,61] It seems evident that protein aggregation can be lethal for a cell but it is unknown what percentage of aggregated intracellular protein kills the ATP-depleted cell. Nevertheless, at some level the aggregation due to ATP decrease is reversible (see chapter 7). Possibly, this characteristic is similar for various types of mammalian cells, and the diversity in cell survival during metabolic stress[25,26,29] may signify only different rates of achievement of the lethal aggregability in different cell lines. The fact that proteins aggregate with different intensities in cells of various lines equally devoid of ATP appears to be influenced by many factors including different expression and activity of molecular chaperones.[26,44]

Although we do not know to date how much protein matter must aggregate to kill the ATP-depleted cell, we hypothesize which cellular proteins play a fatal role. Indeed, high aggregability of the cytoskeleton constituents, especially actin and actin-associated proteins, might explain the correlation between protein aggregation, blebbing and cell death upon metabolic stress. At least blebbing and necrotic death of energy-deprived cells may occur by the above suggested mechanism (see Fig. 3.4) with the addition of actin aggregation as a factor destabilizing the cortical microfilaments.[26,44] This assumption is supported by excellent correlations between the actin insolubility (aggregation), blebbing and necrosis (Fig. 3.2 and Table 3.1).[24-30] Myosin,[15,26,28] α-actinin,[15] vinculin,[15,24] α-crystallin (see ref. 72 and subsection 3.3.2. below) and likely other actin-associated proteins co-aggregate with G- and/or F-actin upon ATP depletion. We do not rule the possibility that accessory cytoskeletal proteins initiate the actin aggregation cascade which ends in necrotic death.

If protein aggregation in ATP-depleted cells does not evoke rapid necrosis, it can have more delayed consequences. Aggregation not exceeding the "lethal" level triggers the heat shock transcription response following the ATP-level restoration (see chapters 1 and 4). Alternatively, a cell avoids necrosis during transient energy starvation, but later the stress-induced aggregation of some proteins in the cytoplasm and within the plasma membrane and compartments results in apoptosis or reproductive death (see chapters 2, 8). We discussed a possible relationship between apoptosis induced in Ehrlich tumor cells by transient ATP depletion and aggregation of cytoskeletal proteins.[77,78] In particular, such a characteristic manifestation of energy deprivation as actin aggregation may be conducive to the post-stress apoptosis. On the one hand, G-actin inhibits DNase I, an enzyme probably involved in apoptotic degradation of nuclear DNA.[79,80] On the other hand, the stress-provoked actin aggregation is usually accompanied by strong depletion of the G-actin pool (see subsection 3.1.1. above). One might speculate that the sequestration of the endogenous inhibitor (G-actin) results in activation of DNase I and genome digestion (reviewed in chapter 8).

In the context of apoptosis, the above mentioned ischemia-induced aggregation of the plasma membrane integral proteins (Fig. 3.5)[62,63] is fascinating because analogous clustering of certain transmembrane receptors may fatally affect the signal transduction pathway and activate apoptotic effectors. However, this issue requires further research. We believe that the level of protein aggregation leading to apoptotic or reproductive death may be intermediate between the severe aggregation causing necrosis and the lower (sublethal) aggregation initiating the heat shock response and the development of cell tolerance. Thus, protein aggregation is a crucial event in the mechanism of cellular reaction to metabolic stress.

3.3. HOW DO HEAT SHOCK PROTEINS RESPOND TO ATP DEPLETION?

The constitutively expressed HSP70, small HSPs, and ubiquitin are able to detect the lack of ATP in vivo, which is manifested in their insolubility resulting from oligomerization, relocalization and targeting to protein substrates (see below). The involvement of molecular chaperones and ubiquitin in cellular reactions to metabolic stress clearly points to the proteotoxicity of ATP depletion. Although we have no data regarding HSP100 and HSP60, we consider both of these ATP-consuming chaperones as potential ATP sensors. Not all mammalian HSPs react to a decrease in the level of ATP by losing their detergent solubility. For example, HSP90 remains Triton-extractable in ATP-depleted murine ascites tumor cells[28] and HeLa cells (Kabakov and Bensaude, unpublished). At present, we can only speculate about the behavior of other HSPs upon ATP deprivation.

3.3.1. HEAT SHOCK PROTEIN 70

It has been shown in various mammalian cells that the constitutive HSP70 is translocated into Triton-insoluble cellular fractions in response to ATP depletion (Table 3.1).[26,28,59,60] This effect is reversible and the resolubilization takes place following the restoration of cellular ATP (Fig. 3.7) or the addition of exogenous ATP in cell-extracting Triton buffer.[26,28,60] According to the concepts of "negative regulation" and "proteotoxicity," the free (free of protein targets) HSP70 seems to be a suppressor of heat shock gene expression (see chapter 1), and as a result the sequestration of soluble HSP70 in cells lacking ATP may trigger the heat shock transcription response under metabolic stress. A link between HSP70, ATP depletion and activation of heat shock genes is considered in detail in chapter 4 but here we wish to analyze the HSP70 insolubility within the context of protein aggregation and proteotoxicity.

The insolubility of HSP70 can be due to its multioligomerization (autoaggregation) and/or its association with other proteins (co-aggregation). Is HSP70 oligomerized under ATP depletion? Using native polyacrylamide gel electrophoresis Kim et al[81] revealed that in vitro oligomerization of constitutive HSP70 occurs in an ATP-dependent manner. Briefly, HSP70 purified

from Chinese hamster ovary cells can run in native electrophoresis as three protein bands of molecular mass 70, 153, and 200 kDa corresponding to monomer, dimer and trimer, respectively. The relative amount of the oligomers is dependent upon the ATP concentration. Following the hydrolysis of residual ATP in the sample, HSP70 forms dimers and trimers which dissociate into monomers in the presence of ATP (1-10 mM). Only monomeric HSP70 was found in the complex with ATP, while both monomers and oligomers were able to associate with heat-denatured proteins.[81] The in vitro studies suggest that the equilibrium between mono- and oligomeric forms of the chaperone can be shifted in ATP-depleted cells toward the oligomerization. Interestingly, the constitutive (HSP73) and stress-inducible (HSP72) forms are co-precipitated from ATP-depleted lysates of HeLa cells as heterodimers sensitive to ATP,[81] which also represents the chaperone dimerization as a likely event in ATP-deprived mammalian cells. Although the stress-induced dimerization and even trimerization of HSP70 does not yet prove its insolubility but indicates this phenomenon may play an important role in ATP-mediated regulation

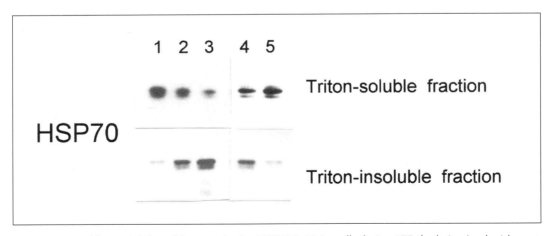

Fig. 3.7. Reversible insolubility of the constitutive HSP70 in HeLa cells during ATP depletion/replenishment. Triton-soluble and insoluble fractions were obtained from equal numbers of the control cells (1) and from the cells exposed to 20 µM CCCP (an uncoupler) in the absence of glucose for a half hour (2) and 1 hour (3). In parallel experiments, the ATP-depleted cells were washed and placed in normal growth medium; then their fractionation was performed at a half hour (4) and at 1 hour (5) of the recovery period. Aliquots of the fractions were analyzed by immunoblotting (see ref. 59).

of the chaperone function and heat shock transcription response.

Thus, co-aggregation of HSP70 with other proteins, rather than oligomerization, explains the chaperone insolubility in cells devoid of ATP. The question then becomes: what are the other proteins? Before the stress, a significant part of constitutive HSP70 is in transient complexes with its natural substrates such as unfolded proteins (immature polypeptides and proteins transported across membranes), clathrin, certain constituents of the cytoskeleton, p53, topoisomerase I, calnexin, HSF, etc. (see chapters 1 and 4). Some of the targets, namely unfolded polypeptides,[83-85] clathrin,[85] p53,[85] actin,[86] cap32/34 actin-binding protein,[87] and keratin intermediate filaments[88] interact with HSP70 in an ATP-dependent manner in which ATP binding but not ATP hydrolysis is necessary to free the chaperone.[70] Therefore, in vivo ATP depletion can inhibit the release of HSP70 from the preexisting substrates and shift the dynamic equilibrium toward complex formation. Beckmann et al[84] have indeed shown the inhibition of HSP70 release from the complexes with maturing proteins in ATP-depleted HeLa cells. In reality, the mechanism of inhibition appears to be more complicated than a simple lack of ATP in HSP70 environment. The problem lies in that even sustained metabolic stress or acute ischemic insult lead to severe, but not total, depletion of cellular ATP. By blocking both glycolytic and mitochondrial ATP generation, the level of ATP can be decreased to 2-7% of the initial (3-5 mM);[24-30] hence the ATP concentration inside such metabolically stressed cells should be a minimum of 60 µM (see chapter 2). Meanwhile, the nucleotide-HSP70 dissociation constant was determined for ATP as 9.5 ± 3.9 µM[69] and K_m for ATP of HSP70 ATPase in the absence and presence of a protein substrate were 1.37 and 1.44 µM, respectively.[85,89] The data suggest ATP availability for HSP70 upon blocking energy metabolism. Thus, ATP deficiency per se is not the principal in vivo factor connecting HSP70 to its

protein substrates under metabolic stress. It seems more likely that the chaperone release is inhibited by ADP, whose concentration in the cytosol increases during ATP depletion up to 1 mM (reviewed in chapter 2). The nucleotide-HSP70 dissociation constant for ADP is 1.6 ± 0.5 µM,[69] which is ~ 6 times lower than that for ATP (see above). As a result, HSP70 binds ADP considerably stronger than ATP. In a cell-free system, ADP does inhibit the ATP-induced dissociation of the HSP70-protein complex,[69] while binding of HSP70 to ADP is tighter than to ATP,[69,85,89] and the HSP70 bound to ADP has the higher affinity toward protein substrates than the free chaperone.[69,70] Taken together these facts allow us to conclude that in the case of ATP-depleting metabolic stress, an increase in the levels of ADP, rather than the lack of ATP, preserves the preexisting HSP70-protein substrate complexes from the ATP-induced dissociation.

Besides the ADP elevation, depletion of cellular ATP is usually accompanied by K^+ efflux (see subsection 3.1.3 and chapter 2). This stress-induced loss of intracellular potassium may be critical for HSP70-substrate interactions as well as the decrease in the ATP/ADP ratio. The HSP70 ATPase is known to be K^+-dependent. Moreover, the ATP-induced HSP70-protein complex dissociation requires K^+.[70] Hence, the inhibition of the chaperone release observed in ATP-depleted cells may be partially due to lack of cytosolic potassium. Certainly, this speculation requires experimental substantiation.

Nascent polypeptides, being the most abundant HSP70 substrate in an unstressed cell,[82] might sequester many of the HSP70 molecules in ATP deprivation. The failure in the free chaperone pool restoration has been indicated as a main cause of heat shock gene activation by metabolic stress.[84] It is poorly understood, though, how the enhanced association of immature proteins with HSP70 can result in its insolubility during ATP depletion, since polypeptide chains complexed with the chaperone should be well protected from aggregation.

Some ribosomes carrying nascent polypeptides can be anchored on the cytoskeleton[90] but the majority of immature (pulse-labeled) proteins remain Triton X-100-extractable upon depletion of cellular ATP.[84]

To explain the phenomenon of rapid HSP70 insolubility in cells losing ATP, we hypothesize that even short-term energy deprivation "creates" new chaperone targets that combine with free HSP70 to form Triton-insoluble complexes. So far, the targets have not yet been identified but the cytoskeleton, chromatin and nuclear matrix seem to be those low-soluble structures which sequester the cytosolic HSP70. The cytoskeleton, on the one hand, can interact with HSP70 in an ATP-dependent fashion.[86-88] On the other hand, the cytoskeleton is strongly damaged because of ATP deprivation (see previous section). Therefore, expected targeting of HSP70 to the affected cytoskeleton or its debris should be considered as a possible cause of the HSP70 insolubility in ATP-depleted cells.

Earlier we suggested actin as the major target[44] because this abundant protein is very sensitive to shifts in the ATP/ADP ratio (see subsection 3.1.1.), its interaction with HSP70 can be ATP-dependent,[86,87] and, finally, actin and HSP70 have clear similarity in their three-dimensional structure[91] that might facilitate binding of one to the other. Nevertheless, our experiments on immunolocalization of the constitutive HSP70 within in vitro ATP-depleted 3T3 mouse fibroblasts (Kabakov and Bensaude, unpublished data) and human vascular endothelial cells (Fig. 3.8) have not revealed its association with either actin microfilaments or other components of the cytoskeleton. However, the absence of apparent fibrillar patterns after immunostaining does not yet exclude HSP70-cytoskeleton associations. The epitope of the sequestered HSP70 may be inaccessible for the monoclonal antibodies used. Furthermore, the positive diffuse extranuclear staining in ATP-depleted endothelial cells (see Fig. 3.8) may be partially due to the stress-induced binding of HSP70 to the fine network of the cortical cytoskeleton but this

is, of course, a subject for electron rather than light or fluorescent microscopy. Hence, the question of HSP70 targeting to the cytoskeleton under ATP depletion remains unresolved.

In the case of heat shock, the cytosolic HSP70 is mainly accumulated within the nucleus where a majority of heat-induced protein aggregation occurs.[73,92] Using immunofluorescent staining, Tuijl et al[93] have found migration of the constitutive HSP70 into nuclei of isolated rat cardiomyocytes in vitro exposed to heating or ischemia-like conditions. It seems likely that ATP depletion as well as thermal stress affects intranuclear structures (e.g., chromatin and matrix) and/or promotes protein aggregation in the nucleoplasm of the ischemic cells. In fact, such events might be a responsible for the nuclear compartmentalization of HSP70. However, we did not observe that ATP deprivation caused marked insolubility of the constitutive HSP70 within nuclei of 3T3 fibroblasts (Kabakov and Bensaude, unpublished) or vascular endothelial cells (Fig. 3.8). In both cases, it was difficult to catch the possible increase in intranuclear HSP70 content, since nuclei of the unstressed cells were already brightly stained. Meanwhile, in contrast to unstressed endothelial cells, the ATP-depleted ones clearly exhibit some of the detergent-insoluble HSP70 outside the nuclei (see Fig. 3.8). Certainly the observed diffuse staining does not allow us to identify any cytostructures or compartments sequestering the cytosolic HSP70 during energy starvation. Perhaps further studies using confocal and electronic microscopy will solve this problem.

At present, we cannot assert with certainty that either cytoskeletal or intranuclear structures become targets for the constitutive HSP70 in cells lacking ATP. In this case, what is real cause of the HSP70 insolubility when cellular ATP is depleted? We hypothesize that the dramatic loss of the HSP70 solubility (see Table 3.1 and Fig. 3.7) is due to its co-aggregation with multiple protein targets randomly distributed in the cytosol rather

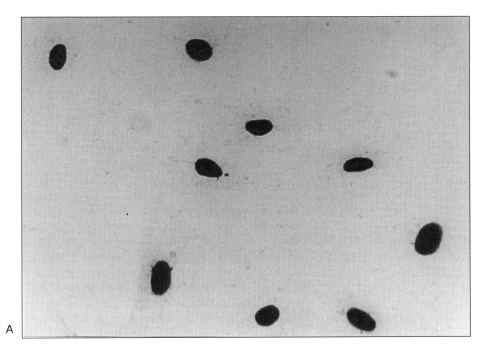

A

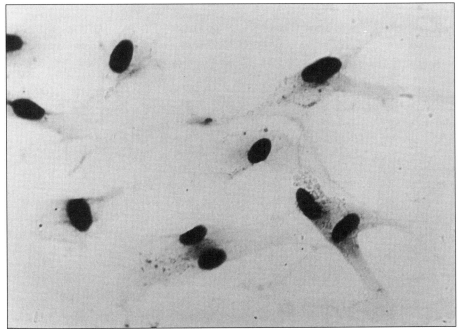

B

Fig. 3.8. Effect of ATP depletion on distribution of the constitutive HSP70 in human aorta endothelial cells. The adherent endothelial cells were subjected to ischemia-simulating exposure in vitro (20 μM CCCP in a glucose-free medium for 2 hours) to deplete cellular ATP.[98] The unstressed (A) and ATP-depleted (B) cells were extracted with 1% Triton X-100 followed by formaldehyde fixation. The Triton-insoluble HSP70 was visualized by means of immuno-histochemical staining with an anti-HSP70 monoclonal antibody N27.[84] Note the clear positive staining outside of the nucleus of ATP-depleted cells (B) that reveals an excess of the insoluble antigen as compared with unstressed cells (A). (x500 magnification) Obtained by Loktionova and Kabakov, previously unpublished.

than its selective binding to preexisting insoluble protein structures (e.g., cyto- and nucleoskeleton). If this is true, HSP70 that is included in large co-aggregates with various proteins may be poorly accessible for antibodies under immunolocalization in situ. Moreover, visualization of HSP70 within the randomly distributed Triton-insoluble deposits should display diffuse images without characteristic patterns (see Fig. 3.8).

Taking into consideration all the above facts and speculations, we propose a scheme (Fig. 3.9) which may, in our opinion, explain the phenomenon of HSP70 insolubility when ATP is depleted. According to this scheme, HSP70 carries out protein refolding via a cyclic K^+, ATP-dependent mechanism, preexisting unfolded proteins, proteins denatured by stress, and spontaneously unfolded (thermolabile) proteins being transient targets for the chaperone. Both DnaJ (HSP40) (see chapter 1) and Hip, a novel cochaperone,[94] most likely act here in cooperation with HSP70, though for simplicity the accessory components are not represented in the scheme. It should be noted that HSP70 may exist in diverse conformational states with different functional activities. Particularly, the ADP-HSP70 complexes formed as a result of ATP hydrolysis or direct binding of ADP to the nucleotide-free HSP70 have the following prominent characteristics: (1) they seem to be dimers[81,82] or trimers[81] which cannot carry ATP;[69,81,85,89] and (2) they possess a high affinity for unfolded proteins.[69,70] Recently discovered Hip, a partner of HSP40 in the HSP70 chaperone machine of eukaryotes, appears to stabilize this highly active "ADP state" of HSP70.[94] Such activated forms of the chaperone rapidly recognize and bind its peptide substrates, including spontaneously unfolded proteins, which leads to the formation of transient and soluble ADP-HSP70-protein complexes. With the assistance of HSP40, the ADP-HSP70-protein interactions typically end in protein refolding (Fig. 3.9). Both ATP and K^+ are essential to dissociate the

chaperone-protein complexes.[70,85] When ATP replaces ADP in the nucleotide-binding domain of HSP70, it is accompanied by the release of the refolded protein and the transformation of the chaperone into a monomeric form with a low affinity for unfolded proteins. Following ATP hydrolysis the active ADP-bound HSP70 dimer (or trimer) is reformed and the reaction cycle can be repeated as long as ATP and K^+ are available (see Fig. 3.9).

However, what will occur upon ATP depletion when both the ATP/ADP ratio and $[K^+]$ drop in the cytosol? Obviously, the multifold increased ADP shifts equilibrium toward the ADP-HSP70 association (upper dotted arrow in Fig. 3.9). In turn, such ADP-mediated "activation" of HSP70 sharply accelerates the chaperone binding to both preexisting and newly formed protein targets. The latter targets are various cellular proteins denatured immediately due to the lack of ATP. Consistent with Bensaude's hypothesis,[59] randomly unfolded thermolabile proteins are also included in the scheme as potential targets for the ADP-HSP70. Hence, a decrease in the ATP/ADP ratio per se may considerably enhance HSP70-unfolded protein complex formation. Furthermore, the ADP-HSP70-protein complexes appear to be more stable and long-lived in the case of ATP depletion, since their dissociation does require ATP and K^{+70} and is strongly inhibited by ADP.[69] It is evident that the dramatic accumulation of unfolded proteins bound to HSP70 will ultimately result in their co-aggregation through intermolecular hydrophobic interactions (lower dotted arrow in Fig. 3.9). In addition, the oligomeric state of the high-affinity HSP70 increases the risk of the co-aggregation, in so far as each dimer and trimer can bind to more than one molecule of a protein substrate. This may lead to the formation of large co-aggregates with high molecular weights. Such co-aggregates can, of course, sediment into Triton-insoluble deposits. This process resembles the precipitation of antigen-antibody network; an

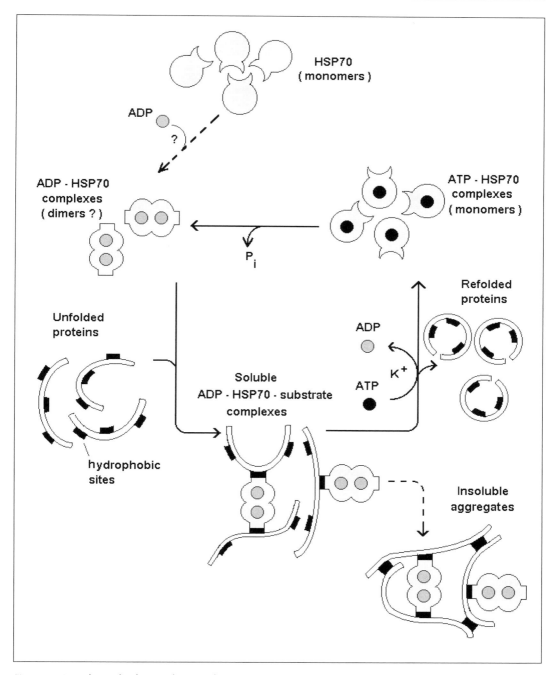

Fig. 3.9. Hypothetical scheme showing how energy deprivation may affect interaction of the constitutive HSP70 with cellular proteins and result in the chaperone insolubility (see the text for details).

analogy between HSP70 and immunoglobulins is clear. We believe that the proposed scheme may explain the phenomenon of HSP70 insolubility during ATP depletion as well as our own failure to identify in situ any concrete Triton-insoluble HSP70 targets within mammalian cells devoid of ATP.

3.3.2. THE SMALL STRESS PROTEINS

The small HSPs are referred to as ATP sensors, since these cytosolic proteins dramatically lose their solubility in response to energy deprivation. In particular, cardiac α-crystallin, a major protein component of the soluble fraction of homogenates from unaffected rat hearts,[95] is rendered irreversibly insoluble during the early period of myocardial ischemia.[96] Chiesi et al, who first documented this phenomenon, explained the loss of solubility during the aggregation of denatured α-crystallin at low pH following ischemic acidification of the cytosol (see chapter 2). Such coagulation resulted in the formation of large amorphous clumps, and occurred only in the presence of other protein components (actin most likely) in the soluble fraction.[96] A considerable increase in the binding affinity of α-crystallin to actin and desmin at slightly acidic pH (6.5) was established later by Chiesi's group.[97] The investigators concluded that the aggregation of α-crystallin in acidic pH was connected but that no association existed between aggregation and ATP depletion since the changes in ATP concentration did not influence the α-crystallin solubility in control heart homogenates.[96] In the above cited work of Barbato et al,[72] the same conclusion was drawn, namely, acidosis but not ATP deprivation evokes the binding of cardiac α-crystallin to myofibrils in ischemic rat hearts. In both studies, however, the effects of ATP were studied in an in vitro system with mechanically-disrupted cells (tissue homogenates and extracts).[72,96] It is quite possible then that fluctuations in ATP levels within structurally intact cardiomyocytes might reversibly affect the solubility of α-crystallin as we

demonstrate with endothelial HSP27 below. In any case, both the rapid, irreversible denaturation of α-crystallin and its probable co-aggregation with cytosolic actin could be involved in the mechanism of ischemic injury of myocardium. It is also possible that at a slightly acidic pH, cardiac α-crystallin associates with actin filaments to protect them against the F-actin aggregation (formation of paracrystals) upon ischemic acidification of the cytosol.[97]

The 27 kDa small stress protein, HSP27, is highly homologous to α-crystallin (see chapter 1) and it exhibits similar behavior under ischemia-mimicking stress. We studied the effect of ATP depletion on the level of Triton X-100-extractable HSP27 in HeLa cells simultaneously treated with carbonyl cyanide m-chlorophenylhydrazone (CCCP), an uncoupler of oxidative phosphorylation, and deoxyglucose, an inhibitor of glycolysis. Immunoblotting of the cellular fractions showed that HSP27 was quickly made insoluble during ATP depletion and its solubility was rapidly restored following ATP recovery (Kabakov and Bensaude, unpublished).

Afterward, we thoroughly explored the same problem on cultured human vascular endothelial cells exposed to metabolic stress in vitro imitating ischemia/reperfusion (incubation with CCCP or rotenone in the absence of glucose with subsequent recovery in a rich growth medium).[98] ATP concentrations and relative amounts of HSP27 in Triton-soluble and insoluble cellular fractions were evaluated during both the ATP-depleting exposure and recovery. In parallel experiments, subcellular localization of HSP27 and its isoform composition were also analyzed. Like HSP27 in HeLa cells, endothelial HSP27 also loses detergent solubility during cellular ATP depletion but becomes soluble again during ATP replenishment (Fig. 3.10). Distribution patterns of antibodies against HSP27[99] reveal an even more intriguing phenomenon: sustained (greater than 30 minutes) ATP deprivation results in the formation of round HSP27-containing granules inside endothelial nuclei (Fig. 3.11).

Moreover, immunoblotting the cell extracts shows that intranuclear granulation is as reversible as its insolubility; when the cellular ATP level is restored, the granular staining completely disappears from nuclei and this process coincides with restored HSP27 solubility. Obviously, both these stress-provoked events are interrelated, since nuclear compartmentalization of a protein and/or its integration in large particles can abolish its extraction with Triton X-100. Although the HSP27-containing granules in the nuclei of ATP-depleted endothelial cells outwardly resemble the above described aggregates of α-crystallin in ischemic rat hearts,[96] we do not correlate the HSP27 granulation with either acidification of the cytosol or co-aggregation of HSP27 with cytosolic actin. In fact, glucose totally abolishes granule formation in the rotenone-treated cells;[98] the exposure does not lead to ATP loss but instead decreases the cytosolic pH through accumulation of glycolytic lactate. Therefore, the observed intranuclear granulation of HSP27 is due to ATP depletion per se rather than pH changes in the cytoplasm or dysfunction of mitochondria. Likewise, actin was not immunodetected in the HSP27-containing granules in situ,[98] which contradicts Chiesi's findings that co-aggregation of α-crystallin and actin under myocardial ischemia occurs.[96]

Why does HSP27 form intranuclear superaggregated particles (granules) in response to ATP depletion? This phenomenon is all the more enigmatic since the small HSPs do not require ATP for their chaperoning actions[100] and HSP27 is probably not an ATP-binding protein at all. At the same time, both the functional activity and supramolecular organization of HSP27 are regulated by phosphorylation/dephosphorylation.[101-103] Taking into account that the protein kinase/phosphatase equilibrium may be shifted in endothelial cells during ATP depletion, we hypothesized that the granule formation is due to the stress-induced changes in the phosphorylation status of HSP27.[98] One may evaluate these changes as shifts in the HSP27 isoform profile. Endothelial HSP27 may be comprised of up to as many as four isoforms: a basic, nonphosphorylated *a* isoform and three more acidic phospho-isoforms *b*, *c* and *d*.[104] Using two-dimensional

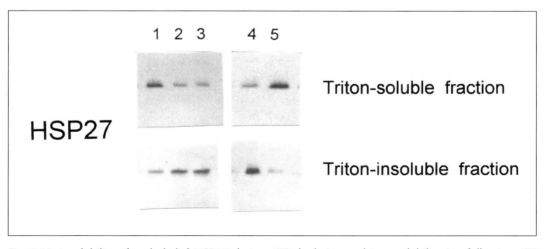

Fig. 3.10. Insolubility of endothelial HSP27 during ATP depletion and its resolubilization following ATP restoration. Equal numbers of adherent human aorta-derived endothelial cells were fractionated with 1% Triton X-100 under normal conditions (1) or at 40 minutes (2) and at 1.5 hours (3) of ATP-depleting treatment (20 μM CCCP in the absence of glucose) or at 0.5 hours (4) and 1.5 hours (5) of the post-stress recovery in a rich growth medium.[98] Aliquots of the Triton-soluble and insoluble cellular fractions were analyzed by immunoblotting with rabbit anti-HSP27 antibodies.[99]

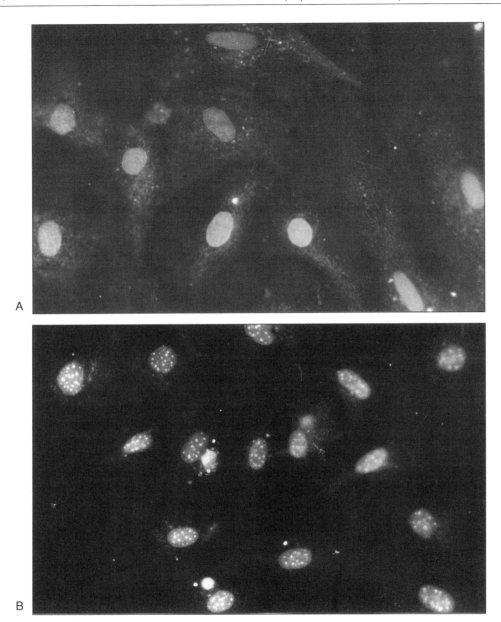

Fig. 3.11. Indirect immunofluorescent staining showing the intranuclear granulation of HSP27 in ATP-depleted endothelial cells. Adherent human aorta endothelial cells were deprived of ATP by incubating with 20 µM CCCP without glucose for 2 hours. The fixed and permeabilized preparations of the unstressed (A) and ATP-depleted (B) cells were labeled with rabbit anti-HSP27 antibodies and anti-rabbit IgG Texas Red-conjugates.[98] (x500 magnification). Note that the formation of intranuclear granules containing HSP27 is accompanied by its insolubility during ATP depletion (see Fig. 3.10 and ref. 98).

immunoblotting, we have visualized the HSP27 isoform patterns in unstressed and ATP-deprived endothelial cells (Fig. 3.12). While all four isoforms are revealed in the unstressed cells, inhibition of energy metabolism for 1 hour resulted in the complete disappearance of most of the phosphorylated *d* isoform and a marked increase in the nonphosphorylated *a* isoform. Only two tracks, a major spot of the nonphosphorylated *a* isoform and a minor spot of the weakly phosphorylated *b* isoform were detected in the cells after 2 hours of ATP depletion (Fig. 3.12) This indicates that

almost a complete removal of phosphates from HSP27 occurred. Thus, the dynamics of the stress-evoked shifts in the isoform spectra clearly indicates the dephosphorylation of endothelial HSP27 during sustained ATP deprivation. As we discussed in ref. 98, this may be the result of an expected inhibition of ATP-consuming protein kinases and/or an activation of phosphatase(s). This seems quite likely since the extracellular signal-regulated protein kinases (ERK) or mitogen-activated protein kinases (MAPK) are involved in the phosphorylation of HSP27 under normal

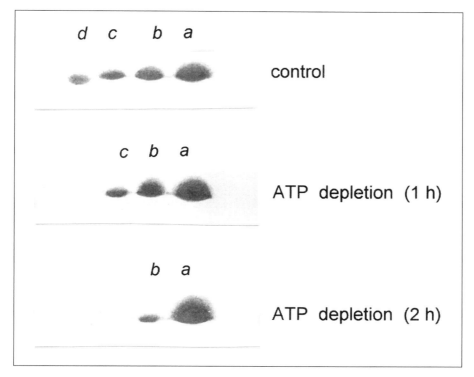

Fig. 3.12. ATP loss-induced shifts in the endothelial HSP27 isoform patterns revealed by two-dimensional immunoblotting. The adherent cells cultured from human aorta endothelium were exposed to sustained ATP depletion (20 μM CCCP in the absence of glucose for 1-2 hours). The antigen material from equal numbers of the unstressed and ATP-depleted cells was immunodetected in each blot. HSP27 from the unstressed endothelial cells is present in four isoforms: basic unphosphorylated a isoform and three more acidic phospho-isoforms b, c and d with a low, middle and high degree of phosphorylation, respectively. It is clearly observed that prolonged ATP depletion provokes dephosphorylation of endothelial HSP27; this may correlate with the stress-induced redistribution of the small stress protein (see Figs. 3.10, 3.11 and ref. 98).

conditions.[101,102] Their activities decrease during ATP depletion as it was shown for Madin-Darby canine kidney epithelial cells.[105]

We suggest a direct relationship between the dephosphorylation and granulation of HSP27 in ATP-depleted endothelial cells.[98] Our suggestion is supported in part by the preceding work of Benndorf et al[103] who demonstrated that unphosphorylated murine HSP25 can form large round particles with characteristic properties. Importantly, researchers isolated these particles from stationary phase Ehrlich carcinoma growing in mice[102] but recently we have established that cells of this tumor line, being in the stationary phase of in vivo growth, experience ATP deficiency (see the next chapter).[30] Therefore, both the HSP27-containing granules in endothelial nuclei (Fig. 3.11) and the purified particles of the murine HSP25[103] may be separate cases of the same phenomenon, namely, multioligomerization (or superaggregation) of the small HSP which is dephosphorylated as a result of ATP depletion.

Dissolution of the intranuclear granules of endothelial HSP27 during post-stress recovery can be accomplished via rephosphorylation of the superaggregated form of HSP27. Indeed, Kato et al[102] have shown in human glioma cells that the activation of cellular protein kinase(s) by various stimulating reagents as well as inhibition of phosphatase(s) by okadaic acid causes dissociation of an aggregated form of HSP27; a correlation between the phosphate incorporation and the dissociation is established. An analogous mechanism for recovering endothelial cells is realized when the ATP/ADP ratio is restored and some of the protein kinases are activated. Dramatic reactivation of the ERK following ATP repletion occurs in rat kidneys subjected to ischemia/reperfusion and in kidney epithelial cells cultured in vitro subjected to transient ATP depletion.[105] Likewise, a burst in the ERK and the MAPK activities was observed in cultured rat cardiomyocytes after a reversible ATP decrease[106] during posthypoxic reoxygenation.[107]

The physiological significance of the HSP27 granulation in nuclei of ATP-deprived cells requires further research. This response may serve to protect the actin skeleton during depletion of cellular ATP (see chapter 7). This seems plausible since actin polymerization is inhibited by monomers of unphosphorylated HSP27, whereas their integration in the multimeric complexes (particles) abolishes the inhibitory effect.[103] Alternatively, HSP27 is one regulator of microfilament dynamics (reviewed in ref. 101) and accordingly, its super-aggregation (granulation) provoked by ATP loss may result in harmful consequences for the stressed cells via a failure in actin regulation. In this respect, the above described actin perturbations in the cytoplasm of ATP-deprived cells (see subsections 3.1.1. and 3.1.3.) may be in part due to the stress-induced dephosphorylation and/or granulation of HSP27. In any case, we consider the disturbance of cell energy balance as one of the factors modulating the phosphorylation status, supra-molecular organization and functional activity of the small HSPs.[98]

3.3.3. UBIQUITIN

Because ubiquitin is a HSP but not a chaperone, in vivo it forms conjugates with aged, defective and damaged proteins, which marks the proteins for subsequent ATP-dependent degradation in 26S proteasomes (see chapter 1). As a result, enhanced ubiquitination of intracellular proteins occurs in neuronal[64-66] and cardiac[67] tissues undergoing transient ischemia. This is considered a hallmark of proteotoxicity intrinsic to ischemic stress. Hayashi et al[64] documented an increase in Triton X-100-insolubility: high molecular weight conjugates appear in the hippocampus after 20 minutes of rat forebrain ischemia followed by 24 hours of reperfusion. Immunostaining with a monoclonal antibody that preferentially recognizes ubiquitin-protein conjugates shows a semiquantitative rise in the level of the conjugates during the early reperfusion period in the cortex of gerbils subjected to short-term forebrain ischemia.[65] The enhancement of protein ubiquitination correlates with the duration of ischemia and the extent of the neuronal damage,[66] which implies an accumulation of defective proteins in ischemic neurons and probably a subsequent activation of proteolysis in the post-ischemic tissue.

Using the same monoclonal antibody,[65] Andres et al showed the transient formation of new ubiquitin-protein conjugates during reperfusion of porcine myocardium stunned by brief ischemia.[67] Certainly, these results strongly resemble the above observations of the post-ischemic brains in gerbils.[65,66] Moreover, the authors' hypothesis that such ischemia-provoked protein ubitiquination acts as a marker for molecular injury[67] fits well with conclusions of the previously cited work.[66]

Kato and colleagues documented the disappearance of ubiquitin immunoreactivity in the gerbil hippocampus 4 hours after 3 minutes of ischemia,[108] suggesting a connection between this effect and the subsequent delayed CA1 neuronal death (see also chapters 6 and 7). Recently, Morimoto et al[109] used antibodies to distinguish the difference between free and conjugated ubiquitin in order to differentially study their distribution in the ischemic gerbil hippocampus. They demonstrated that transient ischemia depletes free ubiquitin but does not deplete conjugated ubiquitin in the hippocampal CA1 neurons.[109] To solve the phenomenon, the researchers suggested the accumulation of denatured proteins in the ischemic hippocampus which then undergo enhanced ubiquitination followed by proteolysis. In that case, if ubiquitinated proteins efficiently degrade, both a decrease in the number of ubiquitin-protein conjugates and the restoration of the free ubiquitin pool should be observed; however, the data of immunostaining does not support this.[109] Therefore, it remains possible that ubiquitination but not proteolysis occurs in the ischemic CA1 neurons. This seems quite probable when ATP deficiency in ischemic cells is taken into account, since the proteolysis in 26S proteasomes does require ATP (see chapter 1) and moreover, 26S protease complexes

themselves are transformed into the inactive 20S form as a result of ATP depletion.[76] The hydrolyzing activity of 26S proteasomes is suppressed after 10 minutes of forebrain ischemia in gerbils.[110] The inhibition of the ATP-dependent 20S →26S conversion through ATP depletion is suggested as one of the causes of the accumulation of ubiquitin-protein conjugates in ischemic tissue. It should be noted, however, that lack of ATP in the ischemic neurons may be responsible for blocking both the ubiquitinated protein degradation and recycling of ubiquitin (i.e., restoration of its free pool).

Thus, a drop in the level of ATP during transient ischemia appears to upset the ubiquitin-mediated proteolysis pathway. Obviously, both the depletion of free ubiquitin and the failed degradation of ubiquitinated proteins can have serious consequences for post-ischemic tissues. On the one hand, it causes the intracellular accumulation of abnormal proteins that by itself may activate the transcription of heat shock genes following ischemia/reperfusion (see chapters 1 and 4). On the other hand, dysfunction of the ubiquitin system affects many cellular processes including proliferation and differentiation, DNA repair, regulation of transcription, organelle biogenesis, apoptosis, etc. (reviewed in ref. 111). Hence, even a transient disturbance of ubiquitin-dependent protein degradation during ischemic episodes may cause severe metabolic derangement leading to apoptosis afterward (discussed in chapter 8). Finally, it seems likely that ubiquitin plays an indispensable role in the reparative processes following ATP replenishment and saves the stressed cell from irreversibly damaged proteins (see chapter 7).

CONCLUDING REMARKS

In completing this long chapter, we would like to underline the fact that ATP loss in mammalian cells causes not only ionic imbalance and the blocking of biosynthesis but such "rapid" proteotoxic effects as disruption of the cytoskeleton, denaturation and aggregation of various proteins in the cytosol, compartments and membranes.

Perhaps some of the damaged and aggregated proteins become targets for molecular chaperones and ubiquitin to be renatured, disaggregated or digested during cell recovery.

The destructive effect of ATP depletion on the cytoskeleton is evinced by the fragmentation of filamentous and microtubular structures, collapse of the cortical actin framework and its detachment from the plasma membrane, side-to-side actin aggregation through bundling of F-actin and misassembling of G-actin, etc. Evidently, the events affect the state of both the cytoplasm and the plasmalemma, which may result in the death of the stressed cells. Nevertheless, the precise mechanism of the cytoskeleton rearrangement upon ATP deprivation remains undefined. Likewise, nothing is known about any of the interactions between cytoskeletal proteins and HSPs during or following ATP depletion.

ATP-binding and non-ATP-binding, phosphorylated and dephosphorylated, thermolabile and thermostable, cytoskeletal, cytosolic, nuclear and membrane proteins aggregated in vivo as a result of ATP depletion. Such diversity of proteins aggregating due to the lack of cellular ATP demonstrates that aggregation is a very complex, multifactor process which can occur through different mechanisms. Besides the ATP loss itself, an increase in the ADP and AMP levels and ionic imbalance in the cytosol, shifts in protein kinase/phosphatase equilibrium, a block in both chaperone machinery and ubiquitin-mediated proteolysis also seem to be the factors promoting protein aggregation in metabolically-stressed cells.

At least HSP70, the small HSPs and ubiquitin are directly involved in the cellular response to ATP depletion. In regard to the former, it appears to be sequestered by its protein substrates as a result of a decrease in the ATP/ADP ratio and the lack of K^+ in energy-deprived cells; importantly, this dissipation of the free HSP70 pool may trigger the heat shock transcriptional response following transient ATP depletion or ischemic episode (see the next chapter).

Undoubtedly, the damage and aggregation of cytoskeletal proteins, as well as the insolubility and inactivation of cytosolic enzymes are responsible for the injury and dysfunction of the ATP-depleted mammalian cells. In turn, the aggregation of actin and actin-associated proteins and the redistribution and dysfunction of membrane ion pumps play the main role in the mechanism of blebbing and rapid necrotic death of cells lacking ATP. If the "aggregability" of cellular proteins actually defines cell sensitivity to energy deprivation, a device preventing fatal aggregation (e.g., molecular chaperones) might protect ATP-depleted mammalian cells from injury and death. We direct our readers to chapter 7 for further discussion of this idea.

RECENT NEWS

Molitoris et al[112] studied the effect of ATP depletion on the stectrin (fodrin) cytoskeletal network which links the actin cytoskeleton to the surface membrane of epithelial cells. Fodrin and ankyrin, being Triton X-100 insoluble under physiological conditions, underwent rapid and duration-dependent increase in solubility during ATP-depletion, whereas myosin solubility was reduced. Additionally, association between fodrin and ankyrin was absent in ATP-depleted cells. The authors concluded that ATP-depletion in epithelial cells leads to dissociation of ankyrin and fodrin from each other and from the actin cytoskeleton.[112] We suggest that similar events could promote bleb formation and plasma membrane breakage in ATP-depleted cells (see subsection 3.1.3).

REFERENCES

1. Ganote C, Armstrong S. Ischaemia and the myocyte cytoskeleton: review and speculation. Cardiovasc Res 1993; 27: 1387-1403.
2. Cooper JA. The role of actin polymerization in cell motility. Annu Rev Physiol 1991; 53: 585-605.
3. Condeelis J. Life at the leading edge: The formation of cell protrusions. Annu Rev Cell Biol 1993;9: 411-444.

4. Janmey PA, Hvidt S, Oster GF et al. Effect of ATP on actin filament stiffness. Nature 1990; 347: 44-49.

5. Pollard TD, Goldberg I, Schwartz WH. Nucleotide exchange, structure, and molecular properties of filaments assembled from ATP-actin and ADP-actin. J Biol Chem 1992; 267: 20339-20345.

6. Orlova A, Egelman EH. Structural basis for the destabilization of F-actin by phosphate release following ATP hydrolysis. J Mol Biol 1992; 227: 1043-1053.

7. Pantaloni D, Carlier M-F. How profilin promotes actin filament assembly in the presence of thymosin β_4. Cell 1993; 75: 1007-1014.

8. Carlier M-F, Jean C, Rieger KJ et al. Modulation of the interaction between G-actin and thymosin β_4 by the ATP/ADP ratio: Possible implication in the regulation of actin dynamics. Proc Natl Acad Sci USA 1993; 90: 5034-5038.

9. Safer D, Nachmias VT. Beta thymosins as actin binding peptides. BioEsseys 1994; 16: 473-479.

10. Ikeuchi Y, Iwamura K, Machi T et al. Instability of F-actin of ATP: A small amount of myosin destabilizes F-actin. J Biochem 1992; 111: 606-613.

11. Laham LE, Way M, Yin HL, Janmey PA. Identification of two sites in gelsolin with different sensitivities to adenine nucleotides. Eur J Biochem 1995; 234: 1-7.

12. Bershadsky AD, Gelfand VI, Svitkina TM, Tint IS. Destruction of microfilament bundles in mouse embryo fibroblasts treated with inhibitors of energy metabolism. Exp Cell Res 1980; 127: 421-429.

13. Sanger JW, Sanger JM, Jockusch BM. Differential response of three types of actin filament bundles to depletion of cellular ATP level. Eur J Cell Biol 1983; 31: 197-204.

14. Bershadsky AD, Gelfand VI. Role of ATP in the regulation of stability of cytoskeletal structures. Cell Biol Int Rep 1983; 7: 173-187.

15. Wang YL. Reorganization of α-actinin and vinculin in living cells following ATP depletion and replenishment. Exp Cell Res 1986; 167: 16-28.

16. Glascott PA Jr, NcSorley KM, Mittal B et al. Stress fiber reformation after ATP depletion. Cell Motil Cytoskeleton 1987; 8: 118-129.

17. Hinshaw DB, Armstrong BC, Beals TF, Hyslop PA. A cellular model of endothelial cell ischemia. J Surg Res 1988; 44: 527-537.

18. Hinshaw DB, Armstrong BC, Burger JM et al. ATP and microfilaments in cellular oxidant injury. Am J Pathol 1988; 132: 479-488.

19. Kuhne W, Besselmann M, Noll T et al. Disintegration of cytoskeletal structure of actin filaments in energy-depleted cells. Am J Physiol 1993; 264: H1599-H1608.

20. Hinshaw DB, Burger JM, Miller MT et al. ATP depletion induces an increase in the assembly of a labile pool of polymerized actin in endothelial cells. Am J Physiol 1993; 264: C1171-C1179.

21. Bacallao R, Gafinkel A, Monke S et al. ATP depletion: a novel method to study junctional properties in epithelial tissues. I. Rearrangement of the actin cytoskeleton. J Cell Sci 1994; 107: 3301-3313.

22. Golenhofen N, Doctor RB, Bacallao R, Mandel LJ. Actin and villin compartmentation during ATP depletion and recovery in renal cultured cells. Kidney Int 1996; 48: 1837-1845.

23. Molitoris BA, Geerdes A, McIntoch JR. Dissociation and redistribution of Na^+,K^+-ATPase from its surface membrane actin cytoskeletal complex during cellular ATP depletion. J Clin Invest 1991; 88: 462-469.

24. Gabai VL, Kabakov AE, Mosin AF. Association of blebbing with assembly of cytoskeletal proteins in ATP-depleted EL-4 ascites tumour cells. Tissue Cell 1992; 24: 171-177.

25. Gabai VL, Kabakov AE. Tumor cell resistance to energy deprivation and hyperthermia can be determined by the actin skeleton stability. Cancer Lett 1993; 52: 3648-3654.

26. Kabakov AE, Gabai VL. Protein aggregation as primary and characteristic cell reaction to various stresses. Experientia 1993; 49: 706-710.

27. Gabai VL, Kabakov AE. Rise in heat-shock protein level confers tolerance to energy

deprivation. FEBS Lett 1993; 327: 247-250.

28. Kabakov AE, Gabai VL. Stress-induced insolubility of certain proteins in ascites tumor cells. Arch Biochem Biophys 1994; 309: 247-253.

29. Kabakov AE, Gabai VL. Heat shock-induced accumulation of 70-kDa stress protein (HSP70) can protect ATP-depleted tumor cells from necrosis. Exp Cell Res 1995; 217: 15-21.

30. Kabakov AE, Molotkov AO, Budagova KR et al. Adaptation of Ehrlich ascites carcinoma cells to energy deprivation in vivo can be associated with heat shock protein accumulation. J Cell Physiol 1995; 165: 1-6.

31. Piper HM. Metabolic processes leading to myocardial cell death. Meth Achiev Exp Pathol 1988; 13: 144-180.

32. Steenbergen CJ, Hill ML, Jennings RB. Cytoskeletal damage during myocardial ischemia: changes in vinculin immunofluorescence staining during total in vitro ischemia in canine heart. Circ Res 1987; 60: 478-86.

33. Armstrong SC, Ganote CE. Flow cytometric analysis of isolated adult cardiomyocytes: vinculin and tubulin fluorescence during metabolic inhibition and ischemia. J Mol Cell Cardiol 1992;24: 149-162.

34. VanWinkle WB, Snuggs M, Buja LM. Hypoxia-induced alterations in cytoskeleton coincide with collagenase expression in cultured neonatal rat cardio-myocytes. J Mol Cell Cardiol 1995; 27: 2531-2542.

35. Ganote CE, Vander Heide RS. Cytoskeletal lesions in anoxic myocardial injury: a conventional and high voltage electron microscopic and immunofluorescence study. Am J Pathol 1987; 129: 327-344.

36. Barry WH, Hamilton CA, Knowlton KU. Regulated expression of a contractile protein gene correlates with recovery of contractile function after reversible metabolic inhibition in cultured myocytes. J Mol Cell Cardiol 1995; 27: 551-561.

37. Kellerman PS, Bogusky RT. Microfilament disruption occurs very early in ischemic proximal tubule cell injury. Kidney Int 1992, 42: 896-902.

38. Leiser J, Molitoris BA. Disease processes in epithelia: the role of the actin cytoskeleton and altered surface membrane polarity. Biochim Biophys Acta 1993; 1225: 1-13.

39. Kellerman PS, Norenberg S, Guse N. Exogenous adenosine triphosphate (ATP) preserves proximal tubule microfilament structure and function in vivo in a maleic acid model of ATP depletion. J Clin Invest 1993; 92: 1940-1949.

40. Maro B, Bornens M. Reorganization of HeLa cell cytoskeleton induced by an uncoupler of oxidative phosphorylation. Nature 1982; 295: 334-336.

41. Nover L. Heat Shock Response. Boca Raton: CRC Press, 1991.

42. Wiegant FAC, Van Bergen en Henegouwen PMP, Van Dongen G, Linnemans WAM. Stress-induced thermotolerance of the cytoskeleton of mouse neuroblastoma N2A cells and rat Reuber H35 hepatoma cells. Cancer Res 1987; 47: 1674-1680.

43. Bellomo G, Mirabelli F, Vairetti M, Malormi W. Cytoskeleton as a target in menadione-induced oxidative stress in cultured mammalian cells: biochemical and immunocytochemical features. J Cell Physiol 1990; 143: 118-128.

44. Kabakov AE, Gabai VL. Heat shock proteins maintain the viability of ATP-deprived cells: what is the mechanism? Trends Cell Biol 1994; 4: 193-196.

45. Steenbergen C, Hill ML, Jennings RB. Volume regulation and plasma membrane injury in aerobic, anaerobic, and ischemic myocardium in vitro. Circ Res 1985; 57: 864-875.

46. Mandel LJ, Doctor RB, Bacallao R. ATP depletion: a novel method to study junctional properties in epithelial tissues. II. Internalization of Na^+,K^+-ATPase and E-cadherin. J Cell Sci 1994; 107: 3315-3324.

47. Atsma DE, Bastiaanse EML, Jerzewski A et al. Role of calcium-activated neutral protease (calpain) in cell death in cultured neonatal rat cardiomyocytes during metabolic inhibition. Circ Res 1995; 76: 1071-1078.

48. Sun FF, Fleming WE, Taylor BM. Degradation of membrane phospholipids in the cultured human astroglial cell line UC-11MG during ATP depletion. Biochem Pharmacol 1993; 45: 1149-1155.

49. Trump BF, Berezesky IK. The role of calcium in cell injury and repair: a hypothesis. Surv Synth Pathol Res 1985; 97: 248-256.

50. Steenbergen C, Murphy E, Watts JA, London RE. Correlation between cytosolic free calcium, contracture, ATP, and irreversible ischemic injury in perfused rat heart. Circ Res 1990; 66: 135-146.

51. Niemenen A-L, Gores GJ, Wray BE et al. Calcium dependence of bleb formation and cell death in hepatocytes. Cell Calcium 1988; 9: 237-246.

52. Jurkowitz-Alexander MS, Altschuld RA, Haun SE et al. Protection of ROC-1 hybrid glial cells by polyethylene glycol following ATP depletion. J Neurochem 1993; 61: 1581-1584.

53. Cantiello HF. Actin filaments stimulate the Na⁺-K⁺ ATPase. Am J Physiol (Renal Fluid and Electrolyte Physiology) 1995; 38: F637-F643.

54. Nguyen VT, Morange M, Bensaude O. Protein denaturation during heat shock and related stress. J Biol Chem 1989; 264: 10487-10492.

55. Bensaude O, Pinto MP, Dubois M-F et al. Protein denaturation during heat shock and related stress. In: Schlesinger MJ and Santoro G, eds. Stress Proteins. Induction and Function. Berlin: Springer Verlag, 1990: 89-99.

56. Dubois M-F, Hovanessian AG, Bensaude O. Heat shock-induced denaturation of proteins. Characterization of the insolubility of the interferon-induced p68 kinase. J Biol Chem 1991; 266: 9707-9711.

57. Bensaude O, Dubois M-F, Legagneux V et al. Early effects of heat shock on enzymes: heat denaturation of reporter proteins and activation of a protein kinase which phosphorylates the C-terminal domain of RNA polymerase II. In: Maresca B, Lindquist S, eds. Heat Shock. Berlin: Springer Verlag, 1991:97-103.

58. Pinto M, Morange M, Bensaude O. Denaturation of proteins during heat shock. J Biol Chem 1991; 266: 13941-13946.

59. Nguyen VT, Bensaude O. Increased thermal protein aggregation in ATP-depleted mammalian cells. Eur J Biochem 1994; 220: 239-246.

60. Gabai VL, Kabakov AE. Induction of heat-shock protein synthesis and thermo-tolerance in EL-4 ascites tumor cells by transient ATP depletion after ischemic stress. Exp Mol Pathol 1994; 60: 88-99.

61. Gabai VL, Mosina VA, Budagova KR, Kabakov AE. Spontaneous over-expression of heat-shock proteins in Ehrlich ascites carcinoma cells during in vivo growth. Biochem Mol Biol Int 1995; 35: 95-102.

62. Musters RJP, Post JA, Verkleij AJ. The isolated neonatal rat-cardiomyocyte used in an in vitro model for "ischemia". I: a morphological study. Biochim Biophys Acta 1991; 1091: 270-277.

63. Schneijdenberg CTWM, Verkleij AJ, Post JA. Aggregation of myocardial sarcolemmal transmembrane proteins is not hindered by an interaction with the cyto-skeleton. Possible implications for ischemia and reperfusion. J Mol Cell Cardiol 1995; 27: 2337-2345.

64. Hayashi T, Takada K, Matsuda M. Subcellular distribution of ubiquitin-protein conjugates in the hippocampus following transient ischemia. J Neurosci Res 1992; 31: 561-564.

65. Hayashi T, Takada K, Matsuda M. Posttransient ischemia increase in ubiquitin conjugates in the early reperfusion. Neuroreport 1992; 3: 519-520.

66. Hayashi T, Tanaka J, Kamikubo T et al. Increase in ubiquitin conjugates dependent on ischemic damage. Brain Res 1993; 620: 171-173.

67. Andres J, Sharma HS, Knoll R et al. Expression of heat shock proteins in the normal and stunned porcine myocardium. Cardiovasc Res 1993; 27: 1421-1429.

68. Jungbluth A, Eckerskorn C, Gerisch G et al. Stress-induced tyrosine phos-phorylation of actin in *Dictyostelium* cells and localization of the phosphorylation site to tyrosine-53 adjacent to the DNase I binding loop. FEBS Lett 1996; 375: 87-90.

69. Palleros DR, Welch WJ, Fink AL. Interaction of hsp70 with unfolded proteins: Effects of temperature and nucleotides on the kinetics of binding. Proc Natl Acad Sci USA 1991; 88: 5719-5723.

70. Palleros DR, Reid KL, Shi L et al. ATP-induced protein-Hsp70 complex dissociation requires K⁺ but not ATP hydrolysis. Nature 1993; 365: 664-666.

71. Palleros DR, Shi L, Reid KL, Fink AL. Hsp70-protein complexes; complex stability and conformation of bound substrate protein. J Biol Chem 1994; 269: 13107-13114.

72. Barbato R, Menabo R, Dainese P et al. Binding of cytosolic proteins to myofibrils in ischemic rat hearts. Circ Res 1996; 78: 821-828.

73. Michels AA, Nguyen VT, Konings AWT et al. Thermostability of a nuclear targeted luciferase expressed in mammalian cells. Destabilizing influence of the intranuclear microenvironment. Eur J Biochem 1995; 234: 382-389.

74. Imamoto N, Matsuoka Y, Kurihara T et al. Antibodies against 70-kD heat shock cognate protein inhibit mediated nuclear import of karyophilic proteins. J Cell Biol 1992; 119: 1047-1061.

75. Gronostajski RM, Pardee AB, Goldberg AL. The ATP dependence of the degradation of short and long lived proteins in growing fibroblasts. J Biol Chem 1985; 260: 3344-3349.

76. Orino E, Tanaka K, Tamura T et al. ATP-dependent reversible association of proteasomes with multiple protein components to form 26S complexes that degrade ubiquitinated proteins in human HL-60 cells. FEBS Lett 1991; 284: 206-210.

77. Gabai VL, Kabakov AE, Makarova YuM et al. DNA fragmentation in Ehrlich ascites carcinoma under exposures causing cytoskeletal protein aggregation. Biochemistry (Moscow, engl transl) 1994; 59: 399-404.

78. Gabai VL, Zamulaeva IV, Mosin AF et al. Resistance of Ehrlich tumor cells to apoptosis can be due to accumulation of heat shock proteins. FEBS Lett 1995; 375: 21-26.

79. Ucker DS, Obermiller PS, Eckhart W et al. Genome digestion is a dispensable consequence of physiological cell death mediated by cytotoxic T lymphocytes. Mol Cell Biol 1992; 12: 3060-3069.

80. Peitsch MC, Polzar B, Stephan H et al. Characterization of the endogenous deoxyribonuclease involved in nuclear DNA degradation during apoptosis (programmed cell death). EMBO J 1993; 12: 371-377.

81. Kim D, Lee YJ, Corry PM. Constitutive HSP70: Oligomerization and its dependence on ATP binding. J Cell Physiol 1992; 153: 353-361.

82. Brown CR, Martin RL, Hansen WJ et al. The constitutive and stress-inducible forms of hsp70 exhibit functional similarities and interact with one another in an ATP-dependent fashion. J Cell Biol 1993; 120: 1101-1112.

83. Beckmann RP, Mizzen LA, Welch WJ. Interaction of hsp70 with newly synthesized proteins: implications for protein folding and assembly. Science 1990; 248: 850-854.

84. Beckmann RP, Lovett M, Welch WJ. Examining the function and regulation of hsp70 in cells subjected to metabolic stress. J Cell Biol 1992; 117: 1137-1150.

85. Hightower LE, Sadis SE. Interaction of vertebrate hsc70 and hsp70 with unfolded proteins and peptides. In: Morimoto RI, Tissieres A, Georgopoulos C, eds. The Biology of Heat Shock Proteins and Molecular Chaperones. New York: Cold Spring Harbor Laboratory Press, 1994: 179-207.

86. Margulis BA, Welsh M. Analysis of protein binding to heat shock protein 70 in pancreatic islet cells exposed to elevated temperature or interleukin 1β. J Biol Chem 1991; 266: 9295-9298.

87. Haus U, Trommler P, Fisher PR et al. The heat shock cognate protein from *Dictyostelium* affects actin polymerization through interaction with the actin-binding protein cap32/34. EMBO J 1993; 12: 3763-3771.

88. Liao J, Lowthert LA, Ghori N, Omary MB. The 70-kDa heat shock proteins associate with glandular intermediate filaments in an ATP-dependent manner. J Biol Chem 1995; 270: 915-922.

89. Sadis S, Hightower LE. Unfolded proteins stimulate molecular chaperone Hsc70 ATPase by accelerating ADP/ATP exchange. Biochemistry 1992; 31: 9406-9412.

90. Hesketh JE, Pryme IF. Interaction between mRNA, ribosomes and the cytoskeleton (a review). Biochem J 1991; 277:1-10.

91. Flaherty KM, DeLuca-Flaherty C, McKay DB. Three dimensional structure of the

ATPase fragment of a 70K heat-shock cognate protein. Nature 1990; 346: 623-628.

92. Kampinga HH. Thermotolerance in mammalian cells. Protein denaturation and aggregation, and stress proteins. J Cell Sci 1993; 104: 11-17.

93. Tuijl MJM, van Bergen en Henegouwen PMP, van Wijk R et al. The isolated neonatal rat-cardiomyocyte used in an in vitro model for 'ischemia'. II. Induction of the 68 kDa heat shock protein. Biochim Biophys Acta 1991; 1091: 278-284.

94. Hohfeld J, Minami Y, Hartl F-U. Hip, a novel cochaperone involved in the eukaryotic Hsc70/Hsp40 reaction cycle. Cell 1995; 83: 589-598.

95. Longoni S, Lattonen S, Bullock G et al. Cardiac alpha-crystallin. II. Intra-cellular localization. Mol Cell Biochem 1990; 97: 121-128.

96. Chiesi M, Longoni S, Limbruno U. Cardiac alpha-crystallin. III. Involvement during heart ischemia. Mol Cell Biochem 1990; 97: 129-136.

97. Bennardini F, Wrzosek A, Chiesi M. αB-crystallin in cardiac tissue. Association with actin and desmin filaments. Circulation Res 1992; 71: 288-294.

98. Loktionova SA, Ilyinskaya OP, Gabai VL, Kabakov AE. Distinct effects of heat shock and ATP depletion on distribution and isoform patterns of human Hsp27 in endothelial cells. FEBS Lett 1996; 392: 100-104.

99. Engel K, Knauf U, Gaestel M. Generation of antibodies against human hsp27 and murine hsp25 by immunization with a chimeric small heat shock protein. Biomed Biochim Acta 1991; 50: 1065-1071.

100. Jakob U, Gaestel M, Engel K, Buchner J. Small heat shock proteins are molecular chaperones. J Biol Chem 1993; 268: 1517-1520.

101. Arrigo A-P, Landry J. Expression and function of the low-molecular-weight heat shock proteins. In: Morimoto RI, Tissieres A, Georgopoulos C, eds. The Biology of Heat Shock Proteins and Molecular Chaperones. Cold Spring Harbor, NY: Cold Spring Harbor Laboratory Press, 1994: 335-373.

102. Kato K, Hasegawa K, Goto S, Inaguma Y. Dissociation as a result of phosphorylation of an aggregated form of the small stress protein, hsp27. J Biol Chem 1994; 269: 11274-11278.

103. Benndorf R, Hayeb K, Ryazantsev S et al. Phosphorylation and supra-molecular organization of murine small heat shock protein HSP25 abolish its actin polymerization-inhibiting activity. J Biol Chem 1994; 269: 20780-20784.

104. Saklatvala J, Kaur P, Guesdon F. Phosphorylation of the small heat-shock protein is regulated by interleukin 1, tumor necrosis factor, growth factors, bradykinin and ATP. Biochem J 1991; 277: 635-642.

105. Pombo CM, Bonventre JV, Avruch J et al. Stress-activated protein kinases are major c-jun N-terminal kinases activated by ischemia and reperfusion. J Biol Chem 1994; 269: 26546-26551.

106. Bogoyevitch A, Ketterman AJ, Sugden PH. Cellular stresses differentially activate c-jun N-terminal protein kinases and extracellular signal-regulated protein kinases in cultured ventricular myocytes. J Biol Chem 1995; 270: 29710-29717.

107. Seko Y, Tobe K, Ueki K et al. Hypoxia and hypoxia/reoxygenation activate Raf-1, mitogen-activated protein kinase kinase, mitogen-activated protein kinases, and S6 kinase in cultured rat cardiac myocytes. Circ Res 1996; 78: 82-90.

108. Kato H, Chen T, Liu XH et al. Immunohistochemical localization of ubiquitin in gerbil hippocampus with induced tolerance to ischemia. Brain Res 1993; 619: 339-343.

109. Morimoto T, Ide T, Ihara Y et al. Transient ischemia depletes free ubiquitin in the gerbil hippocampal CA1 neurons. Am J Pathol 1996; 148: 249-257.

110. Kamikubo T, Hayashi T. Changes in proteasome activity following transient ischemia. Neurochem Int 1996; 28:209-212.

111. Jentsch S. Ubiquitin-dependent protein degradation: a cellular perspective. Trends Cell Biol 1992; 2: 98-103.

112. Molitoris BA, Dahl R, Hosford M. Cellular ATP depletion induces disruption of the spectrin cytoskeletal network. Am J Physiol (Renal Fluid Electrolyte Physiol) 1996; 40: F790-F798.

CHAPTER 4

ATP DEPLETION AS INDUCER OF HEAT SHOCK PROTEIN EXPRESSION

As we described in previous chapters, reduction in the ATP level leads to many disturbances in cellular structures and functions and, if sustained, will eventually result in cellular death. However, transient energy deprivation is well-tolerated by cells; moreover, as in the case of heat shock, the cells subjected to a reversible ATP decrease become tolerant to subsequent and more prolonged energy starvation. This adaptive response may be governed by HSP accumulation. In this chapter, we examine the data on the induction of *HSP* gene expression due to ATP depletion both in vitro and in vivo and discuss the possible mechanisms of such a response.

4.1. ANOXIA AND MITOCHONDRIAL INHIBITORS ACTIVATE HSP GENE EXPRESSION

The first indication that energy deprivation may induce HSP genes was obtained by Ritossa in the same work that discovered the heat shock response.[1] In *Drosophila* salivary glands, Ritossa has found that new puffs in polytene chromosomes appear not only after heating but also after treatment with 2,4-dinitrophenol (DNP), a well-known uncoupler of oxidative phosphorylation. Thereafter, the induction of all *HSP* genes (*HSP82*, *Drosophila* analog of *HSP90*, *HSP70/68*, and *HSP22-HSP27*) in salivary gland was observed upon recovery from anoxia and upon treatment with certain inhibitors of respiration (amytal, rotenone, azide), while other respiratory inhibitors (antimycin A, 2-heptyl-4-oxyquinoline-N-oxide) induced all *HSP* genes except *HSP82* (see ref. 2 for review). No effect on *HSP* induction, however, was observed with cyanide, another respiratory inhibitor, while oligomycin, an inhibitor of mitochondrial ATPase, caused only *HSP82* induction.[3] Since both cyanide and oligomycin resulted in a reduced ATP level, Leenders and coworkers concluded that ATP depletion per se was not responsible for HSP induction.[3] Nonetheless, at the present time this conclusion seems unsubstantiated since the effect of cyanide on the ATP level was not severe (ATP decreased by only 40%); furthermore, the combination of cyanide with oligomycin induced HSP68/70 and HSP22-27 synthesis in the cells. It seems that the regulation of *HSP82* gene expression in *Drosophila* cells may be different from that of other HSPs, but energy deprivation can, at the very least, induce HSP68/70 and HSP22-27 in these cells.

To our knowledge, the first evidence that anoxia induces thermotolerance in mammalian cells (chinese hamster V79 cell line) was obtained in 1981 by Rajaratnam and co-workers.[4] They found that 8 hours of incubation under anoxic conditions increased cell survival (colony forming ability) after hyperthermia about 100-fold, but the mechanism of this protection was not studied. In 1982, Li and Werb found that such incubation under anoxia resulted in enhanced synthesis of HSP90, HSP70, HSP60, and development of thermotolerance coincided with an increase in their synthesis.[5] Later, anoxia was also found to activate the synthesis of the so-called glucose-regulated proteins (GRP78 and GRP95), the HSP homologues located in the endoplasmic reticulum[6] (see ref. 7 for review). In EMT6/Ro mouse mammary tumor cells, Heacock and Sutherland found enhanced synthesis of some proteins which they termed "oxygen-regulated proteins" (ORPs), since their synthesis was activated during anoxia;[8] some of these proteins were later shown to be identical to GRPs.[9] In contrast to HSPs, GRPs cannot confer thermotolerance, but the cells with elevated GRP content are resistant to the actions of topoisomerase II inhibitors, cytotoxic T cells, and tumor necrosis factor (ref. 10-12). In the latter studies, no HSP synthesis was observed under anoxia, but the proteins were rapidly and transiently induced during reoxygenation.[7] This brings up the question: why does anoxia lead to HSP synthesis in some cases and GRP synthesis in others?

As was revealed recently, it is the accumulation of unfolded proteins inside the endoplasmic reticulum that triggers *GRP* gene activation (see ref. 13 for review),

while HSP induction is triggered by unfolded proteins in the cytosol (see chapter 1). Conditions which disturb protein folding in the endoplasmic reticulum such as glucose deprivation or inhibitors of glycosylation (e.g., 2-deoxyglucose, tunicamycin) are powerful inducers of GRPs.[13] When cellular respiration is blocked by anoxia, glycolysis is greatly stimulated. This stimulation reduces the intracellular glucose concentration and thus activates GRP synthesis. Apparently, all cells respond to anoxia by stimulating glucose consumption and activating GRP synthesis, but some of the cells (e.g., cardiomyocytes) are incapable of maintaining the ATP level during anoxia even if glucose is available (see section 2.1. above). Therefore, we suggest that HSPs are induced during anoxia only in those cells that lose ATP under anoxic conditions (see also section 4.5 for further discussion).

In 1986, Havemann and co-workers[14] studied the effects of uncouplers (DNP and carbonyl cyanide m-chlorophenylhydrazone, CCCP) on HSP synthesis in the murine mammary adenocarcinoma M8013 cell line. They found that after 2-8 hours of incubation uncouplers induced thermotolerance but, surprisingly, only CCCP activated HSP110, HSP90, HSP68/70 and HSP27 synthesis. The authors did not measure the ATP level in the uncoupler-treated cells and hypothesized that a nonspecific effect of CCCP (e.g., SH-group oxidation) was responsible for the stimulation of HSP synthesis, since both arsenite and disulfiram, well-known SH-reagents, also induced HSP synthesis and thermotolerance in the tumor cells.[14]

In 1990, a paramount paper by Benjamin et al[15] was published on the issue. In cultured C2C12 mouse myogenic cells, the researchers observed that 2 hours of severe hypoxia (pO_2 <7 torr) induced the DNA-binding activity of the heat-shock transcription factor (HSF) and the synthesis of mRNA of inducible HSP70 (HSP68). Moreover, the induction of HSP68 by hypoxia requires an intact heat-shock element (HSE). Therefore, the authors concluded that both hypoxia and heat shock induce the expression of the *HSP70* gene by a similar, if not an identical, mechanism. Two years later, the data from the above study was confirmed by Giaccia and co-workers who observed activation of HSF in normal human fibroblasts and squamous cell carcinoma following hypoxia ($pO_2 < 5$ torr), with maximal induction at or after 3 hours.[16] Thus, the above studies clearly demonstrate that hypoxia is capable of activating HSF in mammalian cells.

Meanwhile, Tuijl et al[17] found HSP68 transcription and translation following 30 minutes of "in vitro ischemia" (i.e., anoxia in a glucose-free medium) in cultured rat neonatal cardiomyocytes. Discussing their observation, the authors suggested that the ischemia-induced activation of HSP synthesis was associated with intracellular acidification.[17]

Therefore, by the early 1990s when we began our studies on murine tumor cells there was no clear evidence whether ATP depletion alone induces HSP synthesis and thermotolerance in mammalian cells. This question remained open because no measurements of the ATP level were made in previous studies. To address this problem, we investigated the effect of transient ATP reduction by mitochondrial inhibitors on HSP synthesis and thermotolerance in EL-4 thymoma and Ehrlich carcinoma cells. The cells were treated with CCCP or rotenone for 10-20 minutes in a glucose-free medium; this resulted in the rapid loss of ATP (to 4-6% of initial level, see chapter 2). After removal of the inhibitors and recovery of the ATP level in a rich medium for 3 hours, both thymoma and carcinoma cells became thermotolerant (Fig. 4.1).[18,19]

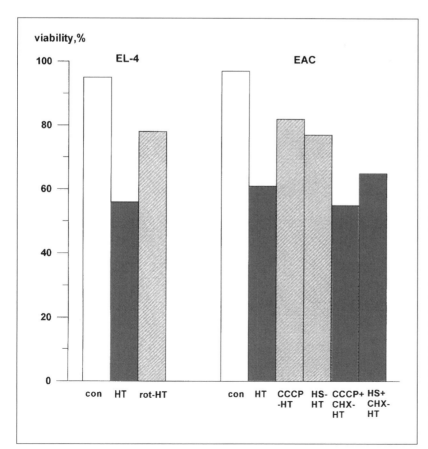

Fig. 4.1. Transient ATP depletion induces thermotolerance in EL-4 thymoma and EAC cells. The cells were first subjected to transient ATP depletion (2 µM rotenone for 15 minutes or 2 µM CCCP for 20 minutes in a glucose-free medium) and recovered for 3 hours in a rich medium to accumulate HSPs; then they were heated for 1 hour at 44°C (HT) and viability was determined by Trypan blue staining. Compared to nonpretreated cells (HT), rotenone or CCCP-pretreated ones (rot-HT or CCCP-HT) exhibit greater viability (p < 0.05), which was prevented by CHX (50 µM). For comparison, in EAC cells, heat shock pretreatment gave almost the same protection (HS-HT). See also refs. 18, 19 for other details.

Importantly, the acquisition of thermotolerance is completely prevented by an inhibitor of protein synthesis, cycloheximide. Moreover, when the cells are exposed to heat shock immediately after energy deprivation, they become, on the contrary, more thermosensitive (Fig. 4.1). The data indicate that the synthesis of some cytosolic protein(s) is necessary for thermotolerance development. To find out what proteins were induced after ATP depletion, the cells were pulse-labeled with ^{35}S-methionine or ^{14}C-leucine. We found that the synthesis of HSP70 and HSP90 was activated in thymoma cells; in addition, the synthesis of HSP27 was observed in Ehrlich carcinoma cells (see refs. 18, 19 and unpublished data). Using immunoblotting with monoclonal antibodies to constitutive HSP70 and inducible HSP68, we observed the accumulation of HSP70 in EL-4 thymoma and HSP68 in the carcinoma cells after ATP deprivation and recovery, while cycloheximide prevented their accumulation (Fig. 4.2). Neither HSP induction nor thermotolerance acquisition are observed when the cells are preincubated in a glucose-free medium without mitochondrial inhibitors or with both the inhibitors and glucose (i.e., when the ATP level was not reduced).[19] In addition to rotenone and uncoupler, HSP induction in the carcinoma cells was observed after oligomycin treatment

(Gabai, Kabakov, unpublished data). Therefore, our data clearly show that transient ATP depletion alone is able to induce a typical stress response (HSP synthesis and thermotolerance) in mammalian cells.

While our study was in progress, a remarkable paper by Benjamin et al was published.[20] When C2C12 myogenic cells were treated with rotenone in a glucose-free medium (i.e., under the same conditions that we used in our work), it reduced the ATP level by 70% and activated HSF after 30 minutes of incubation (Fig. 4.3). At the same time, a 50% decrease in ATP in a glucose-free medium without rotenone did not activate HSF. This work was the first that clearly demonstrated that HSF activation is caused by ATP depletion and some threshold of ATP decrease is necessary for the activation.[20]

Other energy inhibitors, uncoupler CCCP + 2-deoxyglucose, were shown to activate HSP70 synthesis in HeLa tumor cells,[21] while cyanide + 2-deoxyglucose brought about HSP70 mRNA synthesis in neonatal rat cardiomyocytes.[22] In the latter study, Iwaki and coworkers also investigated whether hypoxia-induced HSP70 mRNA synthesis is associated with ATP depletion. Surprisingly, no ATP reduction under hypoxic conditions was observed in the cells up to 3 hours after incubation, while stimulation of HSP70 mRNA

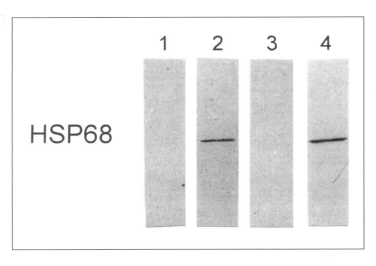

Fig. 4.2. Accumulation of HSP68 in thermotolerant EAC cells following transient ATP depletion (Western blot analysis with monoclonal antibodies). (Lane 1) control cells; (lane 2) cells recovered after treatment with 2 µM of CCCP for 20 minutes; (lane 3) the same as (2) but 50 µM of CHX was present during recovery; (lane 4) heat-shocked cells (44°C, 10 minutes).

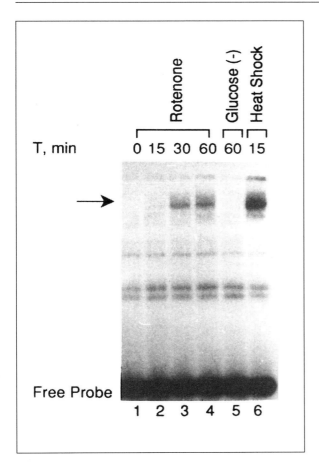

Fig. 4.3. Effect of metabolic inhibitors on DNA-binding activity in protein extracts from cultured C2C12 myogenic cells. Gel retention assays were performed using a synthetic double-stranded oligonucleotide containing a consensus heat shock element as probe. The position of migration of the unbound free probe in the gel is indicated. Stress-inducible DNA protein complexes (arrow) formed after exposure to rotenone (lanes 1-4) or glucose-free medium in the absence of inhibitor (lane 5) and after heat shock for 15 minutes at 42°C (lane 6). Reprinted with permission from Benjamin et al, J Clin Invest 1992; 89: 1685-1689.

synthesis occurred as early as 1.5 hours afterward. The authors suggested that there is another mechanism of HSP induction under hypoxia which is not associated with ATP depletion. In this study no HSF binding was found until after 2 hours of hypoxia, although HSP70 mRNA was already accumulated. As Knowlton hypothesized later in her review,[23] the above effect may be explained by the stabilization of HSP mRNA during hypoxia, similar to the marked increase in HSP70 mRNA stability after the shock in HeLa cells[24] (see also section 4.5 for further discussion).

Finally, Mestril et al studied the transcription of the *HSP70* gene under severe hypoxia (<0.6% O$_2$) in the rat myogenic cell line, H9c2.[25] Transient induction of HSP70 mRNA occurred within 2-4 hours of hypoxia. Two important discoveries were

also made in this work. First, the HSF form involved in the induction of the *HSP70* gene during hypoxia is the same form involved in the heat-shock response, namely HSF1. Second, two heat shock elements (HSEs) present in the HSP70 promoter (see chapter 1) are necessary for activation of chloramphenicol acetyltransferase (CAT) in transfected HSP70-promoter-CAT constructs during hypoxia. In contrast, the presence of either of the two HSEs is sufficient for its inducibility following heat shock. From the data, the authors suggested that the hypoxic induction of the HSP70 gene, although mediated by the same regulatory elements (HSEs) and factor (HSF1) as heat shock, either does not share the same mechanism or that hypoxia may represent a weaker form of stress than heat shock. We believe that the latter suggestion is more correct (see section 4.5 for further discussion).

Most of the findings to date on the induction of heat shock genes under hypoxia or mitochondrial inhibitors in mammalian cells are summarized in Table 4.1. The data clearly show that energy deprivation, at the very least, activates HSF, HSP mRNAs and HSP synthesis in various cells.

4.2. ISCHEMIA-INDUCED HSP SYNTHESIS IN THE HEART

The first indication that ischemia induces HSP synthesis in the heart and other tissues was obtained by Currie and White[26] as early as 1981. They observed that sliced rat tissues, including brain, thymus, lung, spleen, liver, kidney and heart, rapidly synthesize a novel protein with a molecular weight of 71 kDa (P71). The protein is an inducible form of HSP70, as was shown

Table 4.1. Effects of anoxia and mitochondrial inhibitors on the induction of heat shock genes in mammalian cells

Exposure	Cells	HSF Activation	HSP Gene		Ref.
			Transcription	Translation	
1. Anoxia (hypoxia)	Chinese hamster fibroblasts	ND	ND	HSP90, 70, 60	5
+glucose	Mouse myogenic cells (C2C12)	+	HSP70	ND	15
	Human fibroblasts AG1532	+	ND	ND	16
	Human carcinoma JSQ-3	+	ND	ND	16
-glucose	Rat cardiomyocytes	ND	HSP68	HSP68	17
	Rat cardiomyocytes	+	HSP70(68)	HSP70(68)	22
	Rat myogenic cells H9c2	+	HSP68	ND	25
2. Rotenone -glucose	Mouse myogenic cells C2C12	+	–	–	20
	Mouse thymoma EL-4	ND	ND	HSP90, HSP70	19
3. Uncoupler CCCP +glucose	Mouse carcinoma M8013	ND	ND	HSP110, HSP90, HSP70, HSP27	14
+2-deoxyglucose	Human HeLa carcinoma	ND	ND	HSP70	21
-glucose	Mouse Ehrlich carcinoma	ND	ND	HSP90, HSP70(68) HSP27	18
4. Cyanide +2-deoxyglucose	Rat cardiomyocytes	+	HSP70	HSP70	22

ND - not determined

in later studies done by Currie and White. During the following year, the synthesis of a similar protein was found by Hammond and coworkers[27] in the rat heart after 1 hour of aortic banding or heat stress (42°C). The authors suggested that lactic acidosis was the cause of HSP70 induction since increased lactate accumulation was found during both stresses. In 1986, Dillmann et al[28] observed mRNA accumulation of the inducible HSP70 in the ischemic area of dog heart, 6 hours after occlusion of the left anterior descending artery. In the same year, Howard and Geoghegan[29] observed increased transcription of the *HSP70* and *GRP* genes in the mouse heart under hypoxia (hypobaric decompression).

In 1987, Currie conducted the first systematic study of this problem.[30] He used isolated rat hearts and investigated the effect of the ischemia duration and the temperature of incubation on protein synthesis during reperfusion. At 4°C, ischemia for 0.5-4 hours had no effect on HSP synthesis, but after 17 hours, synthesis of a protein with a MW of 71 kDa (an inducible form of HSP70) was stimulated more than 10-fold. At 20°C, as little as 1 hour of ischemia is sufficient to activate HSP70 synthesis, and at 30°C this effect is even more pronounced, resembling that of heat shock (perfusion at 40°C). Therefore, the data clearly indicate that it is not ischemia per se but some ischemia-induced temperature-dependent metabolic event that is responsible for HSP induction in the heart. In particular, it is well-known that a decrease in temperature slows down energy deprivation in ischemic organs. In later studies on rat hearts in situ, Donnelly et al[31] revealed through Western blotting the accumulation of inducible HSP70 (HSP72) at 8 hours of reperfusion after 20 minutes of left coronary artery occlusion, although this effect was not as pronounced as after 20 minutes of hyperthermia at 42°C.

In 1991, Knowlton et al[32] found that in open chest preparation of rabbit heart as little as 5 minutes of ischemia increases inducible HSP70 mRNA synthesis within

the ischemic zone about two-fold. Repetitive ischemia with four 5 minute occlusions/5 minute reperfusions greatly affected HSP70 mRNA and increased HSP70 content as early as 2 hours after ischemia, with maximal accumulation after 24 hours when HSP70 mRNA expression returned to the basal level. Interestingly, four episodes of ischemia also increased HSP70 mRNA in the nonischemic area of the ventricle and, as the authors suggested, this may be due to an increased workload of the nonischemic ventricle, a condition which is capable of inducing HSP synthesis.[33] Using the same experimental rabbit model, Marber et al observed 2.5-fold increase in HSP72 content after 24 hours of four episodes of 5 minute ischemia-reperfusion; in addition, a $1^2/_3$-fold accumulation of HSP60, a mitochondrial stress protein, was also found.[34] Recently, the experiments of Tanaka et al confirmed the HSP72 accumulation in rabbit hearts after similar treatment.[35] Besides HSP72, the induction of HSP27 and HSP90 transcription was found by Das et al in the rat heart following four episodes of 5 minute ischemia/10 minute reperfusion, whereas one episode resulted in the induction of HSP70 only.[36] The protocol of HSP induction (so called "ischemic preconditioning") is usually used in the studies of HSP protective function in the heart (see the next chapter).

In the porcine myocardium, which more closely resembles the human one, Sharma et al[37] reported an increased accumulation of HSP70 mRNA after two cycles of 10 minutes left anterior descending coronary artery occlusion and 30 minutes reperfusion. In the next study of this group, Knoll et al[38] showed by run-on transcriptional assay that this effect is due to increased transcription of the gene rather than RNA stabilization. On the same model, Andres and coworkers[39] studied the effect of two episodes of 10 minute ischemia/30 minute reperfusions on ubiquitin, HSP27 and HSP60 mRNA expression. No increase in HSP60 mRNA was observed whereas about a three-fold accumulation of ubiquitin and HSP27 mRNA was found after 30 minutes of reperfusion with a subsequent decrease to the control level after 210 minutes. The mRNA accumulation, as in the case of HSP70 mRNA, was due to an increase of four to five times the rate of transcription rather than their elevated stability. Ubiquitin mRNA synthesis was associated with the formation of new ubiquitin-protein conjugates. Surprisingly, despite marked activation of transcription, no accumulation of HSP27 after ischemia was found. The authors suggested that the absence of HSP27 accumulation following its gene transcription may be due to either a negative control of the translation or a rapid turnover (degradation) of the protein under ischemia/reperfusion. However, further studies are necessary to resolve the problem.

The question arises as to whether the induction of stress proteins in post-ischemic heart is associated with ATP depletion. Regrettably, in all of the above described studies ATP content was not determined. However, there are some works in which the parameter was measured under similar experimental conditions. For instance, in the isolated rat heart 10-20 minutes of ischemia reduced ATP content by 33-50% (see refs. 40-43). In porcine myocardium, a 50% reduction in blood flow decreased the subendocardial ATP level by approximately two-fold after 10 minutes.[44] In canine heart, even 5 minutes of ischemia resulted in a reliable (20%) fall in the ATP level, and four 5 minute episodes of ischemia/reperfusions (the preconditioning protocol used in the above studies on HSP induction)[32,34,35] caused an approximate two-fold decrease in the ATP content.[45] Therefore, we conclude that even very short-term ischemia reduces ATP content in the heart. Ischemia-induced transcription of the major cardiac *HSP* genes (*HSP70*, *HSP90*, *HSP27*) indicates HSF activation during ATP depletion. This was shown in vitro on myogenic cells by Benjamin et al[20] (see the previous section).

At the same time, in vivo models of ischemia are much more complex than in vitro models of isolated cells, and other factors besides ATP depletion may be important for the induction of stress proteins. In particular, it is well-known that not only ischemia but subsequent reperfusion may be harmful for myocardium because of post-ischemic oxidative stress (see section 2.4). Recently, Kukreja et al[46] observed a marked increase in HSP70 mRNA in isolated rat hearts after the oxidative stress induced by exogenous xantine/xantine oxidase, irradiated rose bengal, or H_2O_2; moreover, the addition of superoxide dismutase (SOD) diminished the increase in HSP70 mRNA after ischemia-reperfusion. As the authors concluded, one of the potential mechanisms of HSP70 expression caused by ischemia-reperfusion may be oxygen radicals. Das and coworkers[36] observed that ischemia-reperfusion activates transcription of not only HSPs but also some antioxidative enzymes, namely catalase and Mn-SOD (a mitochondrial form of SOD), indicating the occurrence of oxidative stress during reperfusion. We believe that oxidative stress plays a certain role in the stress protein induction under particular conditions, but only as a secondary stressing factor which follows ATP depletion (see also section 4.5).

An interesting study has been recently carried out in patients undergoing cardiac operations. Mcgrath et al[47] obtained sequential right atrial biopsy specimens from patients undergoing repair at three intervals: before bypass, after reperfusion and after bypass. Immunoblot analysis of the inducible HSP72 demonstrated high expression of the protein in the human heart compared with other mammalian hearts, but its level was not changed after reperfusion or after bypass. The authors suggested that the high level of HSP72 detected in human hearts reflects preoperative disease, drug therapy, or an inherently high level of this stress protein.[47] In future investigations, it will be interesting to compare the data with the HSP72 level in the hearts of healthy humans, e.g., who died from accidents.

More exciting work has recently been done on young (2-month old) and old (18-month old) rats subjected to left coronary artery occlusion. Nitta et al[48] found that 10 minutes of ischemia markedly induces HSP72 mRNA in young hearts but only slightly in old hearts, while 20 minutes of ischemia had pronounced effects on both young and old hearts. From the data, it was hypothesized that old hearts have a defective sensing mechanism for ischemia. Furthermore, Rowland et al[49] observed that hearts from newborn rats (7-10 days old), when compared to young rat hearts, demonstrate a greater expression of HSP72 after 20 minutes of ischemia. It will be very interesting to compare ischemia-induced ATP depletion in hearts from newborn, young and old rats. Interestingly, old rats showed significantly decreased accumulation of HSP70 in the myocardium[50] and other organs after whole-body hyperthermia (see ref. 51 for review), which may indicate a similar mechanism for HSP induction under ischemia and heating (see section 4.5).

4.3. STRESS PROTEIN EXPRESSION IN POST-ISCHEMIC BRAIN

The brain is the most hypoxia- and ischemia-sensitive organ of mammals. As much as a few minutes of global cerebral ischemia is sufficient to cause irreversible brain dysfunction and death of the whole organism, whereas more prolonged ischemia (about 30 minutes) is tolerated by the heart (see the previous section). The dramatic brain vulnerability to ischemia may be due, on the one hand, to its strong dependence on oxidative metabolism and, on the other hand, to its high activity of ATP-consuming ion transporters.[52] Since several good reviews on ischemia-induced expression of HSPs in the brain have been recently published (see refs. 53-55), we only summarize the data here.

4.3.1. HSP72

Rats and gerbils are the most popular and intensively studied models of stress protein expression under cerebral ischemia. First of all, it is well-established that transient global or focal ischemia leads to a marked increase in mRNA of inducible HSP70 (HSP72) in various brain regions. Importantly, the activation of the heat shock transcription factor (HSF) was shown in gerbil hippocampus following transient ischemia (see ref. 55). Rat hippocampal slices showed that as little as 5 minutes of in vitro ischemia resulted in the stimulation of HSP72 synthesis for the first hour after the stress.[56] Similarly, an early appearance of HSP72 mRNA was detected following 10 minutes of forebrain ischemia in gerbils.[57] When localization of HSP72 mRNA was studied by in situ hybridization in gerbil hippocampus after 5 minutes of global ischemia, Nowak[58] found that the duration of HSP72 mRNA expression was considerably more prolonged (up to 48 hours) in CA1 pyramidal neurons, the most vulnerable cell population, than in CA3 neurons (Fig. 4.4). However, transcription of the *HSP72* gene did not lead to the accumulation of the protein in CA1 neurons, apparently because of the irreversible arrest of translation and subsequent delayed death (within 2-4 days) of the cells. The translationary block is associated with the suppression of eukaryotic initiation factor 2 (eIF-2) activity.[59] At the same time, shorter ischemic insult results in the preferential accumulation of HSP72 in the area, and as a result pharmacological interventions that reduce ischemic damage imitate brief ischemic periods.[55] The expression of HSP72 in the hippocampus following an increased period of global ischemia correlates with the regional and cellular vulnerability to ischemia: CA1 neurons express HSP72 after the briefest periods of ischemia followed by CA4, CA3, dendate granule neurons, glia and lastly, endothelial cells.[61] The data demonstrate the dependence of HSP72 expression on the severity of cell damage. Other studies also showed the transcription of *HSP72* genes both in surviving and dying neurons, but the synthesis and accumulation of the

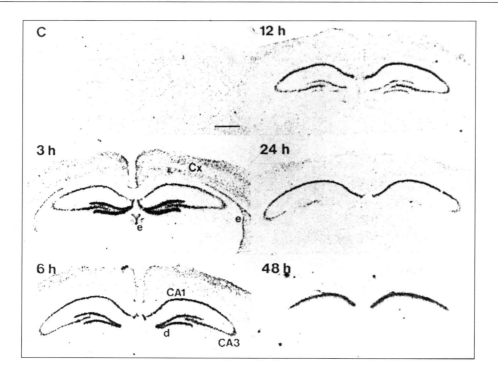

Fig. 4.4. In situ hybridization of HSP72 mRNA expression in gerbil hippocampus after transient ischemia, illustrating prolonged expression in the selectively vulnerable CA1 pyramidal neurons. C, control (nonischemic tissue). The signal is progressively lost from dentate granule cells (d) and CA3 neurons that survive the insult. Ependymal cells (e) that constitutively express HSP72 in the gerbil also show transient induction after ischemia. Reprinted with permission from Nowak TS, J Cereb Blood Flow Metab 1991; 11: 432-439.

protein occurred only in surviving neurons.[55,62,63] As in the case of ischemia-induced stress protein expression in rat heart (see the previous section), a decrease in temperature to 30°C during ischemia markedly diminishes the HSP72 expression in gerbil brain,[64] and hypothermia at 15° had a similar effect in piglet brain.[65]

In contrast to the heart, there is a prolonged lag period in the detection of HSP72 immunoreactivity in the brain, and this delay in detection is well beyond recirculation, at which point translation activity is restored.[55] As an example, after 15 minutes of cerebral ischemia in gerbils, HSP72 is detected in the CA3 region after only 24 hours and in the cerebral cortex after 96 hours.[66] One of the possible explanations of this effect may be the

masking of HSP72 epitopes involved in the protein-protein interaction of functional importance in the stress response.[55] However, this requires further investigation.

As we mentioned above, in comparison with other brain cells, neurons are the most vulnerable cell population. After focal ischemia, neurons rapidly die within the ischemic core, while a major surviving cell type of this area, vascular cells, accumulates HSP72. At the edge of infarct, astrocytes and microglia also express HSP72. Accumulation of HSP72 in neurons is detected only in the regions surrounding a focal insult. However, shorter periods of focal ischemia that allow for the subsequent survival of neurons within the injury focus result in a more striking HSP72 expression throughout the reperfused territory.[55,67]

Therefore, the accumulation of HSP72 after focal ischemia occurs only in those cell types that survive ischemic insult, which coincides with HSP72 accumulation after global ischemia.[61]

4.3.2. OTHER STRESS PROTEINS

Expression of other stress proteins have also been detected after cerebral ischemia. In particular, a moderate increase in the level of constitutive HSC70 mRNA is observed in gerbil hippocampus.[60,68,69] Wang et al[70] found the ischemia-induced transcription of the *GRP78* gene along with HSP72 in rat cerebral cortex. Both GRP78 and GRP95 mRNA synthesis was observed by Lowenstein and collegues[71] in rat hippocampus and cortex. Massa et al[72] studied the mRNA induction of the mitochondrial analog of HSP70 (GRP75) after focal ischemia in rat brain. They observed that when the degree of injury is small, GRP75 mRNA expression in the area of injury is similar to the pattern observed for HSP70 mRNA. However, when the injury is extensive, GRP75 is upregulated only in the neurons outside the ischemic area. Abe et al[68] observed the expression of mitochondrial HSP60 in CA1 neurons after 3.5 minutes of transient ischemia in gerbils. The data indicate that not only cytoplasmic but endoplasmic and mitochondrial chaperones are also induced after cerebral ischemia (see section 4.5 for discussion).

Recently, several studies on HSP27 accumulation following cerebral ischemia have been conducted by Kato and coworkers. A 3 minute period of global ischemia caused a 2.4-fold increase in HSP27 after 3 days in the rat hippocampus but unlike HSP70, glial cells rather than neurons expressed HSP27. No alteration in αB-crystallin content was detected.[73] In rat brain following 1 hour of middle cerebral artery occlusion, HSP27 is induced in the microglia of the ischemic center after 4 hours, and after 1 through 14 days in reactive astrocytes distributed widely throughout the conlateral hemisphere.[67] In gerbil brain, 2-3 minute periods of global ischemia does not result in HSP27 accumulation in

any neurons either, but this protein is induced in a small number of astrocytes in the CA3 regions and in many astrocytes in the dendate hilus.[74] From the data, the authors suggest that HSP27 expression might be a component of glial rather than neuronal reaction to injury.[67,74]

Nowak et al[75] found a 70% increase in the ubiquitin mRNA level 6 hours after transient ischemia in gerbil brain. This may be associated with the post-ischemic activation of protein ubiqitination[76] and thus resembles the effect observed in the porcine heart (see the previous section).

Takeda and coworkers have studied the expression of hemeoxygenases (an inducible form, HO-1, and a constitutive form, HO-2) in rats following 20 minutes of forebrain ischemia. Maximal expression of HO-1 mRNA was observed after 12 hours of reperfusion in both neuronal and glia-like cells distributed in the neocortex, hippocampus and thalamus.[77] The level of HO-1 protein expression was also maximal at 12 hours. The maximum expression occurred in pyramidal neurons and astrocytes of the cortical mantle and astrocytes of hippocampal CA2 and CA3 subfields, whereas no HO-1 protein was detected in CA1 subfield despite an increased level of transcripts.[78] The absence of HO-1 protein accumulation in the CA1 region may be due to an impairment of translation after severe ischemia, as it was demonstrated for HSP72 (see above). The level of HO-2 protein was not noticeably affected by ischemia.[78]

4.3.3. ATP DEPLETION AND STRESS RESPONSE

What is the mechanism of the ischemia-induced expression of stress proteins in the brain? Cerebral hypoxia or ischemia very rapidly (within several minutes) depletes high energy phosphate stores (see refs. 79, 80), and in vitro isolated neurons under anoxia die within 5-10 minutes.[81] However, their death can be markedly delayed at low temperatures (25°C), demonstrating that the suppression of metabolic activity (i.e., ATP consumption by ion transporters) exerts a powerful protective effect.[81] Therefore, short-term ischemia evoking HSP induction substantially reduces the ATP level in the brain, and the inhibitory effect of low temperature on the stress response may be associated with ATP preservation. Prevention of HSP72 mRNA induction in rat hippocampus after global ischemia by K^+-channel openers[82] is also associated with high-energy phosphate maintenance as was observed in the heart (see section 2.3 and ref. 83 for review).

There is a study in which regional alterations of ATP and HSP72 mRNA were examined using ATP-luminescence histochemistry and in situ hybridization. In this work, Kobayshi and Welsh[84] subjected neonatal rats to unilateral carotid artery ligation followed by exposure to hypoxia (8% O_2) for 80 minutes. At the end of hypoxia, ATP levels were decreased in an irregular pattern within the hemisphere ipsilateral to the carotid ligation. Expression of HSP72 mRNA was not detected prior to recovery, except in the ventricular lining of the ipsilateral hemisphere. After 2 hours of recovery, HSP72 mRNA is expressed in a diffuse pattern in the ipsilateral hemisphere, even in the regions in which the distribution of ATP remained patchy. Although the authors have not found close correlation between ATP depletion and HSP72 expression, they conclude that hypoxia-ischemia causes regionally distinct alterations in the ATP and HSP72 mRNA levels, which may be related to nonuniform cell injury in this model (see also section 4.5 for discussion).

One of the popular viewpoints for the mechanism of HSP induction after cerebral ischemia is that it is the result of an elevation of intracellular Ca^{2+} (see refs. 54, 55). Indeed, ischemia evokes rapid influx of Ca^{2+} in neurons due to the membrane depolarization and the opening of voltage-dependent Ca^{2+}-channels.[85] In addition, ischemia in vivo may lead to the accumulation of an excitatory amino acid, glutamate, followed by its binding to the N-methyl-d-aspartate (NMDA) receptors and its opening of large membrane ion channels

through which both sodium and calcium enter into neurons (see ref. 54 for review and chapter 6). However, at the present time there are no clear data on HSP induction in neurons after the glutamate addition and stimulation of NMDA receptors. For instance, Lowenstein and colleagues have not found the accumulation of HSP72 in cerebellar granule cell cultures treated with glutamate, but it has occurred after thermal stress.[86] Moreover, in the rat brain, HSP72 is expressed after injection of NMDA receptor antagonists (e.g., ketamine, MK-801) which block Ca^{2+} influx.[87] Thus, the role (if any) of Ca^{2+} influx in the induction of HSP72 in neurons under ischemia requires further study (see also section 4.5. for discussion).

4.4. ISCHEMIA-INDUCED STRESS PROTEIN SYNTHESIS IN OTHER NORMAL TISSUES AND TUMOR CELLS

Although the heart and the brain are the main organs in which ischemic damage and HSP expression have been intensively studied, there are some data on stress protein synthesis after ischemia in other tissues.

4.4.1. LIVER

The lab of Bernelli-Zazzera investigates HSP synthesis in ischemic liver. The researchers first described the accumulation of mRNA for HSP70 and HSP90 during reperfusion of liver after 60 minutes of ischemia with a maximum at 2 hours after cessation of ischemia. Second, both constitutive HSC73 and inducible HSP70 mRNAs accumulated. Third, run-on experiments demonstrated that the accumulation of these RNAs was due to their increased rate of transcription.[88-90] Shorter periods of ischemia (15-30 minutes) did not activate *HSP70* gene transcription, whereas more prolonged 120 minute ischemia activated its transcription but did not lead to HSP70 accumulation, apparently because of irreversible liver cell damage and death.[91] The similar time-effect dependence is described above for neurons but it takes a significantly longer time for irreversible injury of liver cells than neurons. The most important finding of the group is the activation of HSF at the end of 60 minutes of ischemia without reperfusion; after reperfusion, HSFs binding to HSE increased at 30 minutes but disappeared in the next 30 minutes.[91] During reperfusion, the activation of hemeoxygenase (HO-1) gene expression was also observed at a maximum at 4 hours.[91] In a recent study, this group demonstrated that chemicals causing oxidative stress were not able to promote HSF activation and HSP70 expression; this indicates that ischemia per se rather than reperfusion-induced oxidative stress is responsible for HSP70 synthesis.[92] The authors suggested that, like heat challenge, ischemia can provoke protein misfolding, thus triggering the stress response[91] (see the next section for further discussion).

4.4.2. KIDNEY

There are also several studies on HSP induction after renal ischemia (reviewed in ref. 93). In the first work, Emami and co-workers[94] examined the expression of inducible HSP72 in rat kidney after transient ischemia and heat stress by immunoblotting (42°C, 15 minutes). They found that 15 minutes was sufficient to induce HSP72 accumulation after 4 hours of reperfusion while the maximal effect was achieved at 60 minutes following ischemia and peaked after 12 hours of reperfusion; this HSP72 accumulation persists for up to 5 days. However, prolongation of ischemia to 90 minutes does not bring about any HSP72 accumulation since, apparently, this results in irreversible renal failure (compare with the above data on the brain and the liver). When the intrarenal distribution of HSP72 after ischemia was examined, the researchers found that relative expression was: papilla > cortex > medulla. Interestingly, this cellular distribution of HSP72 accumulation resembled those cells after heat shock. As in the case of heat shock, ischemia-induced HSP72 was localized to the soluble, but not to the membrane fraction. The

authors concluded that transient ischemia-like heat stress in kidney induces HSP72, which may be important in mediating cell repair or increasing resistance to subsequent injuries.[94]

In the second work, Van Why and colleagues[95] also studied the expression of the inducible HSP72 after 45 minutes of renal artery clamping in rats. They first demonstrated the appearance of inducible HSP70 mRNA at 15 minutes of reperfusion; it peaked between 2 and 6 hours and fell by 24 hours. HSP72 accumulated progressively through 24 hours and was found both in soluble and microsomal fractions following ischemia. Within proximal tubules, the protein is initially localized to the apical membrane but after 2-6 hours of reperfusion it had dispersed through the cytosol in a vesicular pattern and migrated from the apical domain within 24 hours. No intranuclear HSP72 was detected, but a portion of the vesicular HSP72 was associated with lysosomes. Although in both of these studies the renal ATP level was not assessed, renal ischemia in rats rapidly exhausts ATP (to 17% after 5 minutes, and to 10% after 15 minutes, see ref. 96).

A very important finding was made recently. Van Why et al[97] examined the relationship between cellular ATP (monitored in vivo by ^{31}P-NMR spectroscopy) and the induction of the stress response (HSF activation and *HSP72* gene transcription) in the renal cortex. Cellular ATP was reduced and maintained at a specific, stable level by partial aortic occlusion for 45 minutes. When the ATP level was maintained above 60% the of control, no HSF activation or HSP72 mRNA induction was detected. However, reduction in the cortical ATP level to 35-50% preocclusion values resulted in HSF activation and a low-level expression of inducible HSP72 mRNA. A greater level of HSF activation and subsequent transcription was detected at 20-25% of the control ATP level, and HSF binding was revealed at the end of 15 minutes of total occlusion. The researchers concluded that a 50% reduction of cellular ATP in the renal cortex occurs before the stress response is detectable, and that reperfusion is not required for the initiation of the heat-shock response in the kidney. Therefore, as in the liver, ischemia per se is able to activate HSF in the kidney. In addition, there is a threshold of ATP depletion for HSF activation, as was observed earlier by Benjamin et al[20] in myogenic cells in vitro (see section 4.1.).

Expression of stress gene mRNAs has also been studied in obstructed rabbit urinary bladder using a semiquantitative reverse transcription-polymerase chain reaction method (RT-PCR). The partial outlet obstruction also leads to ischemic-like conditions as a result of blood vessel occlusion.[98] At 24 hours following obstruction, mRNA of HSP70, HSP60 and HSP27 increased 2.4-, 5.6-, and 4.3-fold, respectively.

4.4.3. TUMORS

Therefore, we conclude that the activation of HSP expression after ischemia is a widespread phenomenon in various organs and tissues. However, the question arises as to whether the event may occur not only under experimental manipulation such as the clamping of blood vessels, but also under natural conditions in vivo. One such situation is tumor growth, which is characterized at later stages by oxygen and nutrient starvation of cancer cells due to insufficient blood flow and may lead to both hypoxic and ischemic conditions within tumor tissues.[99] As we have mentioned above (see section 4.1.), in vitro activation of HSF by hypoxia in tumor cells was first described by Giaccia et al in 1992. In the same paper, researchers described mouse squamous cell carcinoma SCCVII growing in vivo, in contrast to the same cells in vitro which already contain activated HSF; they found that its binding to DNA could be further stimulated by 15 minutes of tumor clamping. From the data, the authors suggested that HSF activation could be a useful marker to monitor tumor hypoxia.[16]

In the next year, Cai and collegues[100] studied the expression of GRPs during in vivo growth of murine radiation-induced fibrosarcoma (RIF). They observed that GRP78 and GRP94 were not synthesized in small tumors (<0.1 g), but began to accumulate in intermediate (0.2-0.8 g) tumors with maximal expression in large tumors (>1.8 g); the elevation of GRP78 mRNA was also demonstrated by Northern blot both in large tumors and RIF cells exposed to anoxic stress in vitro. The authors concluded that RIF tumor cells undergo a glucose-regulated stress response during tumor growth. Interestingly, they did not observe in these tumors the synthesis of either HSP70 or HSP90, whose synthesis might be expected due to the HSF activation reported in the above study of Giaccia et al.[16] Therefore, it is unclear whether HSF activation leads to HSP accumulation and tumor thermotolerance.

We investigated this problem on Ehrlich ascites carcinoma. Many years ago researchers reported the almost total absence of oxygen and glucose in ascites fluid of stationary EAC cells,[101,102] i.e., ischemic-like conditions which occur as a result of rapid growth and high metabolic activity of these cells. Hence, in principle such conditions may lead to the activation of stress protein synthesis. Indeed, Benndorf et al[103] reported the accumulation of some proteins in stationary cells of EAC; the majority of them were later shown to be HSP27.[104] However, researchers have not studied the expression of other HSPs in these cells.

We first observed that stationary EAC cells become considerably more resistant to heating than exponential cells and have decreased levels of heat-induced protein aggregation.[105] Therefore, we suggested that such effects are associated with the accumulation of HSPs in stationary cells. To test this assumption, we took cells from the peritoneal cavity of mice on the fifth day of growth (exponential phase) or on the eighth day (stationary phase) (see Fig. 4.5) and then performed Western blotting with antibodies to different HSPs. As shown in Fig. 4.6, stationary EAC cells accumulated

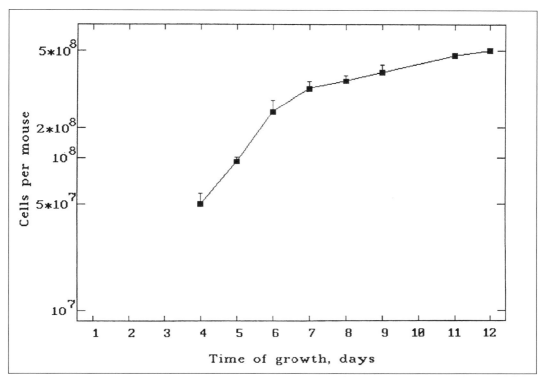

Fig. 4.5. Time course of EAC cell growth in mice. EAC cells were inoculated in peritoneal cavities of C57BL/6 mice (10⁷ cells per mouse), and the number of the cells in isolated ascitic fluids was counted every day starting from the fourth day after inoculation. Each point represent means ± SEM of three to six mice. Reprinted with permission from Kabakov AE et al, J Cell Physiol 1995; 165: 1-7.

all major HSPs, namely HSP70, HSP27, HSP90;[106] this obviously indicates the activation of HSF in these cells. In order to evaluate the possible role of ATP depletion in triggering the stress response in stationary EAC cells, we compared the content of adenine nucleotides in exponential phase with that of stationary cells. Although the total content of adenine nucleotides did not significantly differ in exponential and stationary cells, we observed a marked reduction in ATP content (by 40%), and ATP/ADP ratio, and a reliable decrease in the energy charge of stationary cells compared to exponential ones (Fig. 4.7, Table 4.2). Importantly, when stationary cells were incubated in vitro for several minutes under aerobic conditions with glucose, their ATP content, ATP/ADP ratio and energy charge were rapidly restored

to levels equivalent to exponential cells.[106] In addition, we did not find any significant reduction in the respiratory and glycolytic activity of stationary cells in vitro compared to exponential ones (Gabai, unpublished data). Previously, Skog et al[107] also observed a decrease in ATP content by 60% and a decrease in the ATP/ADP ratio from 3.2 to 1.2 in EAC cells reaching the plateau phase of growth. This energy imbalance is also reversed by the addition of glucose. The observations imply that it is the microenvironment of the stationary cells rather than the impairment of their ATP-generating systems that leads to the energy imbalance within them. Apparently, the energy-depriving conditions in vivo activate stress protein synthesis, which supports the findings that when exponential EAC cells in vitro are subjected to

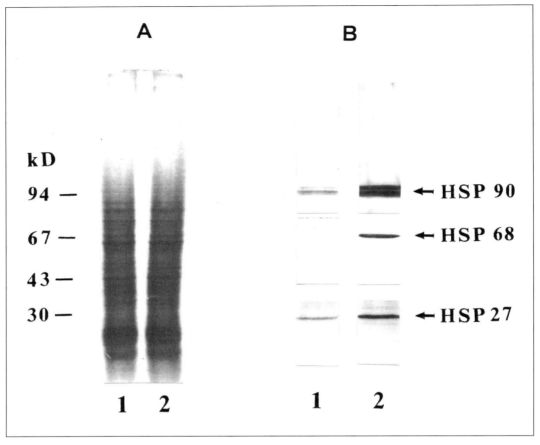

Fig. 4.6. Accumulation of major cytosolic HSPs in stationary EAC cells. One million freshly isolated EAC cells in exponential (lane 1) or stationary (lane 2) phases of growth were lysed and run by PAGE (A). The proteins were blotted onto a nitrocellulose filter and segments of the filter were separately stained by indirect immunoperoxidase reaction with antibodies to HSP90, HSP68 and HSP27 (B). The experiment was repeated three times with similar results. Reprinted with permission from Kabakov AE et al, J Cell Physiol 1995; 165: 1-7.

Table 4.2. ATP/ADP ratio and energy charge in exponential and stationary EAC cells in vivo

Cell Culture State	ATP/ADP Ratio	Energy Charge
Exponential	4.36 ± 0.82	0.886 ± 0.020
Stationary	$2.40 \pm 0.25^*$	0.832 ± 0.003

$^*P < 0.05$ by Student's t-test
ATP/ADP ratio and energy charge ([ATP + 1/2ADP]/ATP+ADP+AMP) were calculated from the data of Fig. 4.7. ATP, ADP, and AMP were measured by HPLC as described in ref. 106.

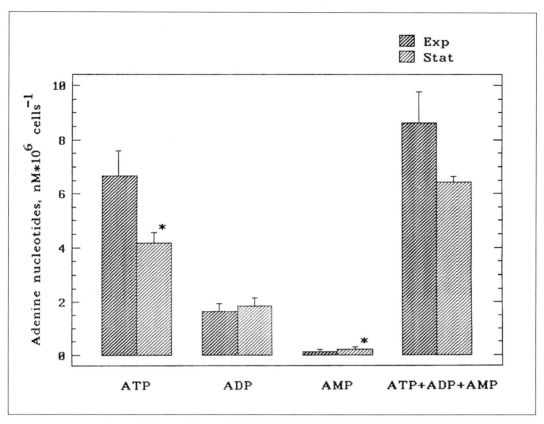

Fig. 4.7. Adenine nucleotide content of EAC cells in exponential and stationary phases of growth. Ascitic fluids were isolated on the fifth day (exponential) or the eighth day (stationary phase) of growth, and the content of ATP, ADP, and AMP was determined by HPLC. * P < 0.05 by Student's t-test. Reprinted with permission from Kabakov AE et al, J Cell Physiol 1995; 165: 1-7.

ATP depletion, they do accumulate HSPs (see section 4.1). This phenomenon is not obviously associated with the genotypic alteration and subsequent selection of resistant genotype, because the inoculation of stationary cells into intact mice after several days reduced their cellular HSP content to that of exponential cells.

It may seem strange that stationary cells down-regulate HSP synthesis as they return to favorable conditions. However, other researchers found that the overexpression of HSP27 markedly reduces the growth rate of EAC cells[104] whereas the overexpression of HSP70 inhibits the proliferation of rat fibroblasts[108] and 10T1/2 mouse cell line.[109] Therefore, tumor cells

probably have to inhibit HSP synthesis to attain maximal growth rate.

Another question is: why have Cai et al[100] observed the accumulation of GRPs but not HSPs in RIF tumors? As we already mentioned in section 4.1., diverse mechanisms exist for the activation of *HSP* genes (HSF-dependent) and *GRP* genes (HSF-independent). In vivo, when tumor cells encounter either oxygen- or glucose-deficient conditions but their ATP level is maintained, this evokes a glucose-regulated response but not a heat shock response. However, combined oxygen- and glucose-deficiency reduce the ATP level and can evoke a HSP response. Although in our study on EAC cells we have not examined expression of GRPs, it is possible that the proteins are expressed some time before HSP expression. There may also be differences between ascites and solid forms of tumors, but obviously further study of this amazing problem is necessary.

4.5. POSSIBLE MECHANISMS OF HSP ACCUMULATION AFTER ATP DEPLETION

The above described results obtained on various cells and organs clearly show that both ATP depletion in vitro and ischemia in vivo stimulate HSP synthesis. Theoretically, the stress response may be governed by several mechanisms, namely the activation of heat shock gene transcription, the stabilization of HSP mRNAs, and the stimulation of their translation. Below we consider the importance of the mechanisms for HSP accumulation after ATP depletion. Another problem we would like to discuss is how a cell can monitor its own energy status and respond to ATP depletion. As we described in previous chapters, ATP depletion leads to ionic imbalance, damage of organelles, disruption of the cytoskeleton, protein aggregation, etc. A priori, all these cellular disturbances may be conducive to an elevated HSP expression, but it is also possible that only one or a few factors are crucial for triggering such a stress response.

4.5.1. pH CHANGES

Historically, the first hypothesis on the mechanism of HSP induction after in vivo ischemia was put forward by Hammond et al, who proposed that the stress response is activated by lactic acidosis.[27] Indeed, the accumulation of inducible HSP70 (HSP68) after 3 hours of extracellular acidification (pH 5.5) was observed by Nishimura et al in cultured rat astrocytes.[110] The researchers also suggested that intracellular acidosis produced during ischemia is the major event triggering the heat shock response in the ischemic brain. The same year that Benjamin and colleagues discovered the activation of HSF by hypoxia[15] (see section

4.1), Mosser et al found that acidification for 1 hour activates HSF in HeLa cytoplasmic extracts, incubation at pH 5.8-6.4 being optimal for the activation.[111]

However, as we have described in section 2.3, ATP depletion cannot decrease cytosolic pH to such low values. To address this issue, Benjamin and coworkers studied whether cytosolic acidification in ATP-depleted C2C12 myogenic cells is responsible for HSF-HSE binding. In their work, 30 minutes of incubation in a glucose-free medium with rotenone (the treatment sufficient for HSF activation) reduced the ATP content by 70% and decreased the intracellular pH (pH_i) from 7.3 to 6.9. However, when the pH_i was artificially reduced to 6.7 for 1 hour by propionic acid and amiloride (an inhibitor of Na^+/H^+-antiporter) but not affecting the ATP level, it did not activate HSF binding to DNA. Conversely, when ATP depletion occurred at normal intracellular pH (in the presence of high $[K^+]$ and nigericin, a K^+/H^+-antiporter), a marked HSF binding to HSE was observed (Fig. 4.8). From the results, the authors concluded that the effect of ATP depletion alone rather than acidification of the cytosol is responsible for HSF activation (see also review of Benjamin and Williams for discussion).[112] Earlier studies by Drummond et al[113] on *Drosophila melanogaster* salivary glands showed that the decrease in pH_i to 6.85 itself did not activate either heat shock gene transcription or translation of HSP mRNA. Therefore, despite the fact that ischemia and ATP depletion result in intracellular acidification, the pH_i drop is not sufficient to trigger HSF activation or stimulate the expression of stress proteins.

4.5.2. CA^{2+} IMBALANCE

Another possible inducer of stress response in ATP-depleted cells is the imbalance of other intracellular ions besides H^+, in particular, Ca^{2+}. In the same work of Mosser et al[111] in which the effect of acidific pH on HSF activation was studied, the researchers also found that the incubation of cytosolic extracts with 1 mM

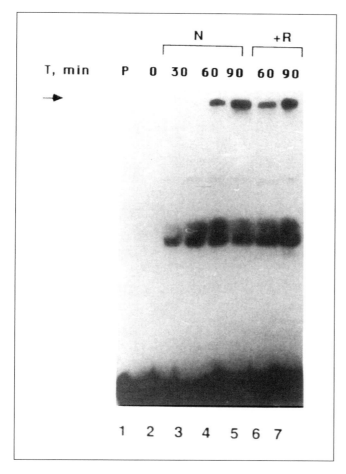

Fig. 4.8. Effects of ATP depletion on HSF-DNA binding activity at physiological pH. To maintain normal pH at 7.4, the cells were incubated at the indicated times in glucose-free DME (pH 7.4) supplemented with high K^+ (120 meq/nigericin), without (N), or with rotenone (+R). Both treatments induce the activation of HSF-binding activity (see lines 4-7) and deplete high-energy phosphate pools (data not shown). Reprinted with permission from Benjamin et al, J Clin Invest 1992; 89: 1685-1689.

of Ca^{2+} rapidly promotes HSF-HSE binding, although lower concentrations of Ca^{2+} (0.01-0.1 mM) had no effect. However, in the permeabilized cell system, in contrast to the cell extracts, no HSF activation by 0.05-1 mM Ca^{2+} was found, though Ca^{2+} was required for the maintenance of HSF in the activated state after heat shock.[114] Likewise, in *Drosophila* salivary glands, an increase in intracellular Ca^{2+} ($[Ca^{2+}]_i$) to 0.2 mM did not stimulate HSP synthesis.[113] Although, as we have discussed in section 2.3, the $[Ca^{2+}]_i$ elevation during ATP depletion was observed in some cells, in viable cells (i.e., in the cells which can accumulate HSPs) it never exceeded 1 micromole, i.e., three orders of magnitude lower than necessary for HSF activation even in cell extracts. As an example, in our

studies on EL-4 thymoma cells, 30 minutes of incubation with rotenone, depleting ATP by 95%, increased $[Ca^{2+}]_i$ to 170 nM only[115] (see section 2.3), while as little as 10 minutes of rotenone treatment is sufficient for stimulation of HSP synthesis[19] (see section 4.1). Likewise, activation of the stress response by uncoupler CCCP in Ehrlich carcinoma cells[18] (section 4.1) is not accompanied by an elevation of $[Ca^{2+}]_i$ above 200 nM (see section 2.3). Therefore, the increase in $[Ca^{2+}]_i$ observed in some ATP-depleted cells is apparently too small for HSF activation.

At the same time, imbalance in intracellular Ca^{2+} which was caused, for instance, by Ca^{2+}-ionophores (A23187, ionomycin) can lead to the increased synthesis of a specific set of stress proteins, namely

GRPs, which are localized within the endoplasmic reticulum (ER). The promoter of *GRP* genes does not contain HSE elements and is not activated by HSF, but it depends on CBF (CCAAT binding transcription factor) which responds to Ca^{2+} imbalance.[116] Another distinction between *GRP* gene activation and that of HSP is its dependence on protein synthesis.[117] As we described in section 4.1, it is the unfolding of proteins within the ER that is responsible for triggering of GRP synthesis.[13] In addition, GRP synthesis is not stimulated by ATP depletion.[118] Thus, although a disturbance in calcium homeostasis can promote the synthesis of GRPs both in vitro and in vivo (e.g., in ischemic brain, section 4.3), apparently it is not responsible for the HSF activation with subsequent HSP synthesis in energy-deprived cells.

In some cases (e.g., in the heart and the brain, see section 4.2, 4.3), ischemia can also stimulate the synthesis of mitochondrial chaperones, HSP60 and GRP75. At the present time, the mechanism of gene activation is unknown. However, it is tempting to speculate that the ischemia-induced damage of mitochondrial proteins may trigger the response. Indeed, as we have described in section 2.4, mitochondria are rather severely damaged during ischemia/reperfusion, especially in the brain and in the heart. Therefore, the unfolding of mitochondrial proteins may somehow activate nuclear *HSP60* and *GRP75* genes, similar to the mechanism described above for *GRP* genes whose products are localized in the ER.

In regard to other intracellular ions, whose imbalance during ATP depletion may be also implicated in stress response, thus far there are no data on the involvement of Na^+, but K^+ decrease may play some role (see below). At the same time, the inhibition of Na^+/K^+-ATPase with ouabain may lead to the transcriptional activation of *c-fos* and *c-jun* protooncogenes.[119] Usually the genes are promptly and transiently activated by a variety of hormones, growth factors and mitogens, but their enhanced transcription is often observed

after ischemic stress (see subsection 4.5.4 below).

4.5.3. PROTEIN UNFOLDING AND AGGREGATION

It is well-known that HSF activation, in principle, is provoked by the accumulation of unfolded (or abnormal) protein in the cytoplasm of the stressed cell; the unfolded proteins bind to HSP70 and decrease its free pool, thus releasing HSF from the inhibitory action of HSP70 (see chapter 1 for details). The close resemblance of HSF activation during ATP depletion to that during heating allowed Benjamin et al[20] in 1992 to hypothesize that the ATP depletion may augment the intracellular load of denatured and/or unfolded proteins, thereby increasing the number of targets of HSP70 and depleting the pool of free HSP70. The same year Beckmann et al[21] also suggested that the reduction in the available level of HSP70 is the trigger by which the stress response is initiated both in heat-shocked and ATP-depleted cells. Initially they found that when HeLa carcinoma cells were pulse-labeled with ^{35}S-methionine for 20 minutes, lysed with a detergent and then immunoprecipitated with antibodies to HSP70 (HSP72/73), a smear of proteins, obviously immature polypeptides, was detected by gel electrophoresis and autoradiography. The polypeptide tracks disappeared within 30-60 minutes of cell incubation in the label-free medium after pulse-labeling, indicating the release of HSP70 from newly synthesized proteins.[120] However, when the pulse-labeled cells were subsequently incubated under energy-depleting conditions (CCCP + 2-deoxyglucose), no release of HSP70 from maturating polypeptides was detected and 2 hours of such incubation caused HSP70 synthesis. Incubating the cells for 4 hours with cycloheximide to block protein synthesis (with subsequent washing) prior to energy deprivation significantly reduced the HSP70 induction. From the data, the authors suggested that the prevention of the HSP70 release from translated polypeptides in ATP-depleted cells is conducive to the promotion of the stress response, and the inhibition of protein synthesis may mitigate this response by reducing the number of HSP70 targets, namely immature polypeptides. According to the above hypothesis, ATP depletion, preventing the release of HSP70 from maturating proteins during translation, leads to a decrease in the pool of free HSP70 and HSF activation (see also section 1.4).

In our opinion, such a mechanism of HSP induction under energy deprivation may indeed take place but is probably not the sole mechanism. Firstly, the hypothesis suggests that the cells with low translational activity would become unresponsive to energy deprivation. Yet, in the same work, Beckmann et al detected some stress response even in the cells with complete suppression of translation.[21] Secondly, the authors suggest that immature proteins are the main and sole target of HSP70 both in unstressed and ATP-depleted cells. However, when the effect of heat shock on release of HSP70 from maturating proteins was studied, Beckmann and coworkers observed no suppression of its release, but they did find a decrease in the Triton X-100 solubility of some proteins together with HSP70. Importantly, after severe heating (45°C, 40 minutes), more than a four-fold increase in insolubility of *mature* proteins was observed and in this case the prevention of translation by cycloheximide did not suppress the stress response.[21] There is much data (discussed in the previous chapter) indicating that the energy imbalance may affect some mature proteins, in particular, constituents of the cytoskeleton. Therefore, ATP depletion leads not only to prevention of HSP70 release from preexisting targets (nascent proteins) but to the appearance of *new* targets for HSP70, namely unfolded and aggregated proteins (see the previous chapter). The newly formed protein targets in ATP-depleted cells may sequester HSP70 like heat-damaged proteins and bring about a stress response.

From the above reasoning, we decided to study the effect of ATP depletion on the solubility of HSP70. Since powerful Triton-insolubility of HSP70 is a well-known phenomenon for heat-stressed cells, one might expect the same effect in energy-deprived cells. Indeed, we observed impressed translocation of HSP70 from the Triton-soluble to insoluble fraction in energy-deprived EL-4 and EAC cells (see section 3.3.1). Besides HSP70, among other proteins similarly responding to ATP depletion were HSP27, myosin, actin, vinculin and a 57 kDa protein of intermediate filaments (apparently, vimentin) while a number of other proteins (e.g., HSP90, MAP kinase, etc.) remained in the Triton-soluble fraction (see refs. 18, 19, 115, 121-124). It is of importance that the HSP insolubility was not obviously associated with the intracellular acidification since incubation of EL-4 cells for 1 hour at pH 6.0 (which decreased pH_i to 6.3) did not affect the HSP70 solubility[19] (compare with the above data on HSF activation). Moreover, the addition of 1 mM ATP but not ADP to cell-extracting Triton buffer made HSP70 in the ATP-depleted cells completely soluble again.[122,123] From the data, we concluded that it is decrease in the ATP level alone that is responsible for HSP70 insolubility (refs. 19, 122-124). The insolubility of HSP70 simultaneously with some other proteins in the energy-deprived cells was later demonstrated by Nguyen and Bensaude, who also suggested the implication of the effect for triggering a stress response under ischemic stress.[125] Recently, Bergeron and coworkers observed on cultured astrocytes enhanced coimmunoprecipitation of some proteins with HSP70 both after heating and ischemia, and this was accompanied by HSF activation and HSP synthesis.[126]

In the previous chapter (subsection 3.3.1), we considered the possible reasons for HSP70 insolubility during ATP deficiency. Briefly, the chaperone may associate with the injured insoluble structures of a cell such as the cytoskeleton or the nucleus and/or co-aggregate with the proteins whose solubility is decreased when ATP declines (see refs. 17, 125-129). In addition, a decrease in the pool of free HSP70 may result from the retardation of its release from the soluble substrates (e.g., immature peptides, clathrin, p53, etc., see ref. 130 for review). The latter effect seems to be connected not only with the ATP decrease but also with the increase in ADP (which stabilizes HSP70-protein substrate complexes[131,132]) and the diminution of $[K^+]_i$ (see chapter 2), since this ion is necessary for dissociation of HSP70-protein complexes.[132]

Although a complete list of the HSP70 targets in the ATP-depleted cells as well as their relative contribution in HSP70 sequestration have not been established, undoubtedly it is a decrease in the free HSP70 pool that is a necessary and sufficient reason for activation of HSF. When the ATP level is increased again, enhanced expression of HSPs occurs and thereby, makes the cells markedly more resistant to subsequent stresses, including ischemia. Thereafter, as the HSP level rises, the HSF is converted into an inactive form and transcription of the *HSP* genes ceases (see Fig. 2.9).

Despite similarity of ATP depletion with heat shock in activation of HSF, there is also a clear distinction between the stresses. First, the spectra of proteins made insoluble during energy deficiency and heating are rather different: in particular, insoluble HSP90 is a marker for heating, while an excess of insoluble myosin is a marker for ATP depletion.[122,123] Second, HSP27, insoluble during both heating and energy starvation, demonstrates an absolutely diverse intracellular localization in endothelial cells: in heat-stressed cells it is associated with microfilaments, while in ATP-depleted ones it is granulated within the nucleus (see section 3.3.2). Finally, Larson et al[133] found that the heat-induced monomer to trimer transition which activates HSF (see chapter 1) occurs in vitro during the preparation of purified HSF.

4.5.4. HYPOXIA

Besides ATP depletion, some other events occurring during ischemia-reper-

fusion also contribute to the promotion of the stress response, hypoxia and oxidative stress being the most important ones. In their recent review, Mestril and Dillman,[134] discussing the mechanisms of HSF activation in the ischemic heart, suggest that there is another pathway which is not associated with ATP depletion. As we have already considered in section 4.1, the researchers observed the stimulation of HSP70 mRNA synthesis in hypoxic rat neonatal cardiomyocytes in the absence of a detectable decline in the ATP level.[25] From the data, Mestril and Dillman proposed that hypoxia activates *HSP* gene transcription independently of energy deprivation. Indeed, studies of gene regulation by oxygen have recently discovered a widely operative system that responds to hypoxia and involves the induction of a DNA-binding complex termed hypoxia-inducible factor 1 (HIF-1). This factor was initially found involved in the transcription of erythropoietin gene, but later HIF-1 was detected in a number of cell lines (e.g., CHO, fibroblasts, HeLa, etc.) in which this gene is not expressed at all.[135] It has been established that HIF-1 activates some genes whose transcription may help a cell withstand hypoxia, such as certain angiogenic growth factors, glucose transporter Glut-1, and some glycolytic enzymes (see refs. 136, 137). In contrast to HSF, HIF-1 requires protein synthesis for activation and is not induced by heat shock.[135] Up to date, there are no data on the involvement of HIF-1 in activation of *HSP* genes.

In addition to HIF-1, another transcription factor, AP-1, is induced by hypoxia, its induction being dependent on protein synthesis.[138-140] Apparently, the DNA-binding activity of this transcription factor depends on the redox state of cysteines located close to the DNA-binding regions of protooncogenes c-fos and c-jun (see above), which are components of the AP-1 complex. Nuclear redox factor Ref-1 is necessary to maintain the reduced state of c-fos and c-jun.[139] Indeed, Webster et al found that hypoxia without ATP depletion results in the induction and nuclear accumulation of c-fos and c-jun protooncogenes in cardiac myocytes.[138] Likewise, hypoxia but not ATP depletion induces AP-1 binding in HeLa cells.[140] The induction of *c-fos* and *c-jun* protooncogenes is frequently associated with the induction of *HSP* genes. For instance, Das et al[36] observed the induction of *c-fos* and *c-jun* mRNA in the preconditioned rat heart (see section 4.2), whereas Knoll and coworkers observed their transcriptional activation by run-on assay in porcine hearts following brief coronary occlusion.[38] In both these studies, the major *HSP* genes were also activated. Using in situ hybridization, Plumier et al recently reported that *c-fos* and *c-jun* transcription coincided with that of *HSP70* and did not occur before reperfusion.[141] In rat brain, Soriano et al[63] also observed the activation of both *c-fos* and *HSP70* gene transcription after nonlethal focal ischemia. However, Takemoto and coworkers[66] demonstrated that 15 minutes of global ischemia in gerbil brain resulted in an earlier accumulation of c-fos and c-jun proteins than HSP70, whereas shorter ischemia for 5 minutes, stimulating c-fos and c-jun accumulation, did not evoke HSP70 synthesis. The authors concluded that c-fos and c-jun induction is an earlier response to ischemic stress than the heat shock response. Likewise, in ischemic-reperfused liver, c-fos and c-jun mRNA accumulation precedes the expression of HSP70 mRNA.[89,91]

Therefore, a cell does possess several transcriptional factors such as HIF-1 and AP-1 which are activated by hypoxia (ischemia) but their participation in the stimulation of *HSP* genes has not yet been documented.

4.5.5. OXIDATIVE STRESS

Another popular viewpoint considers reperfusion-provoked oxidative stress rather than ischemia as an inducer of HSPs (see section 4.2). Indeed, occurring upon reperfusion, oxidative stress is a well-documented phenomenon (see section 2.4 and ref. 142 for recent review). Certainly, oxidants (in particular, H_2O_2, menadione) activate

HSF in cell cultures[143,144] while exogenous xantine/xantine oxidase, irradiated rose bengal or H_2O_2 stimulate HSP70 mRNA synthesis in isolated, perfused rat heart.[46] The question is whether reperfusion-induced oxidative stress is the main reason for HSF activation. To our knowledge, there is one only study where intravascular administration of superoxide dismutase (SOD) during reperfusion after global myocardial ischemia in pigs completely prevented HSP72 mRNA induction.[145] In another work, however, the addition of SOD only diminished this effect, and did not prevent it.[46] It is important to note that the activation of HSF was detected *before* any reperfusion in liver and kidney (i.e., under anoxic conditions when oxidative stress cannot occur, see section 4.4) as well in cell cultures *during* ATP depletion (section 4.1). Therefore, we consider oxidative stress provoked by reperfusion as only an additional factor contributing to HSF activation.

Similar to hypoxia, oxidative stress is able to stimulate transcription of some proteins which can protect cells from oxidative damage; the most important among them are catalase, SOD and HO-1, an inducible form of hemeoxygenase (see ref. 142 for review). There are several studies on various organs where increased expression of the above enzymes is observed after ischemia-reperfusion. Maulik and coworkers found that HO-1 mRNA in rat myocardium is not induced by ischemia but is rapidly expressed following reperfusion; the expression is blocked by SOD and catalase.[146] Likewise, expression of the *HO-1* gene following ischemia-reperfusion was established in porcine myocardium.[147] Induction of catalase and Mn-SOD was observed in the preconditioned rat heart,[36] and hemeoxygenase was expressed in the brain[77,78] and liver[91] (see sections 2.2-2.4). Like c-fos and c-jun oncogenes (see above), expression of these proteins may be independent of *HSP* gene expression, at least in liver.[91,92]

4.5.6. OTHER POSSIBLE MODULATORS OF HSP EXPRESSION

Recently several studies demonstrated that both hypoxia and hypoxia/reoxygenation modulate the activity of some protein kinases (see refs. 148-150). One of these kinases (MAP kinase or ERK1), whose activity is stimulated by the restoration of the ATP level after energy deprivation,[149] was recently shown to act as a negative regulator of the heat shock response through inhibitory phosphorylation of HSF.[151] On the other hand, ATP depletion in cultured astrocytes increased HSF phosphorylation,[152] as was observed during heat shock treatment (see section 1.4). Most likely, protein kinases modulate the activation or inactivation of HSF during and after stresses, but further study is necessary to unravel this intriguing problem.

Another factor modulating HSF activity may be arachidonic acid. This compound is not able to activate HSF per se, but lowers the threshold of its activation during heat shock (see ref. 153 for review). Since the degradation of membrane phospholipids causing the release of arachidonate is often observed during hypoxia and ATP depletion,[22,154,155] it is tempting to speculate that arachidonate promotes HSF activation under ischemia.

Besides activation of transcription, post-transcriptional regulation may also take place during the stress response. In particular, Theodorakis and Morimoto[24] observed a 10-fold increase in HSP70 mRNA half-life during heating of HeLa cells. Knowlton, in her recent review, suggested that the same phenomenon occurs during hypoxia.[23] However, in the study of Stein et al,[156] no stabilization of HSP70 mRNA was observed during hypoxic or hypoglycemic stress although such effect was demonstrated for both the Glut-1 transporter and vascular endothelial growth factor mRNAs.

In addition to HSP70 mRNA stabilization, a number of other factors define preferential translation of stress protein mRNA in heat-shocked cells (see chapter 1 for details). Such phenomena have not yet been observed following ischemic stress, apparently because ATP depletion inhibits the overall translational rate to a much lesser extent than heat shock, as we found in our study on EL-4 cells.[19]

Summarizing, we conclude that the expression of HSPs after ATP depletion is governed mainly (or exclusively) via the HSF-mediated transcriptional activation of heat shock genes. In turn, HSF activation in energy-deprived cells is due to a decrease in the pool of free HSP70, resembling that during heating and other proteotoxic stresses (Fig. 4.9).

CONCLUDING REMARKS

The data on HSP induction in various organs and tissues under ischemia are summarized in Table 2.3. As we described above, duration of ischemia necessary for a stress response varies greatly among different tissues: in some populations of neurons it requires only 3 minutes; in the heart, 5 minutes; in kidney, 15 minutes; in the liver, 60 minutes. The question arises: what determines such great variability in the HSP-inducing ischemic stress?

As it was clearly demonstrated both in cell cultures (see Table 4.1 in section 4.1) and some tissues (Table 4.3), the main mechanism of HSP accumulation after ischemic stress is the transcription activation through HSF binding to heat shock elements (HSE) of *HSP* genes. Benjamin and co-workers found in C2C12 myogenic cells that greater than a 50% decrease in the ATP level is necessary for HSF activation. Likewise, in the kidney Van Why et al established that 50% reduction in cellular ATP must occur in order to activate HSF. Thus, the two studies definitely demonstrate the existence of a threshold of ATP diminution for HSF-HSE binding, at least in myogenic cells and renal cortex cells. Interestingly, Li et al recently demonstrated that chronic impairment of mitochondrial oxidative phosphorylation (in ρ^0 cells lacking mitochondrial DNA) did not lead to the expression of major HSPs.[157] In any

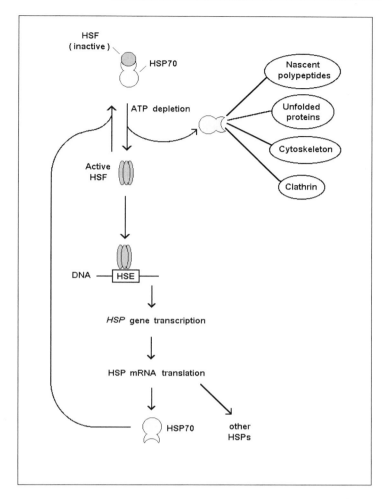

Fig. 4.9. Proposed scheme of stress-response induction following ATP depletion. According to the scheme, the main event is the dissociation of the HSP70-HSF complex due to either the redirection of HSP70 to the newly damaged protein targets (unfolded proteins, cytoskeleton) or the prevention of its release from the constitutive targets (nascent polypeptides, clathrin, etc.). Restoration of the ATP level and disappearance (repair) of damaged proteins elevate a free HSP70 pool again and cease HSF activation.

case, a distinction in the time course of ATP depletion may be one of the reasons for the different times required for triggering the stress response in various tissues.

Another reason for this may be the different level of molecular damage in ischemic cells of various origins. This was clearly established for a different population of cerebral cells (see section 2.3): the greater the vulnerability of cells to ischemia, the briefer the time required for HSP induction. Accordingly, the brain is the most ischemia-sensitive organ, whereas liver is much more resistant (about 2 hours is required for its irreversible damage, see section 4.4). Therefore, the initial mechanisms of stress response triggering and cell death may be the same or very similar, but

the final result (adaptation or death) depends on the intensity of the stress (see also chapter 7).

Considering the different biochemical events within ATP-depleted cells (e.g., ionic imbalance), we came to the conclusion that it is protein unfolding and aggregation that are responsible both for cell death and HSF activation. As we demonstrated in chapter 3, ATP depletion in vivo is a powerful protein-damaging factor, and is thus accompanied by a decrease in the free HSP70 pool and/or HSP70 insolubility. In turn, a decrease in free HSP70 apparently releases HSF from the inhibitory action of HSP70, which leads to the conversion of the inactive monomeric form of HSF to the active DNA-binding trimeric form (Fig. 4.9). At the same time, HSP90 is not made insoluble during ATP depletion,[123] therefore its possible inhibitory influence on HSF (see chapter 1) still continues.

It is important to note that HSF binding is not always sufficient to stimulate

HSP gene transcription (for instance, under salicylate treatment, see ref. 153). However, run-on transcriptional assays clearly demonstrate that the transcriptional activation of HSP genes does occur in ischemic cells. However, an increase in HSP mRNA content may not result in

Table 4.3. Effect of ischemia on the induction of heat shock genes in various organs or tissues

Organ/ Tissue	Region/Cell Type	HSF Activation	HSP Gene Transcription	Translation	Ref.
1. Heart:					
rat	myocytes, endothelium, fibroblasts	+	HSP90; HSP70; HSP27	HSP70	30,36 141, 164
rabbit	ND		HSP70	HSP70;HSP60	32,24
dog	ND		HSP70	HSP70	28
pig	myocytes, endothelium		HSP70;HSP27	ND	37
2. Brain:		+			
rat, gerbil	neurons		HSP70;HSC70; HSP60	HSP70	55
	astrocytes		HSP70	HSP70;HSP27	61,67 74
	microglia		HSP70	HSP70	61,67 74
	endothelium		HSP70	HSP70	61
3. Liver:		+			
rat			HSP70;HSC70	HSP70	88,89
4. Kidney:		+			
rat	cortex, medulla, papilla		HSP70	HSP70	92,95 97
5. Urinary bladder:		ND			
rat			HSP70; HSP60; HSP27	ND	98
6. Tumor:	SSC VII carcinoma (solid)	+	ND	ND	16
mouse	Ehrlich carcinoma (ascites)	ND	HSP27	HSP90; HSP70; HSP27	103, 104, 106

ND - not determined

HSP accumulation after severe ischemia, apparently due to translational failure as was shown in brain, liver and kidney (sections 2.3, 2.4).

Although ATP deficiency is sufficient for HSF activation, other factors may enhance HSF-HSE binding. Among them, the release of arachidonic acid and oxidative stress occurring during reperfusion may be especially substantial. Involvement of protein kinases in the regulation of HSF activity in ischemic cells is also highly probable.

In principle, other mechanisms besides HSF activation may lead to HSP accumulation. For instance, Locke and co-workers recently demonstrated a high content of inducible HSP72 in human and swine hearts in the absence of HSF-HSE binding.[158,159] Likewise, Marcuccilli et al[160] failed to observe HSF activation and HSP synthesis in primary cultured rat hippocampal neurons under hyperthermia, although these cells clearly demonstrated heat shock response to ischemia, at least in vivo (see section 4.3).

Discussing possible reasons for triggering a stress response in post-ischemic brain, Novak and Abe[55] suggested that some secondary messengers (cyclic AMP, Ca^{2+}) may be involved, since several HSF-independent elements are located in the promoter of the *HSP70* gene (see ref. 153). Ebert et al[137] demonstrated the activation of the *Glut-1* gene by mitochondrial inhibitors (rotenone, azide) via serum response elements. At the present time, however, the role of the elements (if any) in the induction of stress proteins under ischemia has not yet been demonstrated. Likewise, there are no data on the existence of post-transcriptional mechanisms (such as the stabilization of mRNA and translation activation) of HSP accumulation in post-ischemic cells.

In hypoxic cells and tissues, transcriptional activation of the *GRP* genes, the chaperones located within the endoplasmic reticulum, is often observed. Their transcription is not mediated by HSF, although the principal mechanism of this is very similar to that of HSF activation, namely an accumulation of unfolded proteins in the endoplasmic reticulum due to a disturbance in protein glycosylation and Ca^{2+} homeostasis in hypoxic cells. Sometimes mitochondrial chaperones (GRP75 and HSP60) are also expressed after ischemic stress in brain and heart. It is tempting to speculate that their expression is also governed by protein unfolding within mitochondria, but further studies are necessary to clarify this enigmatic mechanism.

Although ischemia/reperfusion leads to the activation of some other genes unrelated to heat shock genes, in the next chapters we show that it is HSP accumulation that is mainly responsible for the cytoprotective effect of preconditioning ischemic stress against subsequent severe ischemia.

RECENT NEWS

Winegarden and co-workers[161] demostrated that HSF binding and chromosomal puffing in *Drosophila* under salicilate treatment was associated with ATP decrease; however, this did not lead to HSP gene transcription, apparently due to lack of HSF hyperphosphorylation that is normally associated with HSF activation.

In renal and thyroid cell cultures, transient ATP depletion was shown to induce not only HSP70, but endoplasmic chaperones (GRP78, GRP95) as well.[162] The formation of high molecular weght aggregates of some secretory proteins (e.g., thyroglobulin) with GRPs was demonstrated in ATP-depleted cells. Selective induction of mitochondrial chaperones (HSP60, HSP10) rather than cytosolic HSP70 was observed by Martinus et al[163] in hepatoma cells devoid of mitochondial DNA (ρ^0 cells). The authors suggested that increased aggregation of mitochondrial proteins found in ρ^0 cells may be responsible for the effect.

Nishizawa et al[164] from Nagata's lab revealed activation of HSF (namely HSF1) in isolated rat heart during global ischemia, which was markedly stimulated during reperfusion. The same lab also demonsrated marked activation of HSF1 during ischemia in rat brain.[165] Interestingly, expression of HSP90 mRNA in reperfused myocardium was greater than that after heat shock, in contrast to expression of HSP70 mRNA. The authors suggested the presence of regulatory mechanisms other than HSF in HSP expression.

REFERENCES

1. Ritossa FM. A new puffing pattern induced by a temperature shock and DNP in *Drosophila*. Experientia 1962; 18: 571-573.
2. Ashburner M, Borner JJ. The induction of gene activity in *Drosophila* by heat shock. Cell 1979; 17: 241-254.
3. Leenders HJ, Kemp AJF, Konink JF et al. Changes in cellular ATP, ADP and AMP levels following treatments affecting cellular respiration and the activity of certain nuclear genes in *Drosophila* salivary glands. Exp Cell Res 1974; 86: 25-30.
4. Rajaratnam S, Smith E, Stratford IJ et al. Thermotolerance in Chinese hamster cells under oxic conditions after chronic culture under hypoxia. Br J Cancer 1981; 43: 551-553.
5. Lee SG, Werb Z. Correlation between the synthesis of heat shock proteins and the development of thermotolerance in Chinese hamster fibroblasts. Proc Natl Acad Sci USA 1982; 79: 3918-3922.
6. Sciandra JJ, Subject JR, Hughes CS. Induction of glucose-regulated proteins during anaerobic exposure and of heat-shock proteins after reoxygenation. Proc Natl Acad Sci USA 1984; 81: 4843-4847.
7. Subjeck JR, Shyy TT. Stress protein systems of mammalian cells. Am J Physiol 1986; 250: C1-C17.
8. Heacock CS, Sutherland RM. Enhanced synthesis of stress proteins caused by hypoxia and relation to altered cell growth and metabolism. Br J Cancer 1990; 62: 217-225.
9. Role DE, Murphy KR, Laderoute RM et al. Oxygen regulated 80 kDa and glucose regulated 78 kDa protein are identical. Mol Cell Biochem 1991; 103: 141-148.
10. Shen J, Hughes C, Chao C. et al. Coinduction of glucose-regulated proteins and doxorubicin resistance in Chinese hamster cells. Proc Natl Acad Sci USA 1987; 84: 3278-3282.
11. Shen JW, Subject JR, Lock RB et al. Depletion of topoisomerase II in isolated nuclei during glucose-regulated stress response. Mol Cell Biol 1989; 9: 3284-3291.
12. Sugawara S, Nowicki M, Xie S et al. Effects of stress on lysability of tumor targets by cytotoxic T cells and tumor necrosis factor. J Immunol 1990; 145: 1991-1998.
13. Shamu CE, Cox JS, Walter P. The unfolded-protein-response pathway in yeast. Trends Cell Biol 1994; 4: 56-60.
14. Havemann J, Li GS, Mak JY et al. Chemically induced resistance to heat treatment and stress protein synthesis in cultured mammalian cells. Int J Radiat Biol 1986; 50: 51-64.
15. Benjamin IJ, Kroger B, Williams RS. Activation of heat shock transcription factor by hypoxia in mammalian cells. Proc Natl Acad Sci USA 1990; 87: 6263-6267
16. Giaccia AJ, Auger EA, Koong A et al. Activation of the heat shock transcription factor by hypoxia in normal and tumor cell lines in vivo and in vitro. Int J Radiat Oncol Biol Phys 1992; 23: 891-897.
17. Tuijl MJM, van Bergen en Henegouwen PMP, van Wijk R et al. The isolated neonatal rat—cardiomycyte used in an in vitro model for "ischemia". II. Induction of the 68 kDa heat shock protein. Biochim Biophys Acta 1991; 1091: 279-284.
18. Gabai VL, Kabakov AE. Rise in heat-shock protein level confers tolerance to energy deprivation. FEBS Lett 1993; 327: 247-250.
19. Gabai VL, Kabakov AE. Induction of heat-shock protein synthesis and thermotolerance in EL-4 ascites tumor cells by transient ATP depletion after ischemic stress. Exp Mol Pathol 1994; 60: 88-99.
20. Benjamin IJ, Horie S, Greenberg ML et al. Induction of stress proteins in cultured myogenic cells. Molecular signals for the activation of of heat shock transcription factor during ischemia. J Clin Invest 1992; 89: 1685-1689.
21. Beckmann RP, Lovett M, Welch WJ. Examining the function and regulation of hsp70 in cells subjected to metabolic stress. J Cell Biol 1992; 117: 1137-1150.
22. Iwaki K, Chi S-H, Dillmann WH, Mestril R. Induction of HSP70 in cultured rat neo-

natal cardiomyocytes by hypoxia and metabolic stress. Circulation 1993; 87: 2023-2032.

23. Knowlton AA. The role of heat shock proteins in the heart. J Mol Cell Cardiol 1995; 27: 121-131.

24. Theodorakis NG, Morimoto RI. Post-transcriptional regulation of HSP70 expression in human cells: Effects of heat shock, inhibition of protein synthesis, and adenovirus infection on translation and mRNA stability. Mol Cell Biol 1987; 7: 4357-4368.

25. Mestril R, Chi SH, Sayen MR et al. Isolation of a novel rat heat-shock protein (HSP70) gene and its expression during ischaemia/hypoxia and heat shock. Biochem J 1994; 298: 561-569.

26. Currie RW, White FP. Trauma-induced protein in rat tissues: a physiological role for a "heat shock" protein? Science 1981; 214: 72-73.

27. Hammond GL, Lai YK, Markert CL. Diverse forms of stress lead to new patterns of gene expression through common and essential metabolic pathway. Proc Natl Acad Sci USA 1982; 79: 3485-3488.

28. Dillmann WH, Mehta HB, Barrieux A et al. Ischemia of the dog heart induces the appearance of a cardiac mRNA coding for a protein with migration characteristics similar to heat shock/stress protein 72. Circ Res 1986; 59: 110-114.

29. Howard G, Geoghegan TE. Altered cardiac tissue gene expression during acute hypoxic exposure. Mol Cell Biochem 1986; 69: 155-160.

30. Currie RW. Effects of ischemia and perfusion temperature on the synthesis of stress-induced (heat shock) proteins in isolated and perfused rat hearts. J Mol Cell Cardiol 1987; 19: 795-808.

31. Donnelly TJ, Sievers RE, Vissern FLJ et al. Heat shock protein induction in rat hearts: A role for improved myocardial salvage after ischemia and reperfusion? Circulation 1992; 85: 769-778.

32. Knowlton AA, Brecher P, Apstein CS. Rapid expression of heat shock protein in the rabbit after brief cardiac ischemia. J Clin Invest 1991; 87: 139-147.

33. Locke M, Tanguay RM, Klabunde RE, Ianuzzo CD. Enhanced post-ischemic myo-

cardial recovery following exercise induction of HSP 72. Am J Physiol 1995; 38: H320-H325.

34. Marber MS, Latchman DS, Walker JM, Yellon DM. Cardiac stress protein elevation 24 hours after brief ischemia or heat stress is associated with resistance to myocardial infarction. Circulation 1993; 88: 1264-1272.

35. Tanaka M, Fujiwara H, Yamasaki K et al. Ischemic preconditioning elevates cardiac stress protein but does not limit infarct size 24 or 48 h later in rabbits. Am J Physiol 1995; 36: H1476-H1482.

36. Das DK, Engelman RM, Kimura Y. Molecular adaptation of cellular defences following preconditioning of the heart by repeated ischaemia. Cardiovasc Res 1993; 27: 578-584.

37. Sharma HS, Wunsch M, Brand T et al. Molecular biology of the coronary vascular and myocardial responses to ischemia. J Cardiovasc Pharmacol 1992; 20 Suppl 1: S23-S31.

38. Knoll R, Arras M, Zimmerman R et al. Changes in gene expression following short coronary oclusions studied in porcine hearts with run-on assays. Cardiovasc Res 1994; 28: 1062-1069.

39. Andres J, Sharma HS, Ralph K et al. Expression of heat shock proteins in the normal and stunned porcine myocardium. Cardiovasc Res 1993; 27: 1421-1429.

40. Jennings RB, Reimer KA. Lethal myocardial ischemia injury. Am J Pathol 1981; 102: 241-255.

41. Watts JA, Whipple JP, Hatley AA. A low concentration of nisoldipine reduces ischemic heart injury: enhanced reflow and recovery of contractile function without energy preservation during ischemia. J Mol Cell Cardiol 1987; 19: 809-817.

42. Jeffrey FM, Storey CJ, Nunnally RL et al. Effect of ischemia on NMR detection of phosphorylated metabolites in the intact rat heart. Biochemistry 1989; 28: 5323-5326.

43. Vuorinen K, Ylitalo K, Peuhkurinen K et al. Mechanisms of ischemic preconditioning in rat myocardium. Circulation 1995; 91: 2810-2818.

44. Arai AE, Grauer SE, Anselone CG. Metabolic adaptation to a gradual reduction in

myocardial blood flow. Circulation 1995; 92: 244-252.

45. Cave AC. Preconditioning induced protection against post-ischemic contractile dysfunction: characteristics and mechanisms. J Mol Cell Cardiol 1995; 27: 969-979.

46. Kukreja RC, Kontos MC, Loesser KE et al. Oxidant stress increases heat shock protein 70 mRNA in isolated perfused rat heart. Am J Physiol (Heart Circ Physiol) 1994; 36: H2213-H2219.

47. Mcgrath LB, Locke M, Cane M et al. Heat shock protein (HSP72) expression in patients undergoing cardiac operations. J Thoracic Cardiovasc Surg 1995; 109: 370-376.

48. Nitta Y, Abe K, Aoki M et al. Diminished heat shock protein 70 mRNA induction in aged rat hearts after ischemia. Am J Physiol (Heart Circ Physiol) 1994; 36: H1795-H1803.

49. Rowland RT, Meng X, Ao L et al. Mechanisms of immature myocardial tolerance to ischemia: phenotypic differences in antioxidants, stress proteins, and oxidases. Surgery 1995; 118: 446-452.

50. Kregel KC, Moseley PL, Skidmore R et al. HSP70 accumulation in tissues of heat-stressed rats is blunted with advancing age. J Appl Physiol 1995; 79: 1673-1678.

51. Heidari AR, Takahashi R, Gutsman A et al. Hsp70 and aging. Experientia 1994; 50: 1092-1098.

52. Hochachka PW. Defense strategies against hypoxia and hypothermia. Science 1986; 231: 234-241.

53. Dwyer BE, Nishimura RN. Heat shock proteins in hypoxic-ischemic brain injury: a perspective. Brain Pathol 1992; 2:245-251.

54. Koroshetz WJ, Bonventre JV. Heat shock response in the central nervous system. Experientia 1994; 50:1085-1091.

55. Novak TS, Abe H. Post-ischemic stress response in brain. In: Morimoto RI, Tissieres A, Georgopoulos C, eds. The Biology of Heat Shock Proteins and Molecular Chaperones. Cold Spring Harbor, NY: Cold Spring Harbor Laboratory Press, 1994: 553-576.

56. Raleysusman KM, Murata J. Time course of protein changes following in vitro ischemia in the rat hippocampal slice. Brain Res 1994; 694:94-102.

57. Kumar K, Wu XL. Expression of beta-actin and alpha-tubulin mRNA in gerbil brain following transient ischemia and reperfusion up to 1 month. Mol Brain Res 1995; 30:149-157.

58. Nowak TS. Localization of 70 kDa stress protein mRNA induction in gerbil brain after ischemia. J Cereb Blood Flow Metab 1991; 11: 432-439.

59. Hu BR, Wieloch T. Stress-induced inhibition of protein synthesis initiation: modulation of initiation factor 2 and guanine nucleotide exchange factor activities following transient ccerebral ichemia in the rat. J Neurosci 1993; 13: 1830-1838.

60. Kawagoe J, Abe K, Kogure K. Reduction of HSP70 and HSC70 heat shock mRNA induction by pentobarbital after transient global ischemia in gerbil brain. J Neurochem 1993; 61: 254-260.

61. Sharp FR, Kinouchi H, Koistinaho J et al. HSP70 heat shock gene regulation during ischemia. Stroke 1993; 24: 172-175.

62. Chopp M, Li Y, Zhang ZG et al. p53 expression in brain after middle cerebral artery occlusion in the rat. Biochem Biophys Res Comm 1992; 182: 1201-1207.

63. Soriano MA, Ferrer I, Rodriguezfarre E et al. Expression of c-fos and inducible hsp-70 mRNA following a transient episode of focal ischemia that had non-lethal effects on the rat brain. Brain Res 1995; 670: 317-320.

64. Kumar K, Wu XL, Evans AT et al. The effect of hypothermia on induction of heat shock protein (HSP)-72 in ischemic brain. Metab Brain Disease 1995; 10: 283-291.

65. Shaver EG, Welsh FA, Sutton LN et al. Deep hypothermia diminishes the ischemic induction of heat-shock protein-72 mRNA in piglet brain. Stroke 1995; 26: 1273-1277.

66. Takemoto O, Tomimoto H, Yanagihara T. Induction of c-fos and c-jun gene products and heat shock protein after brief and prolonged cerebral ischemia in gerbils. Stroke 1995; 26: 1639-1648.

67. Kato H, Kogure K, Liu XH et al. Immunohistochemical localization of the low molecular weight stress protein HSP27 following focal cerebral ischemia in the rat. Brain Res 1995; 679: 1-7.

68. Abe K, Kawagoe J, Aoki M et al. Changes of mitochondrial DNA and heat shock protein gene expressions in gerbil hippocampus after transient forebrain ischemia. J Cereb Blood Flow Metab 1993; 13: 773-780.

69. Aoki M, Abe K, Kawagoe J et al. Acceleration of HSP70 and HSC70 heat shock gene expression following transient ischemia in the preconditioned gerbil hippocampus. J Cereb Blood Flow Metab 1993; 13: 781-788.

70. Wang S, Longo FM, Chen J et al. Induction of glucose regulated protein (grp78) and inducible heat shock protein (hsp70) mRNAs in rat brain after kainic acid seizures and focal ischemia. Neurochem Int 1993; 23: 575-582.

71. Lowenstein DH, Gwinn RP, Seren RP et al. Increased expression of mRNA encoding calbindin-D28K, the glucose-regulated proteins, or the 72 kDa heat-shock protein in three models of acute CNS injury. Mol Brain Res 1994; 22: 299-308.

72. Massa SM, Longo FM, Zuo J et al. Cloning of rat grp75, an hsp70-family member, and its expression in normal and ischemic brain. J Neurosci Res 1995; 40: 807-819.

73. Kato H, Liu Y, Kogure K et al. Induction of 27-kDa heat shock protein following cerebral ischemia in a rat model of ischemic tolerance. Brain Res 1994; 634: 235-244.

74. Kato H, Araki T, Itoyama Y et al. An immunohistochemical study of heat shock protein-27 in the hippocampus in a gerbil model of cerebral ischemia and ischemic tolerance. Neuroscience 1995; 68: 65-71.

75. Nowak TS, Bond U, Schlesinger MJ. Heat shock RNA levels in brain and other tissues after hyperthermia and transient ischemia. J Neurochem 1990; 54: 451-458.

76. Hayashi T, Takada K, Matsuda M. Post-transient ischemia increase in ubiquitin conjugates in the early reperfusion. Neuroreport 1992; 3: 519-520.

77. Takeda A, Onodera H, Sugimoto A et al. Increased expression of hemeoxygenase mRNA in rat brain following transient forebrain ischemia. Brain Res 1994; 666: 120-124.

78. Takeda A, Kimpara T, Onodera H et al. Regional difference in induction of hemeoxygenase-1 protein following rat transient forebrain ischemia. Neurosci Lett 1996; 205: 169-172.

79. Ueda H, Hashimoto T, Furuya E et al. Changes in aerobic and anaerobic ATP-synthesizing activities in hypoxic mouse brain. J Biochem 1988; 104: 81-86.

80. Lukkarainen J, Kauppinen RA, Koistinaho J et al. Cerebral energy metabolism and immediate early gene induction following severe incomplete ischaemia in transgenic mice overexpressing the human ornithine decarboxylase gene: evidence that putrescine is not neurotoxic in vivo. Eur J Neurosci 1995; 7: 18401849.

81. Doll CJ, Hochachka PW, Reiner PB. Effect of anoxia and metabolic arrest on turtle and rat cortical neurons. Am J Physiol 1991; 260: R747-755.

82. Heurteaux C, Bertaina V, Widmann C et al. K$^+$ channel openers prevent global ischemia-induced expression of c-fos, c-jun, heat shock protein, and amyloid beta-protein precursor genes and neuronal death in rat hippocampus. Proc Natl Acad Sci USA 1993; 90: 9431-9435.

83. Pierce GN, Czubryt MP. The contribution of ionic imbalance to ischemia/reperfusion-induced injury. J Mol Cell Cardiol 1995; 27: 53-63.

84. Kobayashi S, Welsh FA. Regional alterations of ATP and heat-shock protein-72 mRNA following hypoxia-ischemia in neonatal rat brain. J Cereb Blood Flow Metabol 1995; 15: 1047-1056.

85. Siesjo BK. Calcium and cell death. Magnesium 1989; 8: 223-237.

86. Lowenstein DH, Chan PH, Miles MF. The stress protein response in cultured neurons: characterization and evidence for a protective role in excitotoxicity. Neuron 1991; 7: 1053-1060.

87. Sharp FR, Butman M, Aardalen K et al. Neuronal injury produced by NMDA antagonists can be detected using heat shock proteins and can be blocked with antipsychotics. Psychopharm Bull 1994; 30: 555-560.

88. Cairo G, Bardella L, Schiaffonati L et al. Synthesis of heat shock proteins in rat liver after ischemia and hyperthermia. Hepatology 1985; 5: 357-361.

89. Schiaffonati L, Rappocciolo E, Tacchini L et al. Reprogramming of gene expression in post-ischemic rat liver: induction of protooncogenes and hsp70 gene family. J Cell Physiol 1990; 143: 79-87.

90. Bernelli-Zazzera A. Patterns of RNA and protein synthesis in post-ischemic livers. Free Rad Res Comm 1990; 7: 301-305.

91. Tacchini L, Schiaffonati L, Pappalardo C et al. Expression of HSP70, immediate-early response and hemeoxygenase genes in ischemic-reperfused rat liver. Lab Invest 1990; 68: 465-471.

92. Tacchini L, Pogliaghi G, Radice L et al. Differential activation of heat-shock and oxidation-specific stress genes in chemically induced oxidative stress. Biochem J 1995; 309: 453-459.

93. Lovis C, Mach F, Donati YRA et al. Heat shock proteins and the kidney. Renal Failure 1994; 16: 179-192.

94. Emami A, Schwartz JH, Borkan SC. Transient ischemia or heat stress induces a cytoprotectant protein in rat kidney. Am J Physiol 1991; 260: F479-F485.

95. Van Why SK, Hildebrandt F, Ardito T et al. Induction and intracellular localization of HSP-72 after renal ischemia. Am J Physiol 1992; 263: F769-F775.

96. Zager RA, Bredl CE, Eng MJ et al. Hyperthermia: effects on renal ischemic/ reperfusion injury in the rat. Lab Invest 1990; 63: 360-369.

97. Van Why SK, Mann AS, Thulin G et al. Activation of heat-shock transcription factor by graded reductions in renal ATP, in vivo, in the rat. J Clin Invest 1994; 94: 1518-1523.

98. Zhao Y, Wein AJ, Levin RM. Assessment of stress gene mRNAs (HSP-27, 60 and 70) in obstructed rabbit urinary bladder using a semi-quantitative RT-PCR method. Mol Cell Biochem 1995; 148: 1-7.

99. Vaupel P, Kallinowski F, Okunieff P. Blood flow, oxygen and nutrient supply, and metabolic microenviroment of human tumors: A review. Cancer Res 1989; 49: 6449-6465.

100. Cai JW, Henderson BW, Shen JW et al. Induction of glucose regulated proteins during growth of a murine tumor. J Cell Physiol 1993; 154: 229-237.

101. Kemp A, Mendel B. How does the Ehrlich ascites tumor obtain its energy for growth? Nature (London) 1957; 180: 131-133.

102. Nakamura W, Hosoda S. The absence of glucose in Ehrlich ascites cells and fluids. Biochim Biophys Acta 1968; 158: 212-219.

103. Benndorf R, Nurberg P, Bielka H. Growth phase-dependent proteins of the Ehrlich ascites tumor analyzed by one- and two-dimensional electrophoresis. Exp Cell Res 1988; 174: 130-138.

104. Knauf U, Bielka H, Gaestel M. Over-expression of the small heat-shock protein, hsp25, inhibits growth of Ehrlich ascites tumor cells. FEBS Lett 1992; 309: 297-302.

105. Mosin AF, Gabai VL, Makarova Yu M et al. Damage and interphase death of the Ehrlich ascites carcinoma tumor cells, being at different growth phases, due to energy deprivation and heat shock. Cytology 1994; 36: 384-391.

106. Kabakov AE, Molotkov AO, Budagova KR et al. Adaptation of Ehrlich ascites carcinoma cells to energy deprivation in vivo can be associated with heat shock protein accumulation. J Cell Physiol 1995; 165: 1-7.

107. Skog S, Ericsson A, Nordell B et al. ^{31}P-NMR-spectroscopy measurements of energy metabolism of in vivo growing ascites tumours following addition of glucose. Acta Oncol 1989; 28: 277-281.

108. Li GC, Li L, Liu Y-K et al. Thermal response of rat fibroblasts stably transfected with the human 70 kDa heat shock protein-encoding gene. Proc Natl Acad Sci USA 1991; 88: 1681-1685.

109. Williams RS, Thomas JA, Fina M et al. Human heat shock protein 70 (hsp70) protects murine cells from injury during metabolic stress. J Clin Invest 1993; 92: 503-508.

110. Nishimura RN, Dwyer BE, Cole R et al. Induction of the major inducible 68-kDa heat-shock protein after rapid changes of extracellular pH in cultured rat astrocytes. Exp Cell Res 1989; 180: 276-280.

111. Mosser DD, Kotzbauer PT, Sarge KD et al. In vitro activation of heat shock transcription factor DNA-binding by calcium and biochemical conditions that affect protein conformation. Proc Natl Acad USA 1990; 87: 3748-3752.

112. Benjamin IJ, Williams RS. Expression and function of stress proteins in the ischemic heart. In: Morimoto RI, Tissieres A, Georgopoulos C, eds. The Biology of Stress Proteins and Molecular Chaperones. Cold Spring Harbor, NY: Cold Spring Harbor Laboratory Press, 1994: 533-552.

113. Drummond IAS, McClure SA, Poenie M et al. Large changes in intracellular pH and calcium observed during heat shock are not responsible for the induction of heat shock proteins in Drosophila melanogaster. Mol Cell Biol 1986; 6: 1767-1775.

114. Price BD, Calderwood SK. Ca^{2+} is essential for multistep activation of the heat shock factor in permeabilized cells. Mol Cell Biol 1991; 11: 3365-3368.

115. Gabai VL, Kabakov AE, Mosin AF. Association of blebbing with assembly of cytoskeletal proteins in ATP-depleted EL-4 ascites tumour cells. Tissue Cell 1992; 24: 171-177.

116. Roy B, Lee AS. Transduction of calcium stress through interaction of the human transcription factor CBF with the proximal CCAAT regulatory element of the grp78/BiP promoter. Mol Cell Biol 1995; 15: 2263-2274.

117. Lee AS. Coordinated regulation of a set of genes by glucose and calcium ionophores in mammalian cells. Trends Biochem Sci 1987; 12: 20-23.

118. Resendez E, Ting J, Kim KS et al. Calcium ionophore A23187 as a regulator of gene expression in mammalian cells. J Cell Biol 1986; 103: 2145-2152.

119. Nakagawa Y, Rivera V, Larner AC. A role for the Na/K-ATPase in the control of human c-fos and c-jun transcription. J Biol Chem 1992; 267: 8785-8788.

120. Beckman RP, Mizzen LA, Welch WJ. Interaction of HSP 70 with newly synthesized protein: implications for protein folding and assembly. Science 1990; 248: 850-854.

121. Gabai VL, Kabakov AE. Tumor cell resistance to energy deprivation and hyperthermia can be determined by the actin skeleton stability. Cancer Lett 1993; 70: 25-31.

122. Kabakov AE, Gabai VL. Protein aggregation as primary and characteristic cell reaction to various stresses. Experientia 1993; 49: 706-710.

123. Kabakov AE, Gabai VL. The stress-induced insolubility of certain proteins in ascites tumor cells. Arch Biochem Biophys 1994; 309: 247-253.

124. Kabakov AE, Gabai VL. Heat-shock proteins maintain the viability of ATP-depleted cells: what is the mechanism?. Trends Cell Biol 1994; 4: 193-196.

125. Nguyen VT, Bensaude O. Increased thermal protein aggregation in ATP-depleted mammalian cells. Eur J Biochem 1994; 220: 239-246.

126. Bergeron M, Mivechi NF, Giaccia AJ et al. Mechanism of heat shock protein induction in primary cultured astrocytes after oxygen-glucose deprivation. Neurolog Res 1996; 18: 64-72.

127. Margulis BA, Welsh M. Analysis of protein binding to heat shock protein 70 in pancreatic islet cells exposed to elevated temperature or interleukin 1β. J Biol Chem 1001; 266: 9295-9298.

128. Haus U, Trommler P, Fisher PR et al. The heat shock cognate protein from *Dictyostelium* affects actin polymerization through interaction with the actin-binding protein cap32/34. EMBO J 1993; 12: 3763-3771.

129. Liao J, Lowthert LA, Ghori N, Omary MB. The 70-kDa heat shock proteins associate with glandular intermediate filaments in an ATP-dependent manner. J Biol Chem 1995; 270: 915-922.

130. Hightower LE, Sadis SE. Interaction of vertebrate hsc70 and hsp70 with unfolded proteins and peptides. In: Morimoto RI, Tissieres A, Georgopoulos C, eds. The Biology of Heat Shock Proteins and Molecular Chaperones. Cold Spring Harbor, NY: Cold Spring Harbor Laboratory Press, 1994: 179-207.

131. Palleros DR, Welch WJ, Fink AL. Interaction of hsp70 with unfolded proteins: Effects of temperature and nucleotides on the kinetics of binding. Proc Natl Acad Sci USA 1991; 88: 5719-5723.

132. Palleros DR, Reid KL, Shi L et al. ATP-induced protein-Hsp70 complex dissociation requires K^+ but not ATP hydrolysis. Nature 1993; 365: 664-666.

133. Larson JS, Schuetz TJ, Kingston RE. In vitro activation of purified human heat shock factor by heat. Biochemistry 1995; 34: 1902-1911.

134. Mestril R, Dillmann WH. Heat shock proteins and protection against myocardial ischemia. J Mol Cell Cardiol 1995; 27: 45-52.

135. Wang GL, Semenza GL. General involvement of hypoxia-inducible factor 1 in transcriptional response to hypoxia. Proc Natl Acad Sci 1993; 90: 4304-4308.

136. Stein I, Neeman M, Shweiki D et al. Stabilization of vascular endothelial growth factor mRNA by hypoxia and hypoglycemia and coregulation with other ischemia-induced genes. Mol Cell Biol 1995; 15: 5363-5368.

137. Ebert BL, Firth JD, Ratcliffe PJ. Hypoxia and mitochondrial inhibitors regulate expression of glucose transporter-1 via distinct cis-acting sequences. J Biol Chem 1995; 270: 29083-29089.

138. Webster KA, Disher DJ, Bishoric NH. Induction and nuclear accumulation of Fos and Jun proto-oncogenes in hypoxic cardiac myocytes. J Biol Chem 1993; 268: 16852-16858.

139. Yao K-S, Xanthoudakis S, Curran T et al. Activation of AP-1 and of a nuclear redox factor, Ref-1, in the response of HT29 colon cancer cells to hypoxia. Mol Cell Biol 1994; 14: 5997-6003.

140. Rupec RA, Baeuerle PA. The genomic response of tumor cells to hypoxia and reoxygenation: differential activation of transcription factor AP-1 and NF-κB. Eur J Bioch 1995; 234: 632-640.

141. Plumier JCL, Robertson HA, Currie RW. Differential accumulation of mRNA for immediate early genes and heat shock genes in heart after ischemic injury. J Mol Cell Cardiol 1996; 28: 1251-1260.

142. Steare SE, Yellon DM. The potential for endogenous myocardial antioxidants to protect the mycocardium against ischaemia-reperfusion injury: refreshing the parts exogenous antioxdants cannot reach? J Mol Cell Cardiol 1995; 27: 65-74.

143. Becker J, Mezger V, Courgeon AM et al. Hydrogen peroxide activates immediate binding of a Drosophila factor to DNA heat-shock regulatory element in vivo and in vitro. Eur J Biochem 1990; 189: 553-558.

144. Bruce JL, Price BD, Coleman CN et al. Oxidative injury rapidly activates heat shock transcription factor but fails to increase levels of heat shock proteins. Cancer Res 1993; 53: 12-15.

145. Schoeniger LO, Curis W, Esnaola NF et al. Myocardial heat shock gene expression in pigs is dependent on superoxide anion generated at reperfusion. Shock 1994; 1: 31-35.

146. Maulik N, Sharma HS, Das DK. Induction of the haem oxygenase gene expression during the reperfusion of ischemic rat myocardium. J Mol Cell Cardiol 1996; 28: 1261-1270.

147. Sharma HS, Maulik N, Gho BCG et. Coordinated expression of hemeoxygenase-1 and ubiquitin in the porcine heart subjected to ischemia and reperfusion. Mol Cell Biochem 1996; 157: 111-116.

148. Pombo CM, Bonventre JV, Avruch J et al. Stress-activated protein kinases are major c-jun terminal kinases activated by ischemia and reperfusion. J Biol Chem 1994; 269: 26546-26551.

149. Bogoyevitch A, Ketterman AJ, Sugden PH. Cellular stresses differentially activate c-jun N-terminal protein kinases and extracellular signal-regulated protein kinases in cultured ventricular myocytes. J Biol Chem 1995; 270: 29710-29717.

150. Seko Y, Tobe K, Ueki K et al. Hypoxia and hypoxia/reoxygenation activate Raf-1, mitogen-activated protein kinase kinase, mitogen-activated protein kinases, and S6 kinase in cultured rat cardiac myocytes. Circ Res 1996; 78: 82-90.

151. Mivechi NF, Giaccia AJ. Mitogen-activated protein kinase acts as a negative regulator of the heat shock response in NIH3T3 cells. Cancer Res 1995; 55: 5512-5519.

152. Bergeron M, Mivechi NF, Giaccia AJ et al. Mechanism of heat shock protein 72 induction in primary cultured astrocytes after oxygen-glucose deprivation. Neurol Res 1996; 18: 64-72.

153. Morimoto RI, Jurivich DA, Kroeger PE et al. Regulation of heat shock gene transcription by a family of heat shock factors. In:

Morimoto RI, Tissieres A, Georgopoulos C, eds. The Biology of Stress Proteins and Molecular Chaperones. Cold Spring Harbor, NY: Cold Spring Harbor Laboratory Press, 1994: 417-455.

154. Hagve T-A, Sprecher H, Hohl CM. The effect of anoxia on lipid metabolism in isolated adult rat cardiac myocytes. J Mol Cell Cardiol 1990; 22: 1467-1475.

155. Sun FF, Fleming WE, Taylor BM. Degradation of membrane phospholipids in the cultured human astroglial cell line UC-11MG during ATP depletion. Biochem Pharmacol 1993; 45: 1149-1155.

156. Stein I, Neeman M, Shweiki D et al. Stabilization of vascular endothelial growth factor mRNA by hypoxia and hypoglycemia and coregulation with other ischemia-induced genes. Mol Cell Biol 1995; 15: 5363-5368.

157. Li K, Neufer PD, Williams RS. Nuclear responses to depletion of mitochondrial DNA in human cells. Am J Physiol (Cell Physiol) 1995; 38: C1265-C1270.

158. McGrath LB, Locke M. Myocardial self-preservation: absence of heat shock factor activation and heat shock proteins 70 mRNA accumulation in the human heart during cardiac surgery. J Cardiac Surgery 1995; 10: 400-406.

159. Locke M, Tanguay RM, Ianuzzo CD. Constitutive expression of HSP72 in swine heart. J Mol Cell Cardiol 1996; 28: 467-474.

160. Marcuccilli CJ, Mathur SK, Morimoto RI et al. Regulatory differences in the stress response of hippocampal neurons and glial cells after heat shock. J Neurosci 1996; 16: 478-485.

161. Winegarden NA, Wong KS, Sopta M, Westwood JT. Sodium salicilate decreases intracellular ATP, induces both heat shock factor binding and chromosomal puffing, but does not induce hsp 70 gene transcription in *Drosophila.* J Biol Chem 1996; 271: 26971-26980.

162. Kuznetsov G, Bush KT, Zhang PL, Nigam SK. Perturbations in maturation of secretory proteins and their association with endoplasmic reticulum chaperones in a cell culture model for epithelial ischemia.Proc Natl Acad Sci USA 1996; 93: 8584-8589.

163. Martinus RD, Garth GP, Webster TL et al. Selective induction of mitochondrial chaperones in response to loss of the mitochondrial genome. Eur J Biochem 1996; 240: 98-103.

164. Nishizawa J, Nakai A, Higashi T et al. Reperfusion causes significant activation of heat shock transcription factor 1 in ischemic rat heart. Circulation 1996; 94: 2185-2192.

165. Higashi T, Nakai A, Uemura Y et al. Activation of heat shock factor 1 in rat brain during cerebral ischemia or after heat shock. Mol Brain Res 1995; 34: 262-270.

CHAPTER 5

HEAT SHOCK PROTEINS AND CARDIORESISTANCE TO ISCHEMIA

Being convinced after reading the previous chapter that a transient drop in cellular ATP induces HSP expression, one must wonder whether the induced HSPs help cells survive prolonged ATP deprivation. The answer is yes, such cell tolerance exists in reality and furthermore, it represents a special interest to the medical community because there are a number of pathological states which lead to long-term energy starvation in some human tissues and organs. These states are usually caused by limited blood circulation due to artery spasm, thrombosis or occlusion of the vessel lumen by atherosclerotic plaques. A failure in blood flow results first in the oxygen and nutrient deficiency of a whole organ or in a part of it which is termed global or regional ischemia. It is well-known that acute ischemia causes myocardial infarction as well as mass cell death and necrotic lesions in many other affected tissues. Among biochemical manifestations of ischemic insult, one of the most critical for cell viability is a sharp decrease in the level of ATP and phosphocreatine (i.e., depletion of high energy phosphates).

At the same time, numerous observations indicate that an elevation of the HSP(s) level enhances the survival of mammalian cells exposed to ischemic-like (i.e., ATP-depleting) conditions. As far as we have found, B.S. Polla and J.V. Bonventre (1987)[1] are the first who published data on heat shock-induced cell resistance to an ischemia-simulating treatment and who made the connection between the beneficial effect and the accumulation of HSPs. Their idea was then supported by a variety of experiments with in vitro and in vivo models of ischemia on various cell lines and organs. In this chapter, we wish to review mainly the studies in which cardioprotection from injury under ischemia or ischemia-mimicking exposures was associated with an increase in HSP(s). Although in many of the works the high energy phosphate content was not examined, the depletion of ATP during ischemic state presumably occurs (see chapter 2).

Fortunately, such fatal human pathology as myocardial ischemia can be investigated on quite adequate models using experimental animals or cell cultures. Indeed, ligation of the left anterior descending coronary artery in the heart of an open-chest experimental animal leads to typical myocardial ischemia. Reperfusion following the coronary artery ligation mimics recanalization after thrombosis. Experiments with perfusion of isolated hearts also enable researchers to imitate hypoxia and energy starvation in the myocardium that resembles natural ischemia in a whole organism. Eventually, ischemic-like conditions are rendered in vitro by blocking glycolytic and mitochondrial ATP generation in cultured cardiomyocytes or heart-derived cell lines. Whole body hyperthermia and short-term ischemic episode are the most popular preconditioning treatments to examine the effects of the heat shock response on heart failure during ischemia/reperfusion. However, other stimuli that induce intracardiac HSP accumulation are also applied (Fig. 5.1).

5.1. HEAT SHOCK-INDUCED CARDIORESISTANCE TO ISCHEMIA/REPERFUSION INJURY

Heat pretreatment affects such parameters of ischemia-induced injury in heart tissue as enzyme efflux, contractile dysfunction, size of necrotic zone, etc. (see Fig. 5.1, refs. 2-5). At least in part, the protective effects of thermal pretreatment are connected with intracardiac HSP accumulation (so-called cross tolerance). Moreover, the initial reports about the increase in HSP70 expression in ischemic hearts[6-8] already suggest the cardioprotective functions of HSP(s) against ischemia. Through the manipulation of isolated, perfused rat heart, Currie and his co-workers showed that heat pretreatment of the donor animals results in reduced creatine kinase efflux and improved contractile recovery in the hearts subjected to ischemia/reperfusion.[9] This improved recovery, namely enhanced ventricular function, was not due to changes of energy metabolism.[10] Contrary to that, Wall et al[11] have not found the beneficial effect of heat pretreatment on rat hearts experiencing no flow ischemia, although the

HSP synthesis was indeed enhanced. Later, Donnelly et al[12] described myocardial salvage in vivo in rat hearts exposed to 35 minutes of coronary occlusion and reperfusion after whole body hyperthermia. They suggested that the cardioprotection against an ischemic insult is dependent on the degree of HSP induction in myocardium (i.e., of intracardiac level of the inducible HSP72). In agreement with this supposition, Yellon et al[13] observed in rabbits preserved mitochondrial activity, diminished enzyme release and enhanced functional recovery after ischemia/reperfusion of preheated hearts. However, that same year Yellon and colleagues documented that in spite of a significant HSP72 accumulation in rabbit myocardium 24 hours after hyperthermia, no reduction of infarct size was observed after 45 minutes of regional ischemia and 3 hours of reperfusion.[14] In addition, the next studies of Currie's group revealed that heating simultaneously elevated levels of HSP70 and increased catalase activity, but inhibition of the catalase by 3-amino-1,2,4-triazole abolished the protective effect of heat pretreatment.[15] In fact, the finding might be interpreted as a discovery of an alternative inducible defense mechanism in which catalase, but not HSP70, plays a role. Rises in catalase activity and expression of the inducible form of HSP70 were also found in heat-pretreated rat hearts which are resistant to reperfusion arrhythmias following short-term ischemia.[16] The generation of free radicals during reperfusion following arterial occlusion is one of the ischemia-associated injurious factors (see chapter 2). In conjunction, it seems important that heat pretreatment reduces free radical release in isolated, perfused rat hearts[17] while increases in both catalase activity and HSP content were found again. Although the attempts to distinguish the protective effects mediated by increased HSP70 and elevated catalase activity were unsuccessful,[14-16] it has been suggested that the effects may be interrelated and the heat-induced rise in catalase activity may be due to modulation of the enzyme's

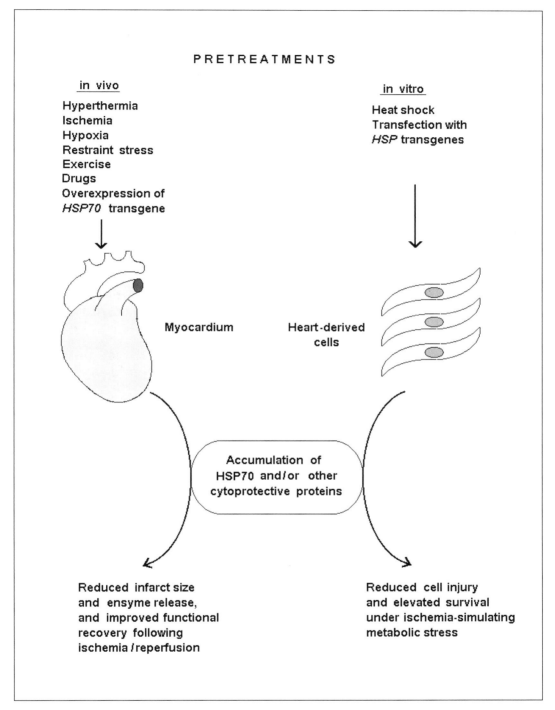

Fig. 5.1. Possible ways to study the role of HSPs in protection of cardiac cells against ischemia/reperfusion injury.

properties by excess HSP70.[18] This problem seems even more difficult to solve after recent data from Auyeung et al[19] that shows catalase inhibition with 3-amino-1,2,4-triazole does not abolish the infarct size reduction in ischemic hearts of heat-shocked rats. Such a result refutes the report by Currie's group[15] and again suggests a major cardioprotective capacity for the newly expressed HSP(s). Thus, the explanation of acquired cardioresistance to ischemia/reperfusion injury is undeniably ambiguous.

Despite the unresolved intrigue with catalase and 3-amino-1,2,4-triazole, studies of the effect of heat pretreatment on infarct size and the contractile recovery of post-ischemic myocardium have continued. Currie et al[20] reported a reduction of the necrotic zone in preheated rabbit hearts exposed to 30 minutes of coronary occlusion. The diminution of infarct size seems to be related to the accumulation of HSP(s), since the protective effect was observed 24 hours after heat pretreatment when the level of HSP70 was increased but not 40 hours after when the HSP70 content in the hearts returns to its initial level. Likewise, Walker and co-workers observed a reduction of infarct size in rabbit hearts that had undergone 30 minutes of coronary occlusion 24 hours post-heat shock.[21] In the case of 45 minute ischemia, they observed no protection.[14] Marber et al[22] established that 15 minute heat shock, as well as a series of short-term coronary occlusions, increased the HSP72 level 2.5- to 2.8-fold and significantly reduced the infarct size in rat hearts subjected to 35 minutes of coronary occlusion. In the study, the degree of cardioprotection following both of the pretreatments correlated with an increase in HSP72 but not with an accumulation of another stress protein, HSP60. Hereafter, the same researchers also found a good correlation between the amount of HSP72 and the recovery of rabbit myocardium contractile function following hypoxia, whereas no correlation was observed between the HSP60 levels and the recovery.[23] Similar results were reported by

Hutter and co-workers,[24] which have quite convincingly demonstrated a good correlation between the content of myocardial HSP72 differentially expressed at various challenging temperatures and infarct size reduction in rats following 35 minutes left coronary artery occlusion (Fig. 5.2). Robinson and colleagues[25] studying recovery of post-ischemic stunned myocardium in dogs found an accelerated recovery of ventricular contractions after heat shock-induced accumulation of myocardial HSP70. Recently, Mccully et al[26] documented the cardioprotective effects of HSP70 induction in a clinically relevant model when 15 minutes retrograde hyperthermic perfusion (42°C) of isolated rat hearts was performed only 5 minutes prior to global ischemia/reperfusion. In contrast to the normothermic perfused hearts, the hyperthermic perfused ones exhibited enhanced myocardial functional recovery. Despite the brief period between heat preconditioning and ischemia/reperfusion, the inducible HSP72 was significantly increased ~1.8-fold by 60 minutes of normothermic post-ischemic reperfusion; the enhanced myocardial recovery correlates with the accumulation of HSP70 mRNA and HSP72.[26] All the data represented in refs. 22-26, of course, strongly support Donnelly's suggestion[12] about the role of myocardial HSP70 expression in cardioprotection against ischemia. Nevertheless, it should be noted that the induction of HSP72 per se may only be a hallmark of the stress response in heart tissue, while the above mentioned protective effects might be mediated by any other stress proteins, excepting perhaps HSP60. Although exact data about the expression of HSP100, HSP90, HSP27, ubiquitin, etc. in preheated myocardium are absent, now it is entirely accepted that the heat shock-induced cardioprotection during ischemia as well as the improved post-ischemic contractile recovery is associated with myocardial HSP70 accumulation (see the next sections).

Additionally, from a clinician's point of view, preconditioning the heart by heat shock prior to cardiopulmonary bypass or

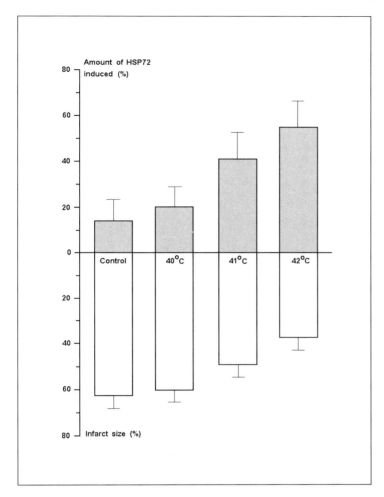

Fig. 5.2. Bar graphs showing the inverse correlation between the amount of heat-induced HSP72 and infarct size after 35 minutes of ischemia and 2 hours of reperfusion in rat hearts. Reprinted with permission from Hutter et al, Circulation 1994; 89: 355-360.

cardioplegia is an important consideration that may be applied in cardiac surgery. The situation resembles the state of myocardium stunned by ischemic insult and hence the effect of heat pretreatment on postcardioplegia recovery has been examined in some studies. In both porcine and rat models of cardioplegia, pretreatment with heat shock results in lesser creatine kinase efflux and better recovery of hemodynamic and endothelial functions after cardiac arrest.[27-29] Since both mechanical and endothelial parameters of rat hearts undergoing cardioplegic ischemia strongly depend upon the level of cardiac HSP70, the researchers associated the beneficial effects of thermal pretreatment directly with the heat-induced increase in HSP70.[28,29] Interestingly, the latter work on the rat

model mimics conditions for heart transplantation; the optimal cardioprotection interval between heat stress and ischemia was determined as 24-30 hours, which correlates with an apparently critical amount of cardiac HSP70.[29]

Finishing the section, we would like to mention that whole body hyperthermia can exert a lot of extracardiac effects, some of which are cardioprotective. In particular, the neutrophil function altered as a result of whole body heating may contribute to cardioprotection against ischemia/ reperfusion injury. Actually, circulating neutrophils attack damaged tissue areas within a post-ischemic heart and generate superoxide radicals which aggravate myocardial injury.[18] As was shown using a model of myocardial ischemia in dogs, infarct size is reduced by means of neutrophil depletion.[30] Human neutrophils respond to heat shock by increasing HSP expression, which results in the inhibition of NADPH oxidase and the suppression of superoxide radical generation.[31] Thus, heat preconditioning to myocardial ischemia in the case of whole body hyperthermia may in part be achieved by affecting neutrophil function.

5.2. THE ROLE OF MYOCARDIAL HSPs IN ISCHEMIC PRECONDITIONING

In addition to heat pretreatment, previous short-term ischemic stress (ischemic preconditioning) confers cardioresistance to repeated severe ischemia (see ref. 32 for review). After the first description of the ischemic preconditioning phenomenon by Murry et al,[33] two distinct phases of the myocardial tolerance have been discovered. The early or classical cardioprotection is observed immediately following brief ischemia (first phase) but then it disappears quickly. However, a delayed (second) phase of the induced cardioprotection is manifested many hours after ischemic preconditioning and is specifically termed the "second window of protection."[22,34] It is generally accepted that the primary heart tolerance conferred by preconditioning ischemia is not dependent on protein synthesis and may be due to multiple stress-provoked effects such as shifts in the high-energy phosphate content, intracellular pH and redox equilibrium, reduced lactate release and slowed metabolism, opening of ATP-dependent potassium channels, stimulation of adenosine A_1 receptors and α_1-adrenoreceptors, and activation of protein kinase C, etc. (reviewed in ref. 32).

The first phase of the cardioprotection is not usually associated with HSPs and molecular chaperones, though the constitutively expressed ones might also be involved in the cytoprotective mechanisms, since priming ischemia appears to affect the distribution and functional activity of some stress proteins. In particular, Tuijl et al[35] described the migration of HSP70 into

nuclei of rat cardiomyocytes undergoing ischemic stress in vitro. Such compartmentalization of the chaperone may perform a protective function toward intranuclear protein structures during ischemia-induced ATP depletion. Moreover, HSP70 translocation to the nucleus of an ATP-deprived cardiomyocyte is directly related to triggering the heat shock response following ischemic episodes (see chapter 4). Likewise, the constitutive HSPs may undergo "activating" modifications in response to preconditioning. For example, the small stress protein, HSP27, is phosphorylated by stress-activated protein kinase(s) that are crucial for cytoprotection (reviewed in detail in chapter 7). Another cardiac small stress protein, α-crystallin, becomes extremely insoluble during myocardial ischemia in rats because of its co-aggregation with cytosolic actin[36] and/or binding to myofibrils;[37] in any case, the reaction may serve in the stabilization of cytoskeletal and contractile proteins in ischemic cardiomyocytes. Although the researchers do not exclude the harmful consequences of α-crystallin insolubility within an ischemic heart,[36] a potential involvement of the small stress proteins as well as other constitutive HSPs in the first phase of cardioprotection should be considered (see subsection 5.5). Eventually, both windows of protection may be tightly interrelated, since the constitutive HSP redistribution and protein kinase cascades triggered within the first early phase may also initiate some later events—for instance, the heat shock transcription response and HSP synthesis that hereafter define the second window (Fig. 5.3).

If the role of constitutive HSPs in ischemic preconditioning remains controversial, the cytoprotective contribution of accumulated inducible HSP(s) in the second window (delayed phase) of ischemia-induced cardioprotection seems more than likely. Indeed, the fact that there is a considerable delay (24 hours or more) in the development of cardioresistance implies the induced synthesis and accumulation of certain effector proteins which mediate the

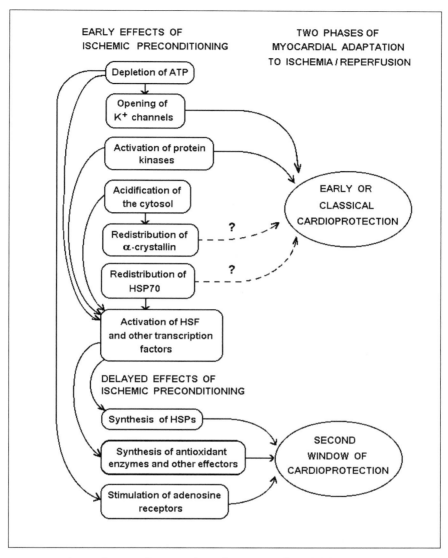

Fig. 5.3. A scheme showing the interrelation between early and delayed cardioprotection after ischemic preconditioning.

protection of heart tissue (Fig. 5.3). Yet in the initial articles on intracardiac HSP expression following ischemia/reperfusion, potential cardioprotective capacity of HSP(s) was discussed.[6-8] The beneficial function of inducible HSPs attenuating heart failure during ischemia has been suggested in studies and reviews.[2-5,11,34,38-41] Likewise, the majority of the above cases of cross tolerance (heat shock-induced cardioresistance to ischemia) lend support to the cytoprotective properties of HSP(s)

under ischemic stress and/or during post-ischemic recovery (reperfusion). Donnelly et al[12] observed in rat heart only mild HSP72 induction by ischemic pretreatment without improved myocardial salvage after ischemia/reperfusion. On the contrary, Marber and co-workers[22] demonstrated that stress induced both marked elevation of cardiac HSP72 and the resistance to myocardial infarction in rabbit hearts subjected to coronary occlusion/reperfusion 24 hours after a series of brief ischemic states (four

5 minute episodes of coronary ligation separated by 10 minutes of reperfusion). Such ischemic preconditioning also increased HSP60 expression in the hearts; however, after examining the effect of heat pretreatment when rise in cardiac HSP60 was insignificant, the authors suggested that, in both cases of preconditioning, HSP72 induced in the myocardium by either stress is responsible for the limitation of infarct size.[22] There is a dissonance described in a work by Tanaka et al[42] in which the same protocol for ischemic preconditioning of rabbit hearts as used by Marber[22] did not result in a reduction of infarct size, although an increase in the expression of HSP72 did occur. The discrepancy seems enigmatic, especially since the majority of findings support the hypothesis concerning the cardioprotective function of the inducible HSP70 under ischemia/reperfusion (see subsection 5.4).

Interestingly, chronic myocardial ischemia alone should accomplish preconditioning of human myocardium. In their reviews,[2,34] Yellon, Marber and Baxter discuss clinicians' observations on patients with myocardial infarction. Briefly, if episodes of unstable angina precede acute myocardial infarction, it results in improved outcome. The time frame of antecedant angina (24-48 hours) suggests the second window of protection (delayed preconditioning). Thus, the improved postinfarction regeneration of heart tissue may be a consequence of HSP-mediated adaptation of myocardium to periodic bouts of ischemia, which induces the synthesis of cardiac stress proteins.

Most likely, the second window of protection is not due only to HSP accumulation (see for example Fig. 5.1). Some scientists quite logically explain the delayed preconditioning phenomenon by such late post-ischemic effects as the rise in endogenous antioxidant (catalase, superoxide dismutase, glutathione peroxidase, etc.) levels, the elevation of protein kinase activities, the stimulation of the adenosine A_1 receptor and the synthesis of other than HSPs effector protein(s) (summarized in

ref. 34). However, HSPs are molecular chaperones that affect the conformation, functional activity and sorting and turnover of cellular proteins, including enzymes and receptors. Increased HSPs may therefore indirectly enhance endogenous protective (perhaps antioxidant) mechanisms which are not related to the heat shock response.

5.3. THE CARDIORESISTANCE INDUCED BY OTHER STIMULI CAN ALSO BE ASSOCIATED WITH HSPs

In addition to the heat stress and ischemic preconditioning, a number of other stimuli induce tolerance to myocardial ischemia. Particularly, a brief hypoxic episode,[43] restraint stress,[44-46] pretreatments with cytokines, interleukin-1 (IL-1)[47] and tumor necrosis factor α (TNFα),[48] endotoxins[49] and such drugs as herbimycin-A[50] or amphetamin[51,52] are able to precondition the myocardium against a subsequent ischemia/reperfusion injury (see Fig. 5.1). The last inducer, amphetamine, causes whole body hyperthermia, probably through the stimulation of endogenous lipolysis.[51] Improved post-ischemic ventricular recovery and increased mRNA synthesis of HSP27, HSP70, and HSP90 were found in hearts of pigs injected with the drug.[51,52] Amphetamine, like heat shock, also elevates catalase and superoxide dismutase activities[52] but the cardioprotective effect of the drug may partially be due to the presence of induced HSP(s).

Previously, herbimycin-A, an antibiotic inhibiting tyrosine kinase, was shown to induce HSP72/73 in various noncardiac cells.[53] Morris et al[50] incubated rat neonatal cardiomyocytes in vitro with herbimycin-A to confer tolerance to heat stress and simulated ischemia upon them. As Western blotting revealed, herbimycin-A induced HSP70 but not HSP90, HSP60, HSP25 or GRP78. The HSP70 induction positively correlated with the herbimycin-A-mediated protection of cultured cardiomyocytes against both heat and ischemia-mimicking stresses: the drug-treated cells accumulated HSP70 and exhibited elevated survival after stresses (on data of trypan blue exclusion and lactate dehydrogenase release tests). The phenomena of drug-induced selective accumulation of HSP70 and cytoresistance to severe stresses are not due to the inhibition of cellular tyrosine kinases by herbimycin-A, since another tyrosine kinase inhibitor, genistein, had no protective effects on the stressed cells.[50] In contrast to amphetamine, herbimycin-A does not cause whole body hyperthermia and therefore the antibiotic would have important clinical implications as a "nonstressful" means of inducing cardiac HSP70 in order to provide myocardial protection from ischemic injury.

Obviously, brief hypoxia as the preconditioning stimulus acts on cardiac tissue like an ischemic pretreatment and the mechanisms of the cardiotolerance appear to be similar in the both cases (see chapter 4). Engelman and co-workers documented intracellular ionic imbalance and augmented expression of catalase and HSP70 in hypoxically-preconditioned rat hearts which exhibit improved post-ischemic function.[43] This allows one to consider HSP induction as one of the adaptive reactions mediating hypoxic preconditioning of the myocardium. In this respect, a "phenomenon of highlanders" may be, at least in part, explained by the naturally occurring hypoxic preconditioning. Indeed, highlanders permanently living under conditions of natural oxygen deficiency usually possess good health; in particular, their prominent features are longer life span and lower frequency of myocardial infarctions. We expect that the elevated level of HSP expression will be recognized among the factors that determine the cardioresistance intrinsic to highlanders.

An interesting experimental approach was applied in Meerson's lab; the researchers showed that the adaptation of rats to intermittent restraint stress is accompanied by inducible HSP70 isoform accumulation in the myocardium and the acquisition of cardiotolerance to reperfusion paradox (post-ischemic myocardial injury during reperfusion).[44,45] It remains unknown why intermittent immobilization of the animals causes an intracardiac increase in HSP70. Perhaps the study provides a good example of the interconnection between the mechanisms of physiological and molecular stress responses. A possible role of the secondary messengers and protein kinases in the adaptive stabilization of heart structures has been previously suggested by Meerson.[44] The recent work of his group reveals that both the accumulation of inducible isoforms of HSP70 and the activation of the inositol triphosphate (IP_3) - diacylglycerol (DAG) circuit in hearts of rats adapted to immobilization stress are equally necessary for the development of cardiotolerance to reperfusion injury.[46] However, it is yet unclear whether the HSP70- and IP_3, DAG-mediated cardioprotective mechanisms act independently of one another or whether they are interrelated.

In regard to IL-1, TNF, and endotoxins, all of them are known to cause free radical bursts within targeted cells. Adaptation of heart tissue to oxidative stress induced by ischemia/reperfusion is the most simple explanation of the cardioprotective effects of the pretreatments.[43-49] Herein, oxidative stress per se provokes the heat shock response and HSP accumulation (see chapter 4). In this connection, the expected induction of HSPs as a result of the cytokine or endotoxin exposures appears to be one of the causes of the post-treatment cardioresistance. Heat shock proteins induced by TNFα delay of cardioprotection against ischemic injury is also associated with an increase in the HSP27 level following cytokine pretreatment.[48] A recent study revealed that TNFα induces cardiac HSP72 as well.[54] Moreover, both IL-1 and TNFα are able to increase HSP27 phosphorylation,[55,56] which in its turn may be significant for acquired cardioresistance.

It is a well known fact that sportsmen have a lesser risk of heart failure. Certainly, preconditioning the human myocardium by athletic exercises includes many factors, among these the improved function of coronary arteries. Interestingly, a contribution of inducible HSPs in the resistance of sportsmen to myocardial infarction may also be real. This speculation arose after a study of Locke et al.[57] They demonstrated a strong correlation between enhanced post-ischemic myocardial recovery and elevated intracardiac content of inducible HSP72 in rats subjected to three bouts of treadmill running. Thus, the myocardial protection associated with exercise, at least in part, may be due to HSP72 induction.[57]

5.4. EVIDENCE FOR THE INVOLVEMENT OF HSP70 IN CARDIOPROTECTION

Obviously, the acquisition of cardiac resistance to ischemia appears to be a multifactor process that includes various inducible mechanisms besides the heat shock response. In addition to the HSP-mediated defense mechanism, at least three alternative ones explain the late cardioprotection induced by preconditioning: (1) activation of endogenous antioxidant systems (adaptation to oxidative stress); (2) stimulation of adenosine A_1 receptors; and (3) protein kinase C activation (see ref. 34 for review). Furthermore, whole body hyperthermia causes some extracardiac effects which might positively influence cardioresistance to ischemia. Although distinct correlations between HSP70 content and cardioresistance to ischemia-induced injury have been documented,[22-26,57] this is not direct evidence for the protective function of HSP70 and may represent only a random temporary coincidence in the dynamics of tolerance development and HSP70 accumulation. Thus, a question arises as to whether myocardial HSP induction is the true defense mechanism against ischemia or if it is only one of the numerous cell reactions to preconditioning stresses. After studies with transfection of the *HSP70* transgene, there is direct evidence for the cardioprotective role of overexpressed HSP70 in the case of ischemic stress.

The first full article about the cytoprotective effect of the *HSP70* transgene

overexpression during in vitro ischemia-mimicking stress was published in 1993 by Williams et al.[58] The researchers introduced a human *HSP70* gene into mouse 10T1/2 cells and subjected the cells to ATP-depleting conditions (incubation with rotenone, an inhibitor of respiration, in the absence of glucose). Cell death and survival were not determined precisely and relative cytotoxicity of the metabolic stress was assessed on the residual activity of a reporter enzyme, firefly luciferase, that is associated with the adherent cells. When compared with the control (nontransfected) cells, the cells constitutively overexpressing human HSP70 displayed elevated (2- to 2.5-fold) "viability" under metabolic stress.[58] The study has indeed demonstrated the protective properties of HSP70 which helps cells sustain themselves during ischemic stress conditions; however, using the firefly luciferase to evaluate cell viability under metabolic stress is not a very appropriate choice because this enzyme is ATP-dependent and it can aggregate (see chapter 3) and lose activity within ATP-depleted mammalian cells.[59,60] The employed assay should, though indirectly, reflect the degree of cell injury or resistance in the case of ATP depletion.[3,58]

A marked step forward was made by Mestril and colleagues[61] when they used a rat embryonic heart-derived cell line H9c2 to generate stably transfected myogenic cells constitutively overexpressing the human inducible HSP70. In their work, ischemic-like conditions were achieved by hypoxia and glucose deprivation; the stress-induced cell injury and post-stress cell survival were monitored on lactate dehydrogenase release and colony formation, respectively. The HSP70 overexpression did confer resistance to the simulated ischemia, which was manifested in lesser efflux of the enzyme and larger colony survival in the transfectants than in the parental cells exposed to the analogous stress (Fig. 5.4). Importantly, in parental H9c2 cells, previous heat shock induced HSP synthesis (i.e., accumulation of the inducible HSP70) and tolerance to the simulated ischemia as well as the transfection with *HSP70* transgene. Summarizing their data, the investigators concluded that the inducible HSP70 plays a major role in protecting cardiac cells against ischemic injury.[61]

A similar research model was applied by Heads, Yellon and Latchman[62] in order to confirm the contribution of certain HSPs in resistance of the heart-derived cells to stresses. These scientists also used the H9c2 cell line for the generation of stable transfectants differentially overexpressing human *HSP90*, *HSP70* and *HSP60* genes.

The cells were subjected to severe heat or ischemic stresses and cell death was evaluated by using trypan blue exclusion. Constitutive overexpression of the human inducible HSP70 conferred powerful cytoprotection against both thermal and ischemic exposures, whereas the overexpressed HSP90 reliably protected only from the former, but not from the latter, treatment. Finally, HSP60 overexpression did not preserve the cells from either heat shock or ischemic stress.[62] Hence, the inducible HSP70 rather than HSP90 or HSP60 appears to be responsible for

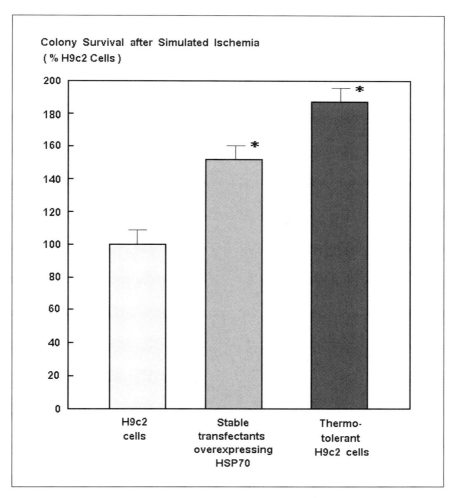

Fig. 5.4. Effects of heat shock-induced thermotolerance and HSP70 transgene overexpression on colony survival of heart-derived H9c2 cells after simulated ischemia. * Significant difference from control, P < 0.05. Redrawn with permission from Mestril et al J Clin Invest 1994; 93: 759-767.

cardioresistance to ischemia. Interestingly, the conclusion is in full agreement with the observations made by Marber et al on in vivo rabbit hearts.[22,23]

The best evidence of the capacity of HSP70 to protect heart tissue from ischemic injury is provided by three research groups which have created and studied transgenic mice overexpressing a foreign inducible HSP70.[63-65] The excellent studies allowed researchers to firmly establish the role of constitutively overexpressed HSP70 in cardioresistance to ischemia in vivo, excluding any accessory effects of preconditioning. Marber and coworkers[63] constructed transgenic mouse lines with a cytomegalovirus enhancer and β-actin promoter driving rat inducible *HSP70* gene expression in heterozygote mice. The unstressed transgene positive animals expressed a very high level of myocardial HSP70 without any detrimental effect. The hearts isolated from transgene positive and transgene negative mice were subjected to 20 minutes of ischemia and up to 120 minutes of reflow while the necrotic zone, contractile recovery and creatine kinase efflux were determined. In transgene positive murine hearts compared with transgene negative ones, the zone of necrosis (infarct size) was diminished by 40%, the contractile function at 30 minutes of reperfusion was two times better, and the release of creatine kinase was decreased by approximately 50%. The conclusion has been made about the anti-ischemic properties of inducible HSP70.[63]

Independently, Plumier et al[64] also created transgenic mice constitutively overexpressing human inducible HSP70 and investigated their resistance to myocardial ischemia. After 30 minutes of ischemia followed by reperfusion, the transgenic mouse hearts evinced significantly improved post-ischemic recovery of the contractile function as compared with the nontransgenic hearts. In particular, a force of the contraction of the myocardium overexpressing human HSP70 more than twice exceeded that of the control tissue; in addition, the parameters such as the rate

of contraction and the rate of relaxation were also improved in the transgenic myocardium. High levels of creatine kinase release, an indicator of cellular injury, were observed at the beginning of reflow after 60 minutes of ischemia in the nontransgenic, but not in the transgenic, hearts (Fig. 5.5).[64] Importantly, it was demonstrated in the same work that the overexpression of human HSP70 does not affect either normal protein synthesis or the stress response in the transgenic animals.

Finally, Williams' group generated and studied transgenic mice constitutively expressing human HSP70 in order to evaluate the metabolic aspects of anti-ischemic cardioprotection.[65] Hearts isolated from the transgenic animals demonstrated the enhanced restoration of high energy phosphate stores and the correction of metabolic acidosis following brief periods of global ischemia. Certainly, the beneficial effects of HSP70 overexpression on recovery of both energy and H^+ balance[65] are responsible for the above described reduction of injury and improvement of functional recovery of the transgenic myocardium during reperfusion.[63,64] Thus, no doubt remains that the inducible HSP70 is directly involved in cardioprotection against ischemia/reperfusion injury.

5.5. ARE OTHER STRESS PROTEINS CARDIOPROTECTIVE?

The question above is still pertinent, in so far as the proved cardioprotective function of HSP70 does not exclude the possibility that other stress proteins contribute to the induction of cardioresistance to ischemia. However, to date nothing is known about the role of other members of the HSP70 family, mitochondrial HSP70 (GRP75) and GRP78, in cardioprotection. Nevertheless, the chaperones localized to mitochondria and the endoplasmic reticulum might locally perform protective and/or reparative functions within the compartments, especially since GRPs are induced by anoxia and glucose deprivation, i.e., by ischemic-like conditions (see ref. 66 and chapter 4). Theoretically, the participation of

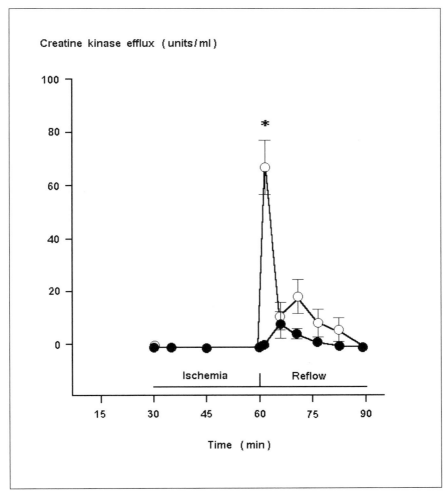

Fig. 5.5. Creatine kinase release in effluent buffer of isolated hearts during reperfusion. The enzyme efflux was measured at the last minute of the pre-ischemic period and during the reperfusion of hearts from nontransgenic (open circles) and transgenic mice (closed circles). * Significant difference between values obtained for transgenic and nontransgenic hearts, $P < 0.05$. Redrawn with permission from Plumier et al J Clin Invest 1995; 95: 1854-1860.

all the reticular and mitochondrial stress proteins in the heart anti-ischemic defensive system is quite possible.

Several independent studies suggest that HSP60 expressed at a high level does not preserve cardiomyocytes from ischemic injury.[22,23,62] The finding is unexpected because the stress protein is induced in the myocardium following ischemic episodes[22] and it is localized to mitochondria which are damaged during anoxia/re-oxygenation (see chapter 2). Therefore, HSP60 might be involved in the protection of cardiac

mitochondria during ischemic insult and/ or their post-stress repair. Although the authors of both articles attribute the protective effects to inducible HSP70, we believe that the heat shock-induced mitochondrial resistance to ischemic injury described for rabbit hearts[13] and cultured rat myocytes[61] might in part be mediated by HSP60 and/ or GRP75 accumulated inside the organelles.

In regard to cardiac HSP90, its expression is also enhanced by ischemic preconditioning.[41] There are no data about correlations between intracardiac HSP90 content and resistance of the myocardium to ischemic insult. The constitutive overexpression of HSP90 in the rat heart-derived H9c2 cells protected them from heat but not from ischemia-simulated stress.[62] In that study, the cell death/survival was determined only by trypan blue staining. Meanwhile, in the case of ischemic stress, the overexpressed HSP90 might positively affect other parameters of cell viability such as the functioning of the cytoskeleton, contractile apparatus and organelles, the ability to form viable colonies, recovery of metabolism, etc. In vitro HSP90 binds to F-actin, interfering with myosin, and ATP induces the dissociation of HSP90 from F-actin.[67] From this it follows that an excess of induced HSP90 might affect the contractile activity of ischemic myocardium and attenuate the rigor myofibril contracture. To understand the role of induced HSP90 and GRP95 in ischemic cells, additional research needs to be done.

The implication of the small stress proteins as myocardial protectors from ischemia/reperfusion injury has not been examined but their potential cardioprotective functions have been discussed.[36,47,68] Increase in the *HSP27* gene expression occurs in a heart as a result of brief ischemia[40] and preconditioning by TNFα.[47] It seems likely that the small HSPs preserve cardiac cells from injury by free radicals during postanoxic reoxygenation (reperfusion), since the ability of these chaperones to protect cells from oxidative stress has been established for the stably transfected cell lines overexpressing HSP27 or αB-crystallin[69-71] and was suggested for the myocardium accumulating HSP27 and becoming tolerant to ischemic injury.[47] Likewise, the small HSPs might be involved in the stabilization of actin filaments in ischemic cardiomyocytes. In addition, expected shifts in the HSP27 phosphorylation status during ischemia/reperfusion or post heat preconditioning are important for the inducible mechanisms of HSP-mediated cardioprotection (reviewed in chapter 7). Activities of protein kinases involved in the

HSP27 phosphorylation increase in rat cardiomyocytes under ATP depletion/repletion[72] or hypoxia/reoxygenation.[73] Another small stress protein, αB-crystallin, is a major water-soluble protein of cardiomyocytes which partially associates with desmin and actin filaments.[68,74] The effects of myocardial ischemia on αB-crystallin and its possible significance to the heart are briefly summarized in Table 5.1. Cardiac αB-crystallin co-aggregates with cytosolic actin[36] and associates with myofibrils[37] in response to ischemic acidification of the cytosol. Obviously, this binding affects contractility of myocardium during ischemia. Likewise, in a cell-free system, α-crystallin was shown to prevent actin aggregation at low pH.[68] This permits us to speculate about the possible chaperoning of cytoskeletal proteins by the small HSPs during ischemic stress, since both the acidification of the cytoplasm and the aggregation of actin occur within ATP-depleted (or ischemic) cells (see chapters 2, 3 and 7). Thus, the small HSPs could actually play an important role in cardioprotection from ischemia/reperfusion injury. However, only further studies with the constitutive transgene overexpression will confirm or reject this supposition. Intriguingly, transgenic mice with heart-specific overexpression of the murine αB-crystallin gene have already been created in Piatigorsky's lab;[75] these animals are excellent models for examining the role of αB-crystallin in cardioresistance to ischemia/reperfusion.

Little is known about how another stress protein, ubiquitin, acts in the case of myocardial ischemia. Andres et al[40] documented increases in the level of ubiquitin mRNA during post-ischemic reperfusion of the porcine myocardium. Furthermore, the researchers found formation of new ubiquitin-protein conjugates within the myocardium reperfused after brief ischemia.[40] The induced ubiquitination suggests the degradation of proteins damaged in cardiomyocytes by ischemic stress. Later, enhanced expression of mRNAs encoding ubiquitin was also found in the post-

Table 5.1. Involvement of cardiac αB-crystallin in myocardial ischemia: Its possible implication in cardioprotection

Effects of Ischemia on αB-Crystallin in Rat Hearts	Proposed Cause of the Response	Expected Consequences for the Ischemic Myocardium
Formation of large, amorphous clumps[36]	Acidosis, co-aggregation with cytosolic actin	Structural damage of cardiomyocytes, promotion of protein aggregation in the cytosol
Association with F-actin and myofibrils[37,68]	Ischemic acidification of the cytosol	Prevention of F-actin aggregation at low pH, modulation of contractility during ischemia
Binding to desmin filaments[68]	Acidification of the cytosol	Stabilization of the cytoskeleton and sarcolemma, protection from post-ischemic oxidative stress

ischemic porcine hearts by Sharma et al.[76] Researchers suspect that ubiquitin participates in the post-ischemic recovery of myocardial tissue, since it can mediate the ATP-dependent proteolysis of damaged proteins in proteasomes and thus accelerate protein turnover in cells that survive the stress (see for review chapters 1 and 7).

Expression of hemeoxygenase-1 is enhanced in heart tissue in response to myocardial ischemia/reperfusion,[76,77] in coordination with the expression of ubiquitin.[76] Since hemeoxygenase is a component of the cellular antioxidant defense system, the augmented post-ischemic expression of hemeoxygenase may contribute to myocardial protection against oxidative stress concomitant with reperfusion. Data suggesting the cardioprotective function for other HSPs (HSP100, HSP47, HSP40, HSP10, etc.) are absent to date.

CONCLUDING REMARKS

After experiments with the constitutive overexpression of the *HSP70* transgene in vitro[58,61,62] and in vivo,[63-65] it is a proven fact that increased HSP70 maintains the viability of cardiac cells exposed to ischemia/reperfusion. What is not yet clear, however,

is the mechanism of HSP70-mediated cytoprotection under myocardial ischemia or in vitro ATP-depleting metabolic stress. So far, HSP70 per se requires ATP for its chaperoning actions, but the beneficial function of this stress protein in the case of ATP deprivation is poorly understood and deserves a more thorough examination. The function of other HSPs and chaperones in an ischemic heart and their possible cooperation with the cardiac antioxidant defense system also requires clarification.

Nevertheless, the discovery of the HSP70-mediated cardioprotection should generate new approaches and strategies for cardiotherapy. Further progress in prophylaxis against fatal myocardial infarction, preservation of the patients' hearts during surgical cardioplegia and the postinfarction regeneration of myocardium might be connected with a search for nonabusive and effective remedies or procedures in vivo inducing HSP70 accumulation in human cardiac tissue. In this respect, encouraging results achieved on experimental models with herbimycine-A,[51] amphetamine[52,53] and "gene therapy"[61-65] are reasons for optimism (see chapter 9 for further discussion).

RECENT NEWS

Since this manuscript was written, several new reports concerning cardioprotective functions of HSPs have been published. A mini-review on the role of stress proteins in cardiotolerance to ischemia/reperfusion has been recently published by Plumier and Currie.[78] Speculating about mechanisms of HSP70-mediated cardioprotection, the authors propose that excess HSP70 may (I) play a role in renaturation of cellular proteins damaged as a result of ischemia/reperfusion, (II) interact with cytoskeletal constituents to prevent collapse of the cytoskeleton due to ischemic stress, (III) increase import of newly synthesized proteins into mitochondria, thus facilitating recovery of mitochondrial function after ischemia/reperfusion.[78] Although the contribution of HSP70 to cardioresistance to ischemic injury is unchallengeable, the authors do not exclude other inducible mechanisms of cardioprotection, in particular, expression of other HSPs and anti-oxidative enzymes.[78]

Hutter et al[79] provide more evidence for in vivo cardioprotective potential of HSP70: overexpression of rat HSP72 in heart tissue of transgenic mice significantly decreases infarct size under myocardial ischemia. Earlier Marber, with colleagues using the same research model, also observed the HSP72-mediated reduction of infarct size but on nonworking, buffer-perfused transgenic hearts.[63] Therein the latter, Hutter's study is more advanced, since the marked diminution of infarct zone was achieved on the working, blood-reperfused transgenic hearts.[79] In contrast to the former model with buffer perfusion, the latter one reveals an additional component of the cardioprotection, namely attenuation of the injurious action which activated neutrophils exert on an ischemic heart (see refs 18,30).

The most important news seems to be the finding of antiischemic cytoprotective properties intrinsic to the cardiac small stress proteins.[80] When our manuscript was in preparation, only overexpressed HSP70 but not HSP90 or HSP60 was shown to protect cardiac cells from ischemia/reperfusion injury or ischemia-simulating metabolic stress. While the role of HSP27 and αB-crystallin was uncertain, we predicted that an excess of these small chaperones may protect cardiomyocytes from ischemic injury (subsection 5.5 and Table 5.1). Actually, this was confirmed by Martin et al who examined how the small HSPs affect cardiomyocyte suspectibility to simulated ischemia.[80] The researchers have constructed adenoviral vectors expressing human HSP27 and rat αB-crystallin in both the sense and antisense orientation. The infected rat cardiomyocytes were in vitro exposed to ischemia-like conditions (hypoxic and hypotonic without glucose). The creatine kinase release was determined as an indicator of cell damage. In adult cardiomyocytes both αB-crystallin and HSP27 overexpression decreased the stress-provoked enzyme efflux; therefore excess small HSPs do protect cells against ischemic injury. Contrary to that, cardiomyocytes infected with the antisense vectors displayed an increase of the enzyme release after simulated ischemia. The conclusion was made that overexpression of the small HSPs is cytoprotective upon ischemia whereas decreasing the level of endogenous small HSPs elevates sensitivity of cardiomyocytes to ischemic insult.[80] This study quite convincingly argues the antiischemic feature of both αB-crystallin and HSP27; however, the mechanisms of cardioprotection mediated by the small HSPs remain to be defined (discussed in subsection 5.5 and chapter 7). In any event, like HSP70, the small HSPs are an endogenous defensive factor that helps cardiomyocytes to survive ischemic stroke. This hopefully provides novel targets for development of antiischemic therapeutic strategies.

REFERENCES

1. Polla BS, Bonventre JV. Heat shock protects cells dependent on oxidative metabolism from inhibition of oxidative phosphorylation. Clin Res 1987; 35: 555A.

2. Yellon DM, Marber MS. Hsp70 in myocardial ischaemia. Experientia 1994; 50: 1075-1084.

3. Benjamin IJ, Williams RS. Expression and function of stress proteins in the ischemic heart. In: Morimoto RI, Tissieres A, Georgopoulos C, eds. The Biology of Stress Proteins and Molecular Chaperones. Cold Spring Harbor, NY: Cold Spring Harbor Laboratory Press, 1994: 533-552.

4. Mestril R, Dillmann WH. Heat shock proteins and protection against myocardial ischemia. J Mol Cell Cardiol 1995; 27: 45-52.

5. Knowlton AA. The role of heat shock proteins in the heart. J Mol Cell Cardiol 1995; 27: 121-131.

6. Dillmann WH, Mehta HB, Barrieux A et al. Ischemia of the dog heart induces the appearance of a cardiac mRNA coding for a protein with migration characteristics similar to heat shock/stress protein 72. Circ Res 1986; 59: 110-114.

7. Currie RW. Effects of ischemia and perfusion temperature on the synthesis of stress-induced (heat shock) proteins in isolated and perfused rat hearts. J Mol Cell Cardiol 1987; 19: 795-808.

8. Mehta HB, Popovich BK, Dillmann WH. Ischemia induces changes in the level of mRNAs coding for stress protein 71 and creatine kinase M. Circ Res 1988; 63: 512-517

9. Currie RW, Karmazyn M, Kloc M, Mailer K. Heat shock response is associated with enhanced post-ischemic ventricular recovery. Circ Res 1988; 63: 543-549.

10. Currie RW, Karmazyn M. Improved post-ischemic ventricular recovery in the absence of changes in energy metabolism in working rat hearts following heat shock. J Mol Cell Cardiol 1990; 22: 631-636.

11. Wall SR, Fliss H, Kako KJ et al. Heat shock does not improve of function after no-flow ischemia in isolated rat hearts. J Mol Cell Cardiol 1990; 22 (suppl 1): 44-55.

12. Donnelly TJ, Sievers RE, Vissern FLJ et al. Heat shock protein induction in rat hearts: A role for improved myocardial salvage after ischemia and reperfusion? Circulation 1992; 85: 769-778.

13. Yellon DM, Pasini E, Cargnoni A et al. The protective role of heat stress in the ischaemic and reperfused rabbit myocardium. J Mol Cell Cardiol 1992; 24: 342-346.

14. Yellon DM, Iliodromitis E, Latchman DS et al. Whole body heat stress fails to limit infarct size in the reperfused rabbit heart. Cardiovasc Res 1992; 26: 342-346.

15. Karmazyn M, Mailer K, Currie RW. Acquisition and decay of heat-shock-enhanced post-ischemic ventricular recovery. Am J Physiol 1990; 259: H424-H431.

16. Steare SE, Yellon DM. The protective effect of heat stress against reperfusion arrhythmias in the rat. J Mol Cell Cardiol 1993; 25: 1471-1481.

17. Mocanu MM, Steare SE, Evans MCW et al. Heat stress attenuates free radical release in the isolated perfused rat heart. Free Rad Biol Med 1993; 15: 459-463.

18. Kukreja RC, Hess ML. The oxygen free radical system: from equations through membrane protein interactions to cardiovascular injury and protection. Cardiovasc Res 1992; 26: 641-655.

19. Auyeung YY, Sievers RE, Weng D et al. Catalase inhibition with 3-amino-1,2,4-triazole does not abolish infarct size reduction in heat-shocked rats. Circulation 1995; 92: 3318-3322.

20. Currie RW, Tanguay RM, Kingma, JG. Heat-shock response and limitation of tissue necrosis during occlusion/reperfusion in rabbit hearts. Circulation 1993; 87: 963-971.

21. Walker DM, Pasini E, Kucukoglu S et al. Heat stress limits infarct size in the isolated perfused rabbit heart. Cardiovasc Res 1993; 27: 962-967.

22. Marber MS, Latchman DS, Walker JM, Yellon DM. Cardiac stress protein elevation 24 hours after brief ischemia or heat stress is associated with resistance to myocardial infarction. Circulation 1993; 88:1264-1272.

23. Marber MS, Walker JM, Latchman DS, Yellon DM. Myocardial protection after whole body heat stress in the rabbit is dependent on metabolic substrate and is related to the amount of the inducible 70-kD heat stress protein. J Clin Invest 1994; 93: 1087-1094.

24. Hutter MM, Sievers RE, Barbosa V, Wolfe CL. Heat-shock protein induction in rat hearts. Circulation 1994; 89: 355-360.

25. Robinson BL, Morita T, Toft DO, Morris JJ. Accelerated recovery of post-ischemic stunned myocardium after induced expression of myocardial heat-shock protein (HSP70). J Thorac Cardiovasc Surg 1995; 109: 753-764.

26. Mccully JD, Lotz MM, Krukenkamp IB et al. A brief period of retrograde hyperthermic perfusion enhances myocardial protection from global ischemia: Association with accumulation of Hsp 70 mRNA and protein. J Mol Cell Cardiol 1996; 28: 231-241.

27. Liu X, Engelman RM, Moraru II et al. Heat shock: A new approach for myocardial preservation in cardiac surgery. Circulation 1992; 86: 11358-11363.

28. Amrani M, Allen NJ, O'Shea J et al. Role of catalase and heat shock protein on recovery of endothelial and mechanical function. Cardioscience 1993; 4: 193-198.

29. Amrani M, Corbett J, Boateng SY et al. Kinetics of induction and protective effect of heat shock proteins after cardioplegic arrest. Ann Thorac Surg 1996; 61: 1407-1411.

30. Romson JL, Hook BG, Kunkel SL et al. Reduction of the extent of myocardial injury by neutrophil depletion in the dog. Circulation 1983; 67: 1016-1023.

31. Maridonneau-Parini I, Malawista SE, Stubbe H et al. Heat shock in human neutrophils: superoxide generation is inhibited by a mechanism distinct from heat-denaturation of NADPH oxidase and is protected by heat shock proteins in thermotolerant cells. J Cell Physiol 1993; 156: 204-211.

32. Cave AC. Preconditioning induced protection against post-ischemic contractile dysfunction: characteristics and mechanisms. J Mol Cell Cardiol 1995; 27: 969-979.

33. Murry CE, Jennings RB, Reimer KA. Preconditioning with ischemia: a delay of lethal injury in ischemic myocardium. Circulation 1986; 74: 1124-1136.

34. Yellon DM, Baxter GF. A "second window of protection" or delayed preconditioning phenomenon: future horizons for myocardial protection? J Mol Cell Cardiol 1995; 27: 1023-1034.

35. Tuijl MJM, van Bergen en Henegouwen PMP, van Wijk R, Verkleij AJ. The isolated neonatal rat-cardiomyocyte used in an in vitro model for 'ischemia'. II. Induction of the 68 kDa heat shock protein. Biochim Biophys Acta 1991; 1091: 278-284.

36. Chiesi M, Longoni S, Limbruno U. Cardiac alpha-crystallin. III. Involvement during heart ischemia. Mol Cell Biochem 1990; 97: 129-136.

37. Barbato R, Menabo R, Dainese P et al. Binding of cytosolic proteins to myofibrils in ischemic rat hearts. Circ Res 1996; 78: 821-828.

38. Yellon DM, Latchman DS. Stress proteins and myocardial protection. J Mol Cell Cardiol 1992; 24: 113-124.

39. Iwaki K, Chi S-H, Dillmann WH, Mestril R. Induction of HSP70 in cultured rat neonatal cardiomyocytes by hypoxia and metabolic stress. Circulation 1993; 87: 2023-2032.

40. Andres J, Sharma HS, Ralph K et al. Expression of heat shock proteins in the normal and stunned porcine myocardium. Cardiovasc Res 1993; 27: 1421-1429.

41. Das DK, Engelman RM, Kimura Y. Molecular adaptation of cellular defences following preconditioning of the heart by repeated ischaemia. Cardiovasc Res 1993; 27: 578-584.

42. Tanaka M, Fujiwara H, Yamasaki K et al. Ischemic preconditioning elevates cardiac stress protein but does not limit infarct size 24 or 48 h later in rabbits. Am J Physiol 1995; 36: H1476-H1482.

43. Engelman DT, Chen C, Watanabe M et al. Improved 4- and 6-hours myocardial preservation by hypoxic preconditioning. Circulation 1995; 92 [suppl. II]: II-417-II-422.

44. Meerson FZ. Adaptive protection of the heart: Protecting against stress and ischemic damage. Boca Raton: CRC Press, 1991.

45. Meerson FZ, Malyshev IYu, Zamotrinsky AV. Differences in adaptive stabilization of structures in response to stress and hypoxia relate with the accumulation of hsp70 isoforms. Mol Cell Biochem 1992; 111: 87-95.

46. Meerson FZ, Malyshev IYu, Zamotrinsky AV, Kopylov YuN. The role of hsp70 and IP_3-DAG mechanism in the adaptive stabi-

lization of structures and heart protection. J Mol Cell Cardiol 1996; 28: 835-843.

47. Brown JM, White CW, Terada LS et al. Interleukin 1 pretreatment decreases ischemia/reperfusion injury. Proc Natl Acad Sci USA 1990; 87: 5026-5030.

48. Maulik N, Das DK. Cardioprotective effect of TNFα is mediated through the myocardial adaptation to oxidative stress. J Mol Cell Cardiol 1993; 25 [suppl.III]: S43.

49. Bensard DD, Brown JM, Anderson BO et al. Induction of endogenous tissue antioxidant enzyme activity attenuates myocardial reperfusion injury. J Surg Res 1990; 49: 126-131.

50. Morris SD, Cumming DVE, Latchman DS, Yellon DM. Specific induction of the 70-kD heat stress proteins by the tyrosine kinase inhibitor herbimycin-A protects rat neonatal cardiomyocytes—A new pharmacological route to stress protein expression? J Clin Invest 1996; 97: 706-712.

51. Maulik N, Wei Z, Engelman RM et al. Improved post-ischemic ventricular recovery by amphetamine is linked with its ability to induce heat shock. Mol Cell Biochem 1994; 137: 17-24.

52. Maulik N, Engelman RM, Wei Z et al. Drug-induced heat-shock preconditioning improves post-ischemic ventricular recovery after cardiopulmonary bypass. Circulation 1995; 92 [suppl II]: II-381-II-388.

53. Murakami Y, Uehara Y, Yamamoto C et al. Induction of Hsp72/73 by Herbimycin A, an inhibitor of tyrosine kinase oncogenes. Exp Cell Res 1991; 195: 338-344.

54. Nakano M, Knowlton AA, Yokoyama T et al. Tumor necrosis factor-alpha-induced expression of heat shock protein 72 in adult feline cardiac myocytes. Am J Physiol 1996; (Heart and Circulatory Physiology) 39: H1231-H1239.

55. Kaur P, Welch WJ, Saklatvala J. Interleukin 1 and tumor necrosis factor increase phosphorylation of the small heat shock protein: Effects in fibroblasts, Hep62 and U937 cells. FEBS Lett 1989; 258: 269-273.

56. Saklatvala J, Kaur P, Guesdon F. Phosphorylation of the small heat-shock protein is regulated by interleukin 1, tumor necrosis factor, growth factors, bradykinin and ATP. Biochem J 1991; 277: 635-642.

57. Locke M, Tanguay RM, Klabunde RE, Ianuzzo CD. Enhanced post-ischemic myocardial recovery following exercise induction of HSP 72. Am J Physiol 1995; (Heart and Circulatory Physiology) 38: H320-H325.

58. Williams RS, Thomas JA, Fina M et al. Human heat shock protein 70 (hsp70) protects murine cells from injury during metabolic stress. J Clin Invest 1993; 92: 503-508.

59. Nguyen VT, Bensaude O. Increased thermal aggregation of proteins in ATP-depleted mammalian cells. Eur J Biochem 1994; 220: 239-246.

60. Barry WH, Hamilton CA, Knowlton KU. Regulated expression of a contractile protein gene correlates with recovery of contractile function after reversible metabolic inhibition in cultured myocytes. J Mol Cell Cardiol 1995; 27: 551-561.

61. Mestril R, Chi S-H, Sayen MR et al. Expression of inducible stress protein 70 in rat heart myogenic cells confers protection against simulated ischemia-induced injury. J Clin Invest 1994; 93: 759-767.

62. Heads RJ, Yellon DM, Latchman DS. Differential cytoprotection against heat stress or hypoxia following expression of specific stress protein genes in myogenic cells. J Mol Cell Cardiol 1995; 27: 1669-1678.

63. Marber MS, Mestril R, Chi SH et al. Overexpression of the rat inducible 70-kD heat stress protein in a transgenic mouse increases the resistance of the heart to ischemic injury. J Clin Invest 1995; 95: 1446-1456.

64. Plumier J-CL, Ross BM, Currie RW et al. Transgenic mice expressing the human HSP70 have improved post-ischemic myocardial recovery. J Clin Invest 1995; 95: 1854-1860.

65. Radford NB, Fina M, Benjamin IJ et al. Cardioprotective effects of 70-kDa heat shock protein in transgenic mice. Proc Natl Acad Sci USA 1996; 93: 2339-2342.

66. Subjeck JR, Shyy T-T. Stress protein systems of mammalian cells. Am J Physiol 1986; 250: C1-C17.

67. Kellermayer MSZ, Csermely P. ATP induces dissociation of the 90 kDa heat shock protein (hsp90) from F-actin: Interference with the binding of heavy meromyosin. Biochem Biophys Res Commun 1995; 211: 166-174.

68. Bennardini F, Wrzosek A, Chiesi M. αB-Crystallin in cardiac tissue: association with actin and desmin filaments. Circ Res 1992; 71: 288-294.

69. Mehlen P, Preville X, Chareyron P et al. Constitutive expression of human hsp27, or human αB-crystallin confers resistance to TNF- and oxidative stress-induced cytotoxicity in stably transfected murine L929 fibroblasts. J Immunol 1995; 154: 363-374.

70. Mehlen P, Kretzrremy C, Preville X, Arrigo AP. Human hsp27, Drosophila hsp27 and human alpha B-crystallin expression-mediated increase in glutathione is essential for the protective activity of these proteins against TNF alpha-induced cell death. EMBO J 1996; 15: 2695-2706.

71. Huot J, Houle F, Spitz DR, Landry J. HSP27 phosphorylation-mediated resistance against actin fragmentation and cell death induced by oxidative stress. Cancer Res 1996; 56: 273-279.

72. Bogoyevitch MA, Ketterman AJ, Sugden PH. Cellular stresses differentially activate c-jun N-terminal protein kinases and extracellular signal-regulated protein kinases in cultured ventricular myocytes. J Biol Chem 1995; 270: 29710-29717.

73. Seko Y, Tobe K, Ueki K et al. Hypoxia and hypoxia/reoxygenation activate Raf-1, mitogen-activated protein kinase kinase, mitogen-activated protein kinases and S6 ki-nase in cultured rat cardiac myocytes. Circ Res 1996; 78: 82-90.

74. Longoni S, Lattonen S, Bollock G et al. Cardiac alpha-crystallin II. Intra-cellular localization. Mol Cell Biochem 1990; 97: 123-128.

75. Gopalsrivastava R, Haynes JI, Piatigorsky J. Regulation of the murine alpha B-crystallin/small heat shock protein gene in cardiac muscle. Mol Cell Biol 1995; 15: 7081-7090.

76. Sharma HS, Maulik N, Gho BCG et al. Coordinated expression of hemeoxygenase-1 and ubiquitin in the porcine heart subjected to ischemia and reperfusion. Mol Cell Biochem 1996; 157: 111-116.

77. Maulik N, Sharma HS, Das DK. Induction of the haem oxygenase gene expression during the reperfusion of ischemic rat myocardium. J Mol Cell Cardiol 1996; 28: 1261-1270.

78. Plumier J-C L, Currie RW. Heat shock-induced myocardial protection against ischemic injury: a role for Hsp70? Cell Stress & Chaperones 1996; 1: 13-17.

79. Hutter JJ, Mestril R, Tam EKW et al. Overexpression of heat shock protein 72 in transgenic mice decreases infarct size in vivo. Circulation 1996; 94: 1408-1411.

80. Martin JL, Mestril R, Hilal-Dandan R, Dillmann W. Small heat shock proteins and protection against ischemic injury in cardiomyocytes. Abstracts of 1996 Meeting on Molecular Chaperones & The Heat Shock Response (May 1-5, 1996, Cold Spring Harbor, NY), p. 342.

CHAPTER 6

INVOLVEMENT OF HEAT SHOCK PROTEINS IN PROTECTION OF VARIOUS NORMAL AND TUMOR CELLS FROM ISCHEMIC INSULT

The heart is the first organ where the participation of HSPs in protecting against ischemia has been demonstrated directly by overexpressing HSP70 in transgenic animals (see the previous chapter). For other organs and tissues, adequate experimental evidence is absent. However, there is are numerous data indicating the possible involvement of HSPs in anti-ischemic protection. Below we discuss the main results obtained on this issue to date.

6.1. HSPS AND ISCHEMIC TOLERANCE IN THE BRAIN

Besides the heart, the brain is the other main organ where the role of HSPs in protection against ischemia is intensively studied (see also refs. 1-3 for review). Before considering the numerous data, we would like to underline the main differences in the responses of these organs to ischemia. First of all, in contrast to the heart, the brain consists of a number of heterogenous cell populations with marked distinction in each cell population's vulnerability to ischemia. So, the most vulnerable cell population is the hippocampal CA1 neurons, which are irreversibly damaged after even a few minutes of complete ischemia. It is important to note that death of these cells after brief ischemic insult occurs in a delayed fashion (i.e., it takes several days to completely develop, see ref. 1). Second, there are no uncomplicated indices of functional activity (similar to the contractile activity in the heart) as well as cell death (like cytosolic enzyme efflux) after ischemic insult in the brain. Therefore determination of cellular damage and death in most of the research is based mainly on the histological examination of the affected brain regions. In addition, immunostaining of certain proteins (in particular, microtubule-associated protein, MAP-2) is frequently employed as a marker of neuronal viability (see below).

6.1.1. HSP70 AND THE TOLERANCE TO GLOBAL ISCHEMIA

The main contributions to the study of ischemic tolerance in the brain have been made by three independent Japanese researcher groups. In 1990, Kitagawa and co-workers from Osaka University were the first to discover this phenomenon.[4] The researchers subjected gerbils to global ischemia by occlusion of the bilateral common carotid arteries for 5 minutes and studied the death of CA1 hippocampal neurons after 7 days by hematoxylin-eosin (H&E) staining and immunohistochemical analysis of MAP-2. The ischemic insult resulted in a marked reduction in neuron numbers (Fig. 6.1). However, pretreatment of the gerbils with 2 minute ischemia 1~2 days before 5 minute ischemia remarkably protected the neurons from death: their number increased about 10-fold as compared to the control. Moreover, when the animals were preconditioned with two episodes of 2 minute ischemia at 1 day intervals, neurons were completely protected from death (Fig. 6.1). Importantly, no protection was observed when the interval between preconditioning and challenging exposures was only 1 hour (Fig. 6.1), indicating that some slow process (possibly protein synthesis) may be necessary for protection. The last result is in great contrast to the preconditioning phenomenon in the heart where only 1 hour is required for the development of classic tolerance (early stage of protection; see the previous chapter). The authors also demonstrated that achieving some threshold during the first insult is necessary, since 1 minute of ischemia is insufficient for tolerance development (Fig. 6.1).[4] Thus, the results clearly show the existence of an ischemic tolerance in the brain, although HSP synthesis was not evaluated in this study.

In 1991, Kirino et al[5] from Teikyo University (Tokyo) confirmed the results using the same model and showed that a maximal protective effect of 2 minute ischemia was observed after 4 days (Fig. 6.2); more importantly, the researchers definitely demonstrated by immunohisto-

chemical analysis the accumulation of inducible HSP70 in the CA1 region after 2 minute ischemia, with the maximum at 2-4 days. Meanwhile, Kitagawa and co-workers[6] found ischemic tolerance not only in CA1 hippocampal neurons but also in the neurons of other brain regions. To induce death of these neurons, prolonged (10 minutes) ischemia was used. The researchers discovered that 2 minute ischemic insult 2 days before 10 minute ischemia provided significant protection to the CA3 region of the hippocampus, the cerebral cortex, the caudoputamen and the thalamus, whereas no protective effect was observed after 1 minute episodes of ischemia insult. In addition, HSP72 (inducible HSP70) accumulation was observed in the hippocampus by immunoblotting after 2 minute ischemia, but not after 1 minute ischemia. The authors concluded that ischemic tolerance occurs widely in the brain and that ischemic pretreatment severe enough to cause HSP72 synthesis might be necessary for this phenomenon. Later, the researchers from this lab detected that prolonged (for 30 days) sublethal mild hypoperfusion by carotid occlusion induced HSP72 synthesis and tolerance to 5 minute global ischemia in the hippocampal neurons, whereas after 1 day of hypoperfusion neurons became more susceptible to the ischemic insult.[7]

The third research group, Aoki and co-workers from Tohoku University, also observed the protection of CA1 hippocampal neurons from 3.5 minute ischemia by 2 minute ischemic pretreatment. They studied the expression of HSP70 mRNA and found that both 2 minute and 3.5 minute ischemia induced the transcription of the HSP70 gene, but after 3.5 minute ischemia only a low amount of HSP70 protein was produced in neurons, apparently due to an irreversible inhibition of translation, and the cells were almost lost by 7 days (see also section 4.3). In preconditioned neurons, the peak time of the HSP70 mRNA accumulation after 3.5 minute ischemia shifted to an earlier period of reperfusion (from 1 day to 3 hours) and the cells

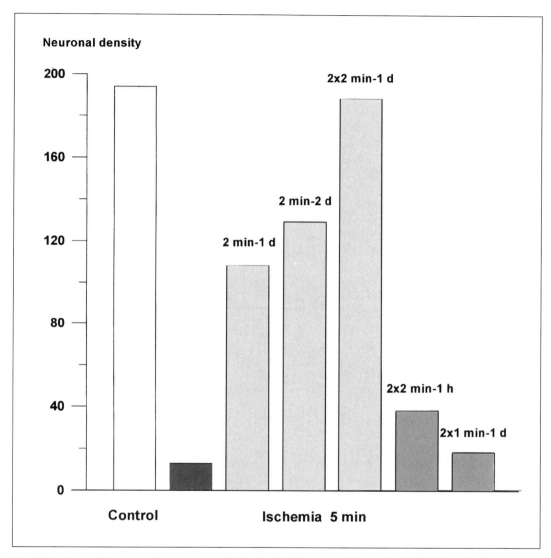

Fig. 6.1. Ischemic tolerance phenomenon in the brain. Gerbils were subjected to global ischemia for 5 minutes either without pretreatment (second column) or preconditioned as indicated above the columns, and neuronal density (number of survived pyramidal neurons per 1 mm of CA1 region) was determined by light microscopy 7 days following the challenging insult. The maximal protection (acheiving control) was observed after two episodes of 2 minute ischemia at 1 day intervals (fifth column). The graph is depicted from the data of Kitagawa et al.[4]

accumulated a large amount of HSP70 beginning from 2 hours.[9] The authors concluded that preconditioning ischemia accelerated HSP70 transcription and ameliorated the disturbance of HSP70 mRNA translation, and that this event could play an important role in the acquisition of ischemic tolerance.[9] It is interesting to note that HSP70 overexpression in rat fibroblast

cell culture was shown to accelerate the recovery from transcriptional and translational inhibition following heating.[10]

In the next study, Aoki and coworkers subjected gerbils to three 2 minute forebrain ischemic insults spaced at 1 hour intervals and to a single 6 minute period of ischemia.[11] Neuronal damage following repeated insults was more drastic than after

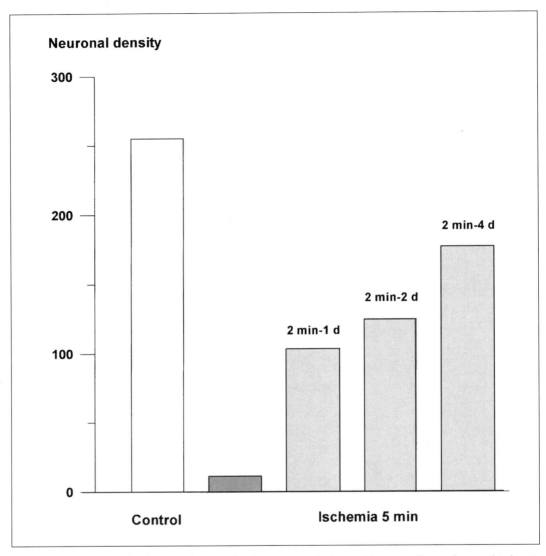

Fig. 6.2. Dependence of ischemic tolerance development on the interval between first and second ischemic insult. Gerbil treatment and assay were the same as in Fig. 6.1. Maximal protection was observed following 4 days after preconditioning 2 minute ischemia. The graph is depicted from the data of Kirino et al.[6]

the single equivalent period of ischemia. No HSP70 immunostaining was observed in CA1 neurons which are destined to die after repeated insults, while mild to moderate staining was observed in neurons after single ischemia. Thus, the data are in accord with the above results of Kitagawa et al[4] who did not observe a protective effect when two ischemic insults were repeated at brief intervals, and in contrast to the results on the myocardium where

similar treatment lead to HSP induction and protection (see section 4.2 and chapter 5). In addition, CA3 neurons, which are resistant to ischemia, showed intensive staining with HSP70 antibody following 6 minutes of ischemia, but only slightly stained after three 2 minute ischemic insults, showing less stress after repeated ischemia.[11] Thus, repeated sublethal insults at short intervals do not yield HSP70 and exert cumulative damage in CA1 neurons, but not in CA3 neurons where such cumulative damage was absent. Interestingly, ischemic tolerance can be induced repeatedly: after 4 weeks of the first preconditioning tolerance disappears, and a second preconditioning confers resistance again.[12]

Similar results were obtained by Tohoku's group on another (rat) model of cerebral ischemia. In Wistar rats, 3 minute ischemia increases HSP70 immunoreactivity in CA1 neurons with the maximum at 3 days of reperfusion and protected them from second 6-8 minutes of ischemia. However, no protection was detected after more severe ischemia (10 minutes).[13,14] The researchers also detected HSP70 in preconditioned neurons after a second 6 minute ischemia accumulated faster, indicating an accelerated translational recovery like that observed in gerbils (see above).

Therefore, all the above data clearly demonstrate an existence of tolerance to global ischemia in hippocampal neurons both in gerbils and rats; this tolerance is induced by short-term ischemia and is apparently associated with HSP70 accumulation. The question arises as to whether other stresses inducing HSPs can confer ischemic tolerance.

Heating is the most powerful inducer of HSPs and it is widely applied in the study of other organs (e.g., heart). There are some studies where hyperthermia-induced protection against global ischemia is demonstrated in rats and gerbil;[15,16] a maximal protection being detected at 18 hours after heat shock (42°C, 15 minutes).[16] However, HSP synthesis after heating is not significant in hippocampal neurons either in vitro or in vivo but is much more pronounced in astrocytes (see refs. 2, 17, 18).

Another well-known inducer of HSP synthesis is oxidative stress (see section 4.5). Othsuki and co-workers from Osaka University studied whether oxidative stress provoked by the administration of superoxide dismutase inhibitor, diethyldithiocarbamate (DEDC), confers tolerance to 5 minute global ischemia in gerbil hippocampal CA1 neurons. DEDC treatment 2 days before ischemia failed to show either

protection of CA1 neurons or HSP70 accumulation. However, this treatment 4 days before ischemia had a significant protective effect associated with an increase in HSP70.[19]

There are also some studies where HSP70 expression was not associated with ischemic protection. When rats were subjected to graded global ischemia, Simon et al[20] detected HSP70 expression preceding neuronal death. Also, the expression of HSP70 in CA1 and CA3 neurons is not sufficient to guarantee survival after prolonged (more than 5-10 minutes) ischemia.[21,22] Certainly, the data do not exclude the protective function of HSP70; they indicate only that the HSP70-expressing cells cannot withstand an extremely severe stress. This finding is not that strange considering the fact that HSP70 in the heart cannot confer tolerance under very prolonged ischemia conditions (see the previous chapter).

In other studies, ischemic protection was found to be independent of HSP70 synthesis. In particular, Kawagoe et al[23] detected that pentobarbital pretreatment prevented CA1 neuronal death after 5 minute ischemia and greatly reduced the induction of HSP70 mRNA. The protective effect may be associated with the hypothermic effect of pentobarbital. Later, Shaver[24] and co-workers demonstrated the hypothermia-induced reduction of HSP70 mRNA expression in ischemic piglet brain. Kumar and co-workers[25] subjected gerbils to 10-minute global ischemia with concurrent moderate hypothermia (30°C). Hypothermic treatment, while protecting hippocampal neurons, completely abolishes HSP70 induction. The authors concluded that a mechanism other than HSP70 induction is responsible for the neuroprotective effect of hypothermia. In accordance with this, a recent study by Yager and Asselin[26] demonstrated that moderate hypothermia (31°C) completely prevents depletion of ATP during brain ischemia in immature rats. Likewise, the neuroprotective effect of K^+-channel openers which is not associated with *HSP70* expression[27] may be due to an alleviation of ischemia-induced ATP loss, a well-known effect on the heart (see section 2.3).

6.1.2. OTHER STRESS PROTEINS AND GLOBAL ISCHEMIA

Ischemia/reperfusion in the brain is often accompanied by the expression of other stress proteins (see section 4.3). Below we consider their possible participation in ischemic tolerance.

Aoki et al[9] found that 2 minute preconditioning ischemia in gerbils accelerated transcription of *HSC70* following a second 3.5 minute ischemia, an effect similar to that observed for HSP70 mRNA. Possibly, accelerated HSC70 expression coupled with HSP70 promotes neuronal survival.

HSP90 is expressed in great amounts in unstressed neuronal tissue and is not further elevated following hyperthermia.[28] To date, there are no data on its participation in ischemic tolerance (see also subsection 6.1.5).

HSP27, in contrast to HSP90, is not expressed in unstressed neurons but is a constitutive protein of astrocytes.[29] Following 2 minute preconditioning ischemia, no increase in HSP27 content in CA1 neurons is detected, but its accumulation is observed in a small number of astrocytes in the CA3 region and in many astrocytes in the dentate hilus.[30,31] From the data, Kato et al concluded that HSP27 is not involved in the protection of hippocampal neurons afforded by preconditioning but may participate in the post-ischemic recovery of reactive astrocytes.[30,31] Interestingly, in astrocyte cultures heating, as well as some other exposures (interleukin-1, tumor necrosis factor), increases the phosphorylation of HSP27[32] that is a well-known phenomenon for other cell types (see chapters 1 and 3), and this post-translational modification may play some role in protecting the cell against stresses (see chapter 7 for discussion).

Besides cytoplasmic chaperones, chaperones of organelles (HSP60, GRPs) are also expressed after ischemia/reperfusion in the brain (see section 4.3). Their protective function has not yet been evaluated.

Ischemia induces the expression of ubiquitin mRNA in CA1 and CA3 regions of rat hippocampus, which reaches a peak at 4-6 hours and persists up to 24 hours of reperfusion.[33] Kato and co-workers[34] studied the localization of ubiquitin in gerbil hippocampus following 3 minutes of ischemia with or without pretreatment with 2 minutes of sublethal ischemia and 3 days of reperfusion. Ubiquitin immunoreactivity in the hippocampus disappeared 4 hours after 3 minute ischemia in both cases. In the resistant cell populations (CA3 and dendate gyrus neurons), ubiquitin recovers by 24 hours. In CA1 neurons, it is recovered by 48 hours after preconditioned treatment, but never recovered after 3 minute ischemia, which only leads to the delayed death of the cells. In studies of other research groups, it was found that short-term ischemia in hippocampus leads to an increase in ubiquitin-protein conjugates (especially in the mitochondrial fraction)[35,36] that is accompanied by a depletion of free ubiquitin.[37] The absence of free ubiquitin recovery in CA1 neurons may be due, in particular, to the inhibition of 26S-proteasome-mediated proteolysis in ATP-depleted cells.[38] Therefore, ubiquitination but not proteolysis occurs in the ischemic CA1 neurons destined to die. Thus, the data indicate that ischemia activates ubiquitination of proteins and the protective effect of preconditioning is associated with the recovery of ubiquitin-dependent proteolysis.

Since ischemia/reperfusion is a well-known inducer of oxidative stress (see section 2.4, 4.5), protection of neurons may also depend upon the level of anti-oxidative enzymes. Uyama et al[39] demonstrated that 5 minute ischemia in gerbils leads to the enhanced transcription of the *Cu,ZnSOD* gene (cytosolic superoxide dismutase). After prolonged (10-minute) ischemia, CA1 neurons showed reduced immunoreactivity of Cu,ZnSOD and total absence of MnSOD (mitochondrial SOD), while reactive glial cells showed increased accumulation of both enzymes.[40] In the work of Ohtsuki et al[19] where ischemic tolerance induced by prior oxidative stress was detected (see above), the researchers also found the accumulation of MnSOD together with HSP70 in the hippocampus of resistant animals. Likewise, Gorgias and co-workers[41] observed that the protective effect of mild hypoxic pretreatment (10-15% O_2) as compare to severe (3% O_2) hypoxia-induced death in gerbil hippocampus was closely related to the increased SOD level. The addition of SOD to cultures of hippocampal neurons significantly decreases their death after hypoxia-reoxygenation;[42] also the administration of SOD to gerbils ameliorates delayed death of CA1 neurons after 5 minute ischemia.[39] Therefore, the expression of anti-oxidative enzymes such as SOD together with HSPs may be important for neuronal survival.

6.1.3. STRESS PROTEINS AND FOCAL ISCHEMIA

Because the brain contains various cell populations, it is of interest to find out whether HSP-mediated protection of other cell types (astrocytes, microglial, and endothelial cells) also occurs. Short-term global ischemia is not fully suitable to address this question because nonneuronal cells of the brain have a much greater resistance to ischemia than neurons. Instead, long-lasting ischemic insult will be used—mainly through the focal model of ischemia—because it results in the injury of almost all cell types. In the models, the middle cerebral artery (MCA) is occluded for a prolonged period of time, and both cell survival and HSP expression are assessed by H&E staining and immunohistochemistry, respectively.

Brief periods of MCA occlusion induce HSP70 primarily in neurons, which survive the insult.[43-46] After 1~2 hours of MCA occlusion, immunoreactive HSP70 within the ischemic core is observed exclusively in the endothelial cells of cerebral blood vessels which are a major surviving cell type in the region; other cell populations which are destined to die do not express inducible HSP70. At the edge of the infarct zone, survival and expression of HSP70 are observed in microglia and sometimes in astrocytes, and in the

surrounding region HSP70 immunostaining is found in neurons that have survived.[1,43,44,47,48] Expression of endoplasmic GRP78 and mitochondrial GRP75 in neurons under focal ischemia is similar to that of HSP70.[49,50]

Kato and co-workers[48] studied the expression of HSP27 following 1 hour of MCA occlusion in the rat. Like global ischemia, no HSP27 synthesis is detected in neurons, but the protein is induced in microglia of the ischemic center after 4 hours and later in the reactive astrocytes. In the same model, Liu et al[51] detected enhanced expression of Cu,ZnSOD and MnSOD in cortical neurons and glial cells in the boundary zone between normal and infarcted areas, while both proteins rapidly diminished or disappeared in the neurons of the ischemic center.

Recently, Nimura et al[52] studied the expression of hemeoxygenase-1 (HO-1) by immunocytochemistry in rats after MCA occlusion. One day following 30 minutes of ischemia, HO-1 staining in striatum occurs mainly in the endothelial cells in infarct and in glial cells surrounding the areas of infarct; no HO-1 is induced in cortex whereas HSP70 is induced in cortical neurons in MCA distribution. One day following 2 hours of ischemia, both HO-1 and HSP70 are induced in neurons of the cortex. The authors suggested that the induction of HO-1 may augment the anti-oxidative defense mechanism compromised by cerebral ischemia, since the protective effect of HO-1 against oxidative stress is established (see chapter 1).

The results indicate that the survival of neuronal, glial and endothelial cells after focal ischemia is associated with HSP70 accumulation; SOD and HO-1 may play a certain role in the protection of the cells. Moreover, in glial cells, HSP27 also probably participates in some protective function, since the protein can rescue cells from oxidative stress (see chapter 7). Certainly, this does not mean that HSP synthesis is specifically responsible for the survival of the cells; moreover, if the insult is severe enough, some HSP70-expressing neurons eventually die (as was also found for global ischemia, see above). However, the data on focal ischemic models indicate that HSP70 is involved in anti-ischemic protection of not only neurons, but other cell types as well.

6.1.4. ISCHEMIA, EXCITOTOXICITY AND HSP-MEDIATED PROTECTION

Ischemic insult in vivo is accompanied by the release of excitatory amino acids (e.g., glutamate), which acts as neurotransmitters at low concentrations but exert cytotoxic effects at high concentration (see refs. 3, 53 for review). The toxicity of glutamate is the result of a Ca^{2+} overload owing to an uncontrolled influx of Ca^{2+} through agonist- (N-methyl-D-aspartate, NMDA) and voltage-sensitive Ca^{2+} channels of the plasma membrane (see sections 2.3 and 4.3). There are a number of studies where blockade of NMDA-receptors is shown to save neurons from ischemic death. For example, Simon et al[54] observed a marked protective effect of NMDA-antagonist, 2-amino-7-phosphonoheptanoic acid, against neuronal death which was caused in rat hippocampus by 30 minutes of ischemia. An elegant article on this issue has been published recently by Wahlestedt and colleagues.[55] Using antisense oligonucleotides (ODNs) to the NMDA receptor channel, the researchers observed that treatment with the ODNs of cortical neuronal cell cultures reduced the number of the receptors, as well as NMDA-induced Ca^{2+}-influx and cell death. Then they revealed that the injection of antisense ODNs (but not sense ones) into the rat brain 2 days before 24 hours of MCA occlusion reduces the infarct size about two-fold. Furthermore, another NMDA-receptor antagonist, MK-801 (dizocilpine maleate), protects cell cultures of cortical and cerebellar granule neurons from anoxic cell death,[56,57] since ATP depletion in vitro promotes the release of neurotransmitters from nerve terminals.[58] Investigators from Kogure's lab studied (by microdialysis technique) whether ischemic tolerance in the gerbil hippocampus is associated with changes in the extracellular glutamate concentration during ischemic insult. They observed that during and immediately after 3 minute forebrain ischemia, the extracellular glutamate level showed a striking increase. However, pretreatment with 2 minute ischemia 4 days before secondary 3 minute ischemia, which induced ischemic tolerance (see subsection 6.1.1 and Fig. 6.2), did not alter the amount of glutamate released.[59] Thus, the finding obviously indicates that hippocampal neurons after preconditioning become resistant to glutamate toxicity. The activation of NMDA receptors during the first ischemic insult may be important for ischemic tolerance, since MK-801 blocked both thermotolerance development and HSP synthesis.[60] The question arises as to whether HSPs can defend brain cells from excitotoxic insults.

In 1991, two groups of researchers working independently of one another studied the neuroprotective properties of HSPs in a model of glutamate-induced excitotoxicity. Rordorf and co-workers[61] used 2 week old cultures (containing both glia and neurons) grown from the cortex of rat pups. Cell death is caused by exposing the cultures to glutamate for 5-20 minutes and assessed by a lactate dehydrogenase (LDH) release assay. After cell exposure to 100 µM of glutamate for 10 minutes, about 80% of cells die by 24 hours; however, heat shock (42°C, 20 minutes) applied 3 hours prior to glutamate treatment, reduced the cell death about fourfold (to 20%). No protection was detected when glutamate was applied at 30 minutes after heat shock, indicating that some slow process is necessary for tolerance acquisition. The authors demonstrated that heat shock led to increased expression of HSP70, and inhibitors of transcription (actinomycin D) or translation (cycloheximide) completely blocked the protective effect of heat shock. Similar results were obtained by Lowenstein and colleagues in cerebellar granule neuron cultures.[62] Thus, the protective effect of heat shock treatment against excitotoxicity in neuronal cultures is dependent on the accumulation of HSPs, in particular, HSP70.

Although in the above studies the mechanism of HSP-mediated protection has not been evaluated, some suggestions can be made. First, Ca^{2+}-influx through NMDA-receptor is shown to stimulate superoxide generation, and some antioxidants are effective blockers of glutamate-induced neurotoxicicty.[63,64] In other cell cultures, it was shown that HSPs (in particular, HSP70 and HSP27) protect cells from oxidant-induced cytotoxicity (see ref. 65 for review and chapter 7 for further discussion). Second, NMDA treatment damages DNA, which leads to the activation of poly (adenosine 5'-diphosphoribose) synthetase and the depletion of NAD and ATP;[66] furthermore, activation of glycolysis via the overexpression of the glucose transporter protects cultured hippocampal neurons from glutamate toxicity.[67] As we consider below, the HSP70-mediated defense from ATP depletion was recently demonstrated. Thirdly, excitotoxic stress induces cytoskeletal damage, namely depolymerization of actin filaments[68] and calpain-mediated spectrin breakdown.[69] Thus, protection of the cytoskeleton by HSPs is also possible (see section 6.4 and chapter 7).

6.1.5. HSP70 OVEREXPRESSION PROTECTS NEURONS AND ASTROCYTES FROM ISCHEMIA-INDUCED NECROSIS IN VITRO

In vivo models of cerebral ischemia are rather complicated and some other factors besides HSP may participate in ischemic tolerance. Therefore, simpler in vitro models can be more suitable for establishing a direct estimation of HSP's protective function. Recently, a few important studies on this issue appeared. Amin et al[70] from Latchman's lab have established that the pretreatment of dorsal root ganglion (DRG) neurons with heat shock (42°C, 40 minutes) 24 hours before ischemic stress (anoxia + deoxyglucose for 2 hours) reduces their death about two-fold as assayed by Trypan blue staining. The heat shock brought about a drastic accumulation of

inducible HSP70 and the appearance of thermotolerance. In the next work,[71] the researchers demonstrated directly that overexpression by transfection of HSP70 alone is sufficient to save DRG neurons from ischemic stress. In contrast, overexpression of HSP90 had no protective effect similar to the observation made on cardiac cell cultures (see chapter 5). Amin and colleagues[70,71] also concluded from the studies that a certain level of HSP70 overexpression provides greater protection against hyperthermia than against ischemia; this may indicate, among other factors, distinct targets of HSP70 in ischemic and heated cells (see chapter 7 for discussion).

Finally, Papadopoulos et al[72] from Giffard's lab observed the protective effect of heat pretreatment against oxygen-glucose deprivation on primary cultures of astrocytes. The cells are markedly more resistant to ischemia than neurons (see section 6.1.3), and their death (assayed by LDH release) began only after 4 hours of ischemic stress.[73] Using retroviral vectors, the investigators transfected astrocytes with HSP70 and revealed that only 12% of HSP70-overexpressing cells died after 7 hours of ischemia, whereas 82% of control (nontransfected) astrocytes are dead by the same time.

Thus, the above studies from Latchman's and Giffard's labs directly demonstrate the protective function of HSP70 in neurons and astrocytes during ischemic stress. The research is in complete agreement with the results obtained from in vivo models of global and focal ischemia (see subsections 6.1 through 6.3 above).

6.1.6. DELAYED NEURONAL DEATH AND HSPs: PROTECTION FROM APOPTOSIS?

As we considered above, HSP70-mediated protection of neurons and astrocytes from ischemia-induced necrosis is clearly shown. However, an unique feature of hippocampal neurons is that after even short-term ischemic insult (3-5 minutes) they die, but their death proceeds in a greatly delayed manner (i.e., it takes about 1 week to develop). The delayed neuronal death (DND) has apparently little likeness to the classic ischemic necrosis which develops very rapidly. In the early 1990s, when the ischemic tolerance phenomenon was first demonstrated in hippocampal neurons, little was known about the mechanism of DND. Therefore, before discussing the possible protective functions of HSPs, we must first consider the recent data on DND in various models of brain ischemia.

In section 2.2 we mentioned that besides necrosis, ischemia, if not severe, can also stimulate apoptotic (programmed) cell death in various cells and organs. Many recent data indicate that DND of hippocampal neurons occurs via apoptosis (see refs. 74-76 for recent review). In 1990, Goto et al[77] from Kogure's lab demonstrated that a consecutive administration of cycloheximide (CHX), a protein synthesis inhibitor, almost totally prevents cell death of CA1 hippocampal neurons (i.e., the same neurons where ischemic tolerance was first demonstrated, see section 6.1) at 72 hours after 10-minute ischemia in rats. In 1993, Okamoto et al[78] were the first to observe internucleosomal DNA fragmentation (a hallmark of apoptosis, see section 2.2) at 48 hours after 5 minute ischemia in CA1 neurons. Volpe et al[79] observed the 3' termini of endonuclease-generated DNA fragments in the hippocampus 72 hours after reperfusion. In the next studies, typical apoptotic (condensed) nuclei were found in the CA1 area 24 hours and 3 days following ischemia; DNA of the nuclei is broken (as assayed by TUNEL method, see section 2.2.), but nuclear pyknosis is observed only on day 7 after ischemia.[80] It is significant that no DNA fragmentation was found in astrocytes, which are resistant to ischemia, and decapitation-induced ischemia caused a random (but not internucleosomal) DNA degradation.[81]

Similar results were obtained on the models of focal ischemia by Li and co-workers. After 2 hours of MCA occlusion in mice and rats, neurons exhibiting DNA fragmentation are located primarily in the inner boundary zone of the infarct[82,83] and DNA fragmentation peaked at 24-48 hours after focal ischemia.[84] Upon short-term MCA occlusion (10-20 minutes) apoptotic cells increased in the regions of selective neuronal necrosis (the preoptic area and the striatum).[85] Of great importance is that infarction after mild focal ischemia (30 minutes) develops in surprisingly delayed fashion (up to 2 weeks) and the infarction volume is as large as that induced by 90 minutes ischemia after 1 day (when it is fully developed). Moreover, the delayed infarction is markedly reduced by CHX pretreatment.[86] Although there is a report that the TUNEL assay cannot discriminate apoptosis from necrosis, morphological examination can make this clear.[87]

Therefore, DND of neurons after global ischemia, as well as neuronal death following focal ischemia, obviously represent apoptosis, provided that the ischemic stress is not prolonged. Although the molecular mechanism of neuronal apoptosis is not fully understood, some speculations on this issue can be made. First of all, a dependence of apoptosis on protein synthesis indicates that ischemic stress induces some protein that promotes apoptosis. A good candidate for such a protein may be p53, the product of the tumor suppressor gene. This protein was first involved in apoptosis induced by genotoxic agents (see refs. 74, 88, 89 for review), but then its participation in neuronal apoptosis was also revealed. So, focal ischemia in rats induces p53 expression,[47,90] and the attenuation of p53 expression in transgenic mice (p53$^{-/-}$ or p53$^{+/-}$) reduces the infarct size after MCA occlusion by 15-27%.[91] It is still unknown why p53 is induced after ischemia, but DNA damage (single-strand breaks) is observed in CA1 neurons as early as 1 hour after of reperfusion, then disappears by 4 hours,[92] and subsequent double-strand breaks (internucleosomal fragmentation) arise much later (see above).

Excitotoxic stress induced by the glutamate analog, kainate, is also associated with p53 accumulation, and both p53 synthesis and apoptosis are prevented by CHX;[93] furthermore, the loss of p53 in transgenic mice protects neurons from kainate-induced cell death.[94] Like ischemic stress, glutamate provokes necrosis at a high dose or apoptosis at a low dose,[95] but apoptosis, in contrast to necrosis, is attenuated by CHX and actinomycin D, an inhibitor of transcription.[56,96] As we have considered above (section 6.1.4), HSP-mediated protection from glutamate-induced necrosis is demonstrated; therefore the protective function of HSPs against glutamate-induced apoptosis is also quite possible.

Another protein whose expression is increased in post-ischemic hippocampus and which can be implicated in programmed death is Bax protein.[97] The expression of the protein is activated by p53[98] and it belongs to the family of Bcl-2 proteins, which are important modulators of apoptosis under various exposures (see refs. 99, 100 for review). Some proteins of this family such as Bcl-2 itself, Bcl-x_L and Mcl-1 are powerful blockers of apoptosis, whereas Bcl-x_S and Bax promote apoptosis.[100] Hara and colleagues recently demonstrated by immunohistochemical assay that after 5 minute ischemia in gerbils, Bax expression in CA1 neurons increases with time and peaks at 72 hours (i.e., just before apoptosis development) while no expression of anti-apoptotic Bcl-2 is detected. Shimazaki et al[101] also did not find Bcl-2 expression after the same ischemic stress. However, preconditioning 2 minute ischemia (which led to HSP70 synthesis, see subsection 6.1.1) causes Bcl-2 expression in the CA1 region after 30 hours. From these data, the researchers suggested that Bcl-2 can contribute to ischemic tolerance. Surprisingly, Bcl-2 expression after focal ischemia also coincided with HSP expression (subsection 6.1.3) and cell survival.[102] Furthermore, overexpression of Bcl-2 can protect hypothalamic tumor cells GT1-7 from apoptotic death after transient energy deprivation (cyanide in a glucose-free medium)[103] and cultured neurons from glutamate-induced toxicity.[104] After focal

ischemia, Bcl-2 overexpressing neurons showed greater survival,[105] and in Bcl-2 overexpressing transgenic mice infarct size was reduced by 50%.[106]

Although Bcl-2-mediated protection of CA1 neurons from apoptosis remains to be demonstrated, we suggest that this protein may complement HSP70-mediated protection. Indeed, Bcl-2 is localized to various intracellular membranes (mitochondrial, endoplasmic and nuclear)[107] and recent data on various models of apoptosis demonstrate that its protective effect lies in preservation of the mitochondrial and the endoplasmic reticulum (ER) functions, apparently by suppression of oxidative stress and disturbance in Ca^{2+} homeostasis (see refs. 99, 100, 108). During DND, both the ER and mitochondria are rather severely damaged. For instance, the membranes of the ER in CA1 neurons after transient ischemia undergo changes and demonstrate decreased binding of the second messenger, IP_3 (inositol-triphosphate).[109] Mitochondria of the neurons after ischemic stress have reduced activity of cytochrome-oxidase[110] and MnSOD[40] that lead to an increased generation of reactive oxygen species (ROS) within them. Likewise, elevation of ubiquitin-protein conjugates is detected in the mitochondrial fraction of hippocampus, also indicating damage to mitochondrial proteins.[35] Since the inhibition of mitochondrial respiration with rotenone induces apoptosis in PC12 neuronal cell culture,[111] mitochondrial damage occurring after ischemia may be important for apoptosis development. In addition, Ca^{2+} imbalance is demonstrated in hippocampal neurons after ischemic insult in vivo[53,112] and in hippocampal neuron cultures after glutamate exposure.[113] Therefore, all the above data suggest a possible participation of Bcl-2 in protection from DND via the preservation of mitochondria (see Fig. 6.3).

We suggest that HSPs (in particular, HSP70), apart from Bcl-2, preserve cytosolic proteins, especially that of the cytoskeleton. Several data show that the expression of cytoskeletal proteins is

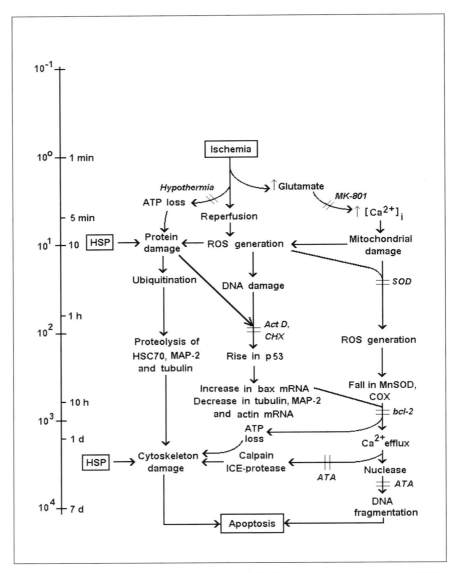

Fig. 6.3. Proposed scheme of delayed neuronal death and protective function of HSPs. Following brief ischemic insult several sequential events (according to log time scale on the left) occurs in the hippocampal neurons destined to die. Protecting proteins from damage as indicated, HSPs can save the cells from apoptosis. Although there are no data on the involvement of ICE-protease in neuronal apoptosis, we expect its role by analogy with other systems (see chapter 8 and text for further explanation).

harshly impaired in the hippocampus after ischemic stress. For instance, Kumar et al[114,115] detected a decline in α-tubulin and β-actin mRNA in the CA1 region as early as 6-24 hours post-ischemia when HSP70 mRNA is peaked. It is important that the decrease in α-tubulin mRNA content is accompanied by a decrease in translation

and a reduction in content of this protein (apparently a result of proteolysis).[116] In addition, expression of MAP-2 is also markedly reduced,[116,117] and distribution of MAP-2 immunoreactivity is markedly disrupted.[118] It is interesting that the level of constitutive HSC70 is also reduced soon after ischemia, although the content of other proteins is not diminished.[116] At 24-48 hours following transient ischemia in the hippocampal CA1 neurons, no gross structural alteration within them is noted, but at 72 hours electron microscopy reveals the accumulation of electron-dense fluffy dark material, possibly aggregates of proteins or internalized plasma membrane fragments.[119] Therefore, all of the above data attest that the damage to the cytoskeleton is intrinsic to neuronal apoptosis. The protective function of HSP (in particular, HSP70) may ensue from the preservation of the cytoskeleton (Fig. 6.3), as we demonstrated in other systems (see section 6.4).

6.2. ISCHEMIC TOLERANCE IN THE KIDNEY

As we have described in section 4.4, the kidney exhibits a rather powerful stress response to ischemia which is accompanied, in part, by a marked accumulation of HSP70. Borkan et al[120] showed that an increase in the HSP70 content protects renal cells from severe heating (50°C), and it was suggested that this protein may also be involved in cellular protection from ischemia/reperfusion. Indeed, there is a report by Perdizet and collegues[121] which concludes that heat shock pretreatment protects the pig kidney allografts against warm ischemia. These researchers heated kidneys in vivo (42.5°C, 15 minute), allowed them to recover for 4-6 hours to accumulate HSP70, and then subjected them to ischemia for 1.5 hours. After kidney removal and transplantation into a littermate, the heat-shocked kidney had a significantly greater survival rate and improved renal functions on the sixth day after transplantation. In such a complicated experimental system, however, some other factors besides ischemic tolerance may determine the transplantant survival and functions.

Joannidis and collegues[122] studied the injuries and functions of post-ischemic kidneys after heat shock pretreatment without their subsequent transplantation. They first subjected rats to whole-body heating

(42°C, 15 minutes) which resulted in a marked stimulation of HSP70, HSP60 and ubiquitin mRNA syntheses in all regions of the kidney, but most prominently in the medulla and papilla. Increased *HSP70* gene transcription is accompanied by its marked accumulation following 24-48 hours (as determined by immunohistochemistry) in inner medullary collecting ducts, medullary thick ascending limbs, and proximal tubules. Forty-eight hours after heat shock when maximal HSP70 accumulation was observed, the animals underwent uninephrectomy and cross clamping of the remaining renal artery for 40 minutes, and 24-72 hours later kidney functional activity (by serum creatinine and urea nitrogen) as well as damage (by histology) were assessed. The experiments failed to detect any difference in renal functions or damage at 24, 48 or 72 hours after ischemia-reperfusion between heat-shocked and control rats.

In the second set of experiments, the researchers removed kidneys following 24 or 48 hours after heat shock and then perfused them in isolation for 1.5 hours. In the experiments, no alleviation was observed in the degree or extent of hypoxic injury to the medullary thick ascending limb (characteristically observed in this system), nor did HSP70 accumulation modify the progressive decline in glomerular filtration rate or fractional reabsorption of glucose seen in perfused kidney. The authors concluded from the results that HSPs do not prevent ischemic renal injuries.

A similar conclusion was reached by Zager[123] who also researched the kidney and the ischemic tolerance phenomenon. Zager[124] subjected rats to extrarenal obstruction (ureteral ligation) and 24 hours later cortical proximal tubule segments (PTS) were isolated and subjected to hypoxia/reoxygenation injury. Cell death (assessed by LDH release) due to the stress was reduced in PTS obtained from the obstructed kidney comparing with the untreated one, and this resistance occurred without increase in HSP70, SOD or catalase. Besides ischemia, the obstruction initiated resistance to Ca^{2+}-ionophore (A23187), lipid peroxidation (induced by Fe^{2+}) and phospholipase A_2 (PLA_2) attack. Zager observed that the PLA_2 treatment of tolerant PTS produced about three-fold less lysophosphatidylcholine and free fatty acids and suggested that the resistant state in PTS is determined by the plasma membrane stability.[124]

In the next study,[125] rats underwent 35 minutes of bilateral renal artery occlusion (i.e., the same ischemic stress which induced HSP70, see section 4.4 and ref. 125). Twenty-four hours later, PTS were subjected to oxidative stress or A23187 exposure. As in the previous work, PTS became resistant to the treatments and lipid peroxidation in them was suppressed, although no increase in SOD, catalase, or hemeoxygenase after preconditioning ischemia was detected.[126]

The above study, however, does not exclude HSP70 involvement in the ischemic tolerance phenomenon. Therefore, the researchers from Zager's lab investigated the dependence of this phenomenon on protein synthesis.[127] They first exposed cultured human proximal tubular cells (HK-2) to sublethal ATP depletion (antimycin A + 2-deoxyglucose), which decreased the ATP level to less than 10% of the initial. After 24 hours of this exposure, HK-2 cells demonstrated a significant resistance (assayed by vital dye exclusion) to A23187/ATP depletion. Of great importance is that protein synthesis inhibitors (CHX and verrucarin A) did not prevent the tolerance development; moreover, treatment with the inhibitors for 24 hours (but not for 6 hours) induced the same cytoprotection as prolonged ATP deficiency. Cytoresistance is dissociated from the extent of free Ca^{2+} loading and ATP depletion but is associated with a decrease in membrane deacylation, indicating reduced phospholipid hydrolysis. Since CHX + ATP depletion did not induce additive benefits, the researchers suggested that a common mechanism operates for the induction of cytoresistant state after either energy deprivation or protein synthesis

inhibition. By immunoblotting with anti-phosphotyrosine antibodies, they observed that the emergence of a resistant state correlated with the dephosphorylation, or possibly the disappearance, of an unidentified 130 kDa tyrosine-phosphorylated protein (denoted pp-130), and prevention of tyrosine dephosphorylation by orthovanadate maintained pp-130 phosphorylation (or expression) and abolished cytoprotection. Thus, the data shows that the ischemic tolerance phenomenon in renal cells, in contrast to myocardial and brain cells, is not dependent on HSP accumulation.

Finally, the researchers from Zager's lab demonstrated that the above cytoresistant state in HK-2 cells, besides ATP depletion or protein synthesis inhibitors, may be induced by prolonged (20 hours) but not short-term incubation with low concentrations of sphingosine, although its higher doses were cytotoxic.[128] They speculated that the protective effect of sphingosine may result from its inhibitory effect on protein trafficking. Thus, if the depletion of specific protein(s) at a particular cellular site is responsible for cytoresistance, then either the impairment of protein trafficking or the blockade of protein synthesis might produce the same result.[128]

Although the proteins whose activation leads to the death of renal cells during hypoxia are not yet known, one such protein may be PLA_2, since the suppression of its activity by arachidonic acid protects proximal tubule epithelial (PTE) cells from hypoxic injury.[129] Likewise, the increased susceptibility of mouse PTE compared with kidney cell culture MDCK to chemical anoxia correlates with an increased free fatty acid accumulation in anoxic PTE cells.[130] Other possible candidates are a Ca^{2+}-dependent protease (calpain)[131] and a 15 kDa endonuclease[132] whose activities are also increased during ischemic stress in PTE cells, since the inhibitors of these enzymes (especially calpain inhibitors) markedly reduce cell death.

The question arises: why does HSP accumulation (in particular, HSP70) not protect kidneys from ischemic stress, as was demonstrated in the heart (chapter 5) or in the brain (section 6.1)? We hypothesize that the main difference between the organs may have ensued from the diverse intracellular targets primarily damaged by ischemia. As we consider below (section 6.4 and chapter 7), HSP-mediated protective action is directed mainly to cytosolic proteins, especially that of the cytoskeleton. During renal ischemia, however, the crucial role in cell death is apparently Ca^{2+}-mediated plasma membrane damage through activation of PLA_2 and proteases like calpain. At this level, the protective effect of HSPs may not be as significant as in other cases. Therefore another protective system which confers resistance primarily to the plasma membrane phospholipids is implicated in kidney tolerance. However, HSP70 may still play some beneficial role. For instance, after transient renal ischemia the chaperone accumulates initially in the apical domain of PTE (ref. 133, see also subsection 4.4.2), i.e., in the same region where morphological changes (loss and internalization of microvilli, large gaps in cytoskeletal terminal web) are most striking (see ref. 134 for review).

Recently, the defensive effect of prolonged incubation with protein synthesis inhibitors (CHX or emetine) against ischemic stress was observed by Lobner and Choi in another cell culture, namely cortical neurons.[135] The protection was not attributable to the reduction of apoptosis (see the previous section), since cell death under these conditions was necrotic. CHX pretreatment also protects various cell lines from another stress, namely hyperthermia, and this protection is associated with reduced heat-induced protein aggregation.[136] Although the precise mechanism of the protective effect of CHX during heating is unknown, Beckmann et al suggests that it may result from an increase in free HSP70 due to the release from its targets, namely immature proteins (see section 4.5 for details). Whether such a mechanism is involved in kidney ischemic tolerance is not yet studied, but the kidney expresses a rather

high level of constitutive HSC70.[137] Furthermore, HSC70, HSP90 and HSP27 are also constitutive HSPs of the kidney;[138,139] their role in anti-ischemic protection remains unevaluated, but HSP27 may be important for anti-oxidant defense during reperfusion (see chapter 7).

6.3. HSPS AND PROTECTION FROM ISCHEMIA OF OTHER ORGANS AND TISSUES

Data on the protective role of HSPs in ischemic tolerance of other organs and tissues are rather scarce, although some of them suggest such a function.

Saad and co-workers[140] studied the effect of heat pretreatment on ischemic tolerance in the rat liver. Rats were first exposed to whole-body heat shock (42°C, 15 minutes) which induced the transcription and translation of HSP72 (determined by Northern and Western blottings, respectively). Two days later the livers in situ were subjected to warm ischemia (occlusion of hepatic artery, portal vein and bile duct) for 30 minutes and reperfusion for 40 minutes and their ATP content and energy charge as well as the levels of liver LDH and alanine aminotransferase (ALT) were determined during reperfusion. The researchers found that during ischemia the ATP level was reduced equally in the control and heat-shocked groups (to less than 15% of the initial) and no enzyme release was detected. However, during reperfusion the heat-shocked group exhibited significantly improved energy metabolism (ATP level and energy charge) and more than four-fold reduced release of LDH and ALT (indicating decreased plasma membrane damage). The effect closely resembles that observed in the HSP70-overexpressing heart (see chapter 5). In addition, the 7 day survival rate in the control group of rats (without heat pretreatment) was 50%, but it increased to 100% in the heat-shocked group. Thus, the authors concluded that heat exposure associated with HSP70 accumulation had a significant protective effect against warm ischemic liver injury. This study also shows that HSPs protecting the liver in turn saves the whole organism from death.

A similar approach taken with the liver was applied by Stojadinovic et al[141] to study the protective function of HSP in ischemia/reperfusion injury of the rat small intestine. Heat shock pretreatment (42°C, 15-20 minutes) accompanied by HSP70 accumulation reduced the mucosal injury (assessed by histology) after 30 minutes of superior mesenteric artery occlusion followed by 60 minutes of reperfusion. The researchers observed in heat-treated intestine significantly reduced leukotriene B-4 production and neutrophil infiltration and proposed that the protective effect of HSP70 accumulation is associated with the prevention of neutrophil activation and chemotaxis (see also chapter 7 for discussion). We believe that such a mechanism for protection may indeed operate but in addition, a direct cytoprotective effect of HSP70 against reperfusion-induced oxidative stress is also highly probable, since Musch et al[143] observed a reduction in cell damage after oxidant monochloramine in HSP70-overexpressing intestinal epithelial cell culture IEC-18.

Recently Hotter and colleagues[144] described the ischemic preconditioning phenomenon on the intestine. They found that increased LDH release after ischemia-reperfusion was prevented after preconditioning. In this study, the role of HSPs was not evaluated. Instead, the researchers suggested the involvement of nitric oxide (NO) in this phenomenon because the inhibition of nitric oxide synthesis abolishes the protective effect of preconditioning, and nitric oxide administration simulates this effect. In addition, increased synthesis of NO is detected after preconditioning. Surprisingly, there is a report by Malyshev et al that heat-induced HSP70 accumulation is attenuated by the NO-synthase inhibitor; heat shock also potentiates the NO production in the intestine, liver, kidney, heart, brain and spleen. However, it remains unknown whether NO generation per se evokes a stress response. In any case, the implication that NO is involved in the preconditioning phenomenon of the intestine does not necessarily exclude the importance of the participation of HSP70.

Finally, there are a few studies where HSP-mediated protection is proposed in skeletal muscle. Garramone et al[145] studied ultrastructure, creatine phosphate and ATP levels of rat limb muscles after ischemia either pretreated with heat shock to stimulate HSP70 accumulation or without pretreatment. Heat-shocked muscles after ischemia demonstrated better preservation of mitochondria (as determined by electron microscopy) and about a two-fold increased level of creatine phosphate, although no conservation of ATP levels was detected. The authors concluded that the heat shock response confers significant biochemical and ultrastructural protection against ischemic injury in rat skeletal muscle.

Liauw and co-workers[146] investigated the effect of sequential ischemia/reperfusion in a paired canine gracilis muscle model. In this model, one gracilis muscle was subjected to 5 hours of ischemia followed by 48 hours of reperfusion; the contralateral (second) muscle was treated in the same way and their ATP content and necrosis were determined. In the second (preconditioned) group of muscles, 60% reduction in necrosis and preservation of the ATP level was observed. The researchers suggested that muscle salvage in this model may be associated with HSP expression judging from preliminary analysis of HSP content.

However, Ianuzzo et al[147] failed to observe any protective effect (assessed by tissue histology and citrate synthase activity) of heat shock pretreatment which associated with two-fold increase in HSP70 content in latissimus dorsi muscle. The reason for the difference between this study and the above works are not yet known, but it is apparent that additional work is necessary to evaluate the role of HSP in skeletal muscle preservation. We suggest that for this tissue, which resembles in many aspects cardial tissue, the protective function of HSP, particularly HSP70, is highly probable.

6.4. HSP-MEDIATED PROTECTION OF TUMOR CELLS FROM ISCHEMIA-INDUCED DEATH

By the early 1990s when we began our studies on tumor cells, there were some indirect indications about the involvement of HSPs in anti-ischemic protection in the heart and the brain, but, surprisingly, almost no data on this subject existed on simpler systems such as cell cultures. To our knowledge, there was only one work by Polla and Bonventre on nontumor cells.[148] They observed heat-induced protection against the cytotoxic effect of oligomycin (an inhibitor of mitochondrial ATPase, see section 2.1) on the glycolysis-deficient mutant chinese hamster fibroblast line, DS7. Prior exposure to heat shock (44°C, 20 minutes) followed by recovery for 5 hours resulted in an increased viability of the cells (determined by Trypan blue exclusion) after oligomycin exposure from 16% to 60%. Additionally, the plating efficiency of cells (i.e., their reproductive capacity, see section 2.2) after low oligomycin concentration increased from 18% in unheated cells to 64% in heat-shocked cells. The authors also observed in heat-pretreated cells the typical pattern of HSPs, including major HSP70, and suggested that the heat shock response may represent intrinsic cellular mechanism for protection against ischemic injury.

6.4.1 HSPs CAN PROTECT ENERGY-DEPRIVED TUMOR CELLS FROM NECROTIC DEATH

Only a few years ago, the protective effect of HSPs (in particular, HSP70) against ischemic injuries looked quite paradoxical since HSP70 apparently requires ATP for its chaperone functions, and the level of ATP drops during ischemia. Therefore, in our first experiments when we observed that heat-shocked cells became resistant to ischemia, we were somewhat amused and began to explore the phenomenon scrupulously.

Our first study was carried out on Ehrlich ascites carcinoma (EAC) cells.[149] As a model of "in vitro ischemia" we used an incubation of cells in a glucose-free medium with mitochondrial inhibitors. We found that heat shock (44°C, 10 minutes) with subsequent recovery for 3 hours significantly reduced rotenone-induced cell death (Fig. 6.4). Then we demonstrated that this protection was indeed dependent on HSP accumulation (in particular, HSP68, an inducible form of HSP70). First, rotenone-resistant cells also became tolerant to hyperthermia (44°C, 30-60 minutes) and this was accompanied by an accumulation of inducible HSP68 as determined by immunoblotting with specific monoclonal antibodies (Fig. 6.4), while no protection from rotenone and thermotolerance was detected immediately after heat shock.[149] Second, besides heat shock, transient ATP depletion also conferred resistance to rotenone, thermotolerance and HSP68 accumulation (Fig. 6.4). Third, an inhibitor of protein synthesis, CHX, completely prevented HSP68 accumulation, thermotolerance development and resistance to rotenone (Fig. 6.4).

We also demonstrated that cells with increased HSP68 content become resistant not only to rotenone but also to other mitochondrial inhibitors (uncoupler CCCP, oligomycin), demonstrating that this tolerance was not drug-specific. To assess whether the resistance to ischemia was associated with the preservation of ATP, we measured ATP content in rotenone-treated preheated and control cells (Table 6.1). Although in heat-shocked cells some delay in ATP depletion was detected after 15 minutes of rotenone treatment, the difference disappeared within 30-60 minutes and no difference was detected after uncoupler pretreatment (Table 6.1) which also induced tolerance to rotenone (Fig. 6.4). Therefore, we concluded that HSP action is associated with the protection of cells from the fatal consequences of ATP depletion rather than to the preservation of ATP. Later, we also failed to observe any preservation of ATP after rotenone treatment in tolerant P_3O_1 myeloma cells (Table 6.1, see also below). The conclusion

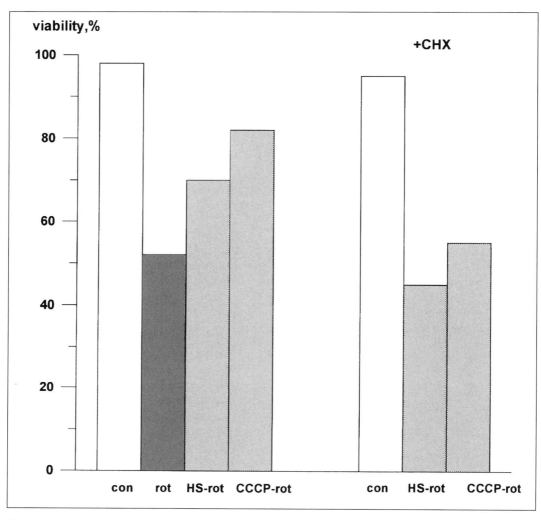

Fig. 6.4. Increased viability of rotenone-treated (rot) EAC cells following heat shock (HS-rot) or uncoupler (CCCP-rot) pretreatment with recovery. The cells were preheated (44°C, 10 minutes) or treated with CCCP (2 µM, 20 minutes), recovered in a rich medium for 3 hours in the absence or presence of CHX (50 µM) and then subjected to rotenone (2 µM) in a glucose-free medium for 2 hours. Viability was determined by the Trypan blue exclusion test. The differences between the nontolerant (second column) and pretreated cells (third or fourth columns) were significant (p < 0.05 by Student's t - test). The graph was depicted from the data of Gabai, Kabakov.[149]

that HSP does not prevent ATP depletion due to simulated ischemia is also supported by the data obtained on the heart (chapter 5) and liver (section 6.3).

In our second study, we demonstrated the above phenomenon on another cell culture, myeloma P_3O_1.[150] We also analyzed by electrophoresis and autography with ^{14}C-leucine labeling what proteins were synthesized after preconditioning heat shock in P_3O_1 and EAC cells. Judging from one-dimensional gel electrophoresis, after heat shock treatment P_3O_1 cells incorporated labeled leucine only in three protein bands, obviously HSP70, HSP90 and HSP27 (Fig. 6.5). In the cells, no inducible HSP70 (HSP68) was detected by either autoradiography or immunoblotting, but within 3 hours of recovery the cells accumulated "constitutive" HSC70 (Fig. 6.5) and be-

Table 6.1. Effect of rotenone on the ATP level in control and tolerant tumor cells

Conditions	Cells	ATP, % of Initial		
		15 Minutes	30 Minutes	60 Minutes
Rotenone	EAC	4.5 ± 0.1	3.6 ± 0.5	3.3 ± 0.2
	P_3O_1	16.5 ± 1.5	11.1 ± 0.9	9.0 ± 0.4
HS - rotenone	EAC	$15.7 \pm 3.0^*$	4.3 ± 0.7	1.4 ± 0.4
	P_3O_1	19.1 ± 0.6	3.9 ± 0.6	3.8 ± 0.1
HS+CHX - rotenone	EAC	8.0 ± 2.3	3.8 ± 0.6	2.4 ± 0.9
	P_3O_1	19.2 ± 2.0	10.0 ± 0.3	6.7 ± 2.6
CCCP - rotenone	EAC	9.3 ± 2.2	3.8 ± 1.6	2.1 ± 1.0

* $P < 0.05$ compared to effect of rotenone in corresponding control cells. Rotenone, 2 µM; HS, 44°C, 10 minutes; CCCP, 2 µM; CHX, 50 µM. After HS or CCCP treatment, the cells recovered for three hours to accumulate HSP, or HSP accumulation was prevented by CHX. The initial ATP levels were 25.5 ± 1.5 nM (EAC) or 34.5 ± 2.4 nM (P_3O_1) per 10^6 cells.

came tolerant to rotenone (Fig. 6.6). The lack of inducible HSP68 despite normal thermotolerance development was demonstrated previously in some of the other tumor cell lines (see, for example, ref. 151), and we failed to observe this chaperone in heat-shocked EL-4 thymoma cells.[152] In contrast to HSC70, no appreciable increase in either HSP27 or HSP90 was detected by immunoblotting.[150] Apparently, the absence of HSP27 and HSP90 elevation is associated with a short-term recovery period since prolongation of recovery to 6 hours brought about HSP27 and HSP90 accumulation as well.[150] Previously absence of HSP27 accumulation despite increased transcription of this gene was observed in post-ischemic myocardium (see section 4.2). Incubation of the cells with CHX during recovery completely inhibited protein synthesis, thus preventing HSC70 accumulation and the acquisition of tolerance to rotenone (Fig. 6.6). Therefore, we concluded that in P_3O_1 cells resistance to ATP depletion coincided with an accumulation of HSC70.

In EAC cells, besides the above-mentioned three HSPs, we found the synthesis of inducible HSP68 as well as a 110 kDa protein (presumably, HSP110), HSP60 and 22 kDa protein (probably, (αB-crystallin)

(Fig. 6.5). By immunoblotting, we observed the accumulation of HSP68/HSC70, but as in P_3O_1 cells, no appreciable increase in HSP90 and only a small increase in HSP27 were detected.[150] Similar changes in HSP expression in EAC cells were detected when tolerance was induced by transient ATP depletion (Kabakov, Gabai, unpublished data). Hence, in EAC cells, resistance to ATP-depletion depended mainly on HSP68/HSC70.

We next became interested in why the cells with increased HSP content became more resistant to ischemic necrosis. Our previous data obtained on various tumor cell cultures (EL-4 thymoma, Ehrlich and HeLa carcinoma cells) established a very good correlation between cell resistance to ATP depletion and their actin skeleton stability (see chapter 3 and refs. 153, 154 for further details). Therefore, we compared the rotenone-induced rise in Triton-insoluble actin (actin aggregation) in control and tolerant cells. Both EAC and P_3O_1 cells after HSP accumulation demonstrated a decreased level of actin aggregation provoked by ATP depletion, and aggregation of other proteins was also diminished; the effects were completely suppressed by CHX (Table 6.2). From these data, we concluded that HSP70-mediated tolerance to energy deprivation in

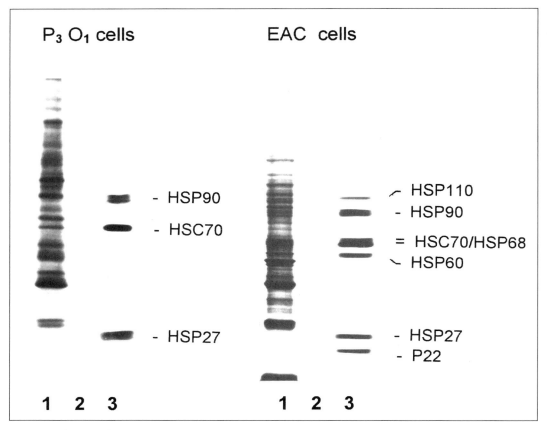

Fig. 6.5. Induction of HSP synthesis in P_3O_1 and EAC cells following heat shock treatment inducing tolerance to ATP depletion. Fluorograms demonstrate the tracks of ^{14}C-labeled polypeptides from the control (lane 1) and heat-shocked (lane 3) cells (43°C or 44°C, 10 minutes for P_3O_1 and EAC cells, respectively). In presence of 50 µM CHX no label incorporation was detected (lane 2).

tumor cells lies in the prevention of "lethal" aggregation of cytoskeletal proteins.[149,150] To our knowledge, this was the first demonstration of a possible mechanism of HSP protective action in ATP-depleted cells.

The question arises as to whether the observed phenomenon of ischemic tolerance in tumor cells is important for their in vivo growth. As we considered in section 4.5, rapidly dividing tumor cells in situ often experience energy-deficiency conditions due to an insufficient supply of nutrients and oxygen. If the cells have sufficient supply of glucose and a high intristic glycolytic rate, they do not die under hypoxia but their proliferation ceases. For instance, Olivetto and Paoletti[155] found in Yoshida ascites hepatoma AH130 that hypoxia and

antimycin A, a respiratory inhibitor, does not diminish the ATP level, but markedly reduces DNA synthesis (assessed by thymidine incorporation). In HL-60 human promyelotic leukemic cells, mitochondrial inhibitors cause the accumulation of cells in G_1-phase at low doses and in G_2/M at higher (but nontoxic) doses, in accordance with the degree of ATP reduction induced by the compounds. As Sweet and Singh[156] suggested from the data, cycling cells must maintain a minimal ATP content to satisfy energy requirements of the checkpoints that allow them to passage through G_1 to S phase and through G_2 to mitosis. Matsunaga et al[157] also observed inhibition of growth and accumulation in G_2/M phase of HL-60 cells treated with 50 nM of rotenone.

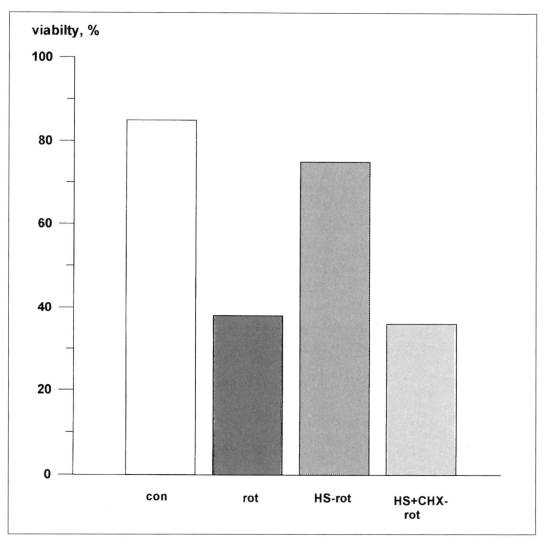

Fig. 6.6. Increased viability of rotenone-treated (rot) P_3O_1 cells following heat shock pretreatment (HS-rot) as in Fig. 6.5. Other conditions were the same as in the legend of Fig. 6.4.

However, when tumor cells underwent joint oxygen-glucose deprivation, they die via necrosis. In particular, Tannock and Kopelyan[158] found the formation of a necrotic center in spheroids of human bladder cancer line MGH-U1 when medium glucose concentration is less than 3 mM. Hlatky et al[159] calculated from their data on the sandwich model of a tumor analog that cells undergo necrosis if their ATP production rate drops lower than one-half of control. Since ascites fluid of EAC cells in late stages of growth contains almost no oxygen and glucose (Gabai, unpublished data), such an environment may in principle be conducive to necrotic death. However, we failed to observe any increase in cell death (assayed by Trypan blue staining) in stationary EAC cells (8 days after inoculation) in comparing with exponential cells (5 days of growth).[160] At the same time, nondividing EAC cells had decreased levels of ATP and energy charge, which apparently caused an accumulation of all

Table 6.2. Increase in rotenone-induced protein aggregation in control (nonpreheated) and tolerant (preheated) cells

	Cells	Increase in Protein Aggregation, %	
		Total	Actin
Control	P_3O_1	89	116
	EAC	74	98
Tolerant	P_3O_1	18*	26*
	EAC	12*	17*
CHX-treated	P_3O_1	104	94
nontolerant)	EAC	75	81

* P < 0.05 compared with control
The aggregation was assessed by measuring the relative increase in the intensities of Coomassie-stained Triton-insoluble polypeptide patterns. The cells were exposed to shock (44°C, 10 minutes) and recovered for three hours without CHX (tolerant) or with 50 µM CHX, and then treated with 2 µM rotenone in a glucose-free medium for one hour.[150]

major HSPs (HSP70, HSP90, HSP27) and appearance of thermotolerance (see section 4.5 and refs. 160-162).

We compared the vulnerability of exponential and stationary cells to energy deprivation and observed that stationary EAC become significantly more resistant to this stress (Fig. 6.7).[160,161] In rotenone-treated stationary cells, the rate of ATP depletion is not delayed, but there are diminished levels of protein aggregation[160] and blebbing (Fig. 6.7), the latter being an indicator of actin skeleton injury (see chapter 3). Likewise, diminished levels of protein aggregation and blebbing are revealed in stationary cells treated with hyperthermia.[161] Besides ATP depletion and hyperthermia, stationary EAC also become resistant to oxidative stress provoked by vikasol or H_2O_2; in this case, we also found decreased protein aggregation in stationary cells compared to exponential ones.[162] Since stationary EAC cells in vivo experience energy starvation we suggested that HSP accumulation in them may represent adaptive response protecting cells from death under energy deficiency in late stages of tumor growth.[160,162]

We have not yet studied, in ATP-depleted cells with elevated HSP content, other parameters whose changes may be implicated in necrotic death. In particular, Arora et al[163] found that hepatocellular carcinoma (HCC) cell lines are much more resistant than hepatocytes to anoxia-induced necrosis. They detected that phospholipid degradation and calpain activation in HCC cells are significantly lower, although no elevation of intracellular Ca^{2+} was detected in either cell type. Mellitin, a PLA_2 activator, increases calpain activity and cell necrosis similarly in HCC cells and hepatocytes. The authors suggested that the adaptive mechanism responsible for hepatomas resistant to anoxia lies in the inhibition of PLA_2-mediated calpain activation. As we have already mentioned, participation of PLA_2 and calpain is highly probable in anoxic death of other cell types, such as brain and renal cells (see section 6.1, 6.2). Although Arora et al did not compare HSP levels in HCC cells and hepatocytes, there was a study by Schiaffonati and co-workers[164] who observed an increased expression of constitutive HSC70 in hepatomas. Since HSP70-mediated prevention of PLA_2 activation is demonstrated in TNF-treated fibrosarcoma cells (ref. 165, see also chapter 7), decreased activation of PLA_2 in HCC may be also associated with an increased level of HSP70 expression.

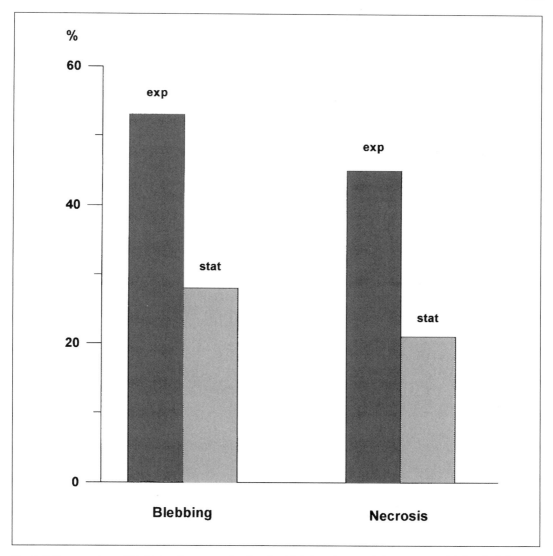

Fig. 6.7. Suppression of blebbing (percent of cells with blebs) and necrosis in stationary EAC cells following ATP depletion. Cells in exponential (5 days of growth) or stationary (8 days of growth) phases of growth were treated with rotenone (2 μM) in a glucose-free medium; the percentage of cells with blebs and Trypan blue-stained cells were determined 1 hour or 3 hours, respectively. The graph was depicted from the data of Kabakov et al.[160]

6.4.2. HSPs AND THE PROTECTION OF TUMOR CELLS FROM ISCHEMIA-INDUCED APOPTOTIC DEATH

Several recent data on various cells and tissues demonstrates that ischemic stress can elicit not only necrosis but apoptotic death as well (see sections 2.2, 6.1). Apoptosis after mild ischemia has also been recently described in tumor cell lines. In

1994, Hartley and co-workers found in PC12 pheochromocytoma that respiratory inhibitors (rotenone or methyl-4-phenyl-pyridium) diminished the ATP level and evoked apoptosis accompanied by DNA fragmentation at low doses of the inhibitors or necrosis at high doses.[111] The same year, Wolvetang et al[166] found that the incubation of various human lymphoblastoid and melanoma cell lines for 18 hours with rotenone, antimycin A or oligomycin increased the number of apoptotic cells from 1-5% in the control to 24-51%. Apparently, the inhibition of respiration by itself was not responsible for this effect since no rise in spontaneous apoptosis was observed in ρ^0 cells lacking mitochondrial DNA and normal respiratory chain. In 1995, hypoxia-induced apoptosis was described by Yao et al[167] in HT29 adenocarcinoma and Shimuzu et al[168] in PC12 cells and hepatoma 7316A. Recently Matsunaga and co-workers[157] found that 50 nM of rotenone inhibited growth of HL-60 cells while its ten-fold higher concentration induced apoptosis. Thus, all of the data including that described above (subsection 6.4.1) indicate that the effect of hypoxia and respiratory inhibitors apparently depends on the intensity of stress: a small perturbation of energy metabolism stops cell proliferation and induces HSP synthesis, a mild perturbation elicits apoptosis and a severe one causes necrosis.

The question arises as to whether stationary EAC cells undergoing energy deficiency become resistant not only to necrosis but to apoptosis as well. To answer this question, we first assessed whether EAC cells undergo apoptosis upon energy deficiency. As possible apoptotic stimuli we used two exposures: chronic starvation resembling in vivo conditions (prolonged incubation in nutrient free medium, Dulbecco phosphate-buffered saline) or transient acute ATP depletion (30-60 minutes with uncoupler CCCP, and subsequent addition of glucose to restore ATP level. To quantify apoptosis, we studied: (1) cell morphology and nuclear changes (chromatin condensation) using conventional microscopy of hematoxylin-eosin stained cells and fluorescent microscopy (acridine orange or Hoechst 33342 staining); (2) DNA fragmentation; (3) cell size and cellular DNA content using flow cytometry (see section 2.2) and (4) plasma membrane permeability, assessed using Trypan blue staining. In a rich medium, EAC cells demonstrated no apoptotic changes by any of the above

criteria within the whole period of incubation (6 hours).[169] However, during starvation, exponential EAC cells demonstrated progressive apoptotic changes including cell shrinkage, blebbing of plasma membrane and chromatin condensation and DNA fragmentation (Fig. 6.8). The same but enhanced changes were observed in the exponential cells under acute ATP depletion (see ref. 169 and Fig. 6.8), although their blebbing was completely reversed after glucose addition.[169] A main distinction of EAC apoptosis from "classical" apoptosis of thymocytes lies in the character of DNA fragmentation: EAC cells demonstrate no distinct internucleosomal patterns in agarose gel and sub-G_1 peak of DNA by flow cytometry, although both assays clearly show DNA degradation.[169] This is significant since such DNA fragmentation is completely absent in EAC cells dying via necrosis (i.e., incubated with mitochondrial inhibitors for 2-3 hours, data not shown). As we have already mentioned in section 2.4, there are data that apoptotic features may greatly vary between different cell types.

In contrast to exponential cells, stationary ones exhibited few apoptotic changes after either exposure (see ref. 169 and Fig. 6.8 for DNA fragmentation). After removing stationary cells (7-9 days of growth) from mice, we failed to observe an increase in the number of apoptotic cells. We concluded that the cells became resistant not only to ischemia-induced necrosis but to apoptosis as well.

One of the possible explanations of the above finding may be the decreased activity of apoptotic machinery, namely proteases and nucleases involved in apoptosis (see ref. 170 for recent review and chapter 8). To test this possibility, we treated cells with a low concentration (0.01%) of a detergent, Triton X-100, which was shown to induce rapid DNA fragmentation and chromatin condensation in carcinoma cells, apparently via direct activation of proteases and nucleases.[171] After 1 hour of incubation with this compound, however, both exponential and stationary cells exhibited almost the same level (50%) of DNA

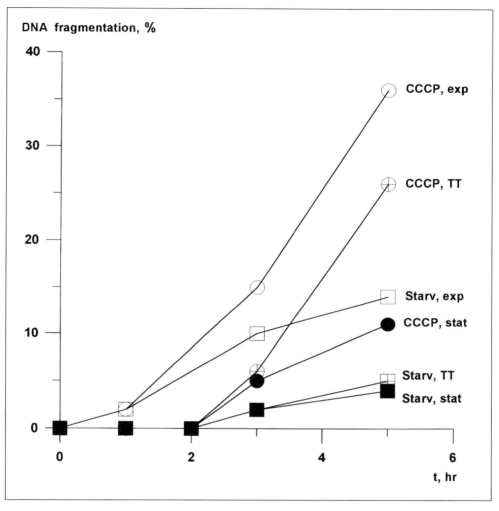

Fig. 6.8. DNA fragmentation in EAC cells in exponential (exp) and stationary (stat) phases of growth following chronic starvation (starv) or transient ATP depletion (CCCP); exponential cells subjected to heat shock (44°C, 10 minutes) with recovery for 3 hours which became thermotolerant (TT) are also shown. Starvation: incubation in a phosphate-buffered saline; transient ATP depletion: 2 μM of CCCP for 1 hour with the subsequent addition of 10 mM glucose. DNA fragmentation was assessed as described in refs. 169, 172.

fragmentation.[169] Furthermore, more prolonged incubation with CCCP of stationary EAC cells induced marked DNA fragmentation within them as well.[161] In terminal stages of tumor growth in vivo (12-14 days), the number of apoptotic cells (with small size and decreased DNA content) are also elevated (Gabai, Zamulaeva, Mosin, unpublished data). Hence, these results indicate that stationary cells by themselves are not defective in apoptotic machinery, but their resistance may be

associated with a decreased level of stress-induced damage activating this machinery.

Therefore, we next investigated whether resistance to apoptosis is induced in exponential cells by heat shock pretreatment leading to an accumulation of HSP68. Employing the same protocol as we used for study of resistance to ischemic necrosis (see subsection 6.4.1), we revealed that heat-pretreated exponential cells became more resistant to apoptotic DNA fragmentation following starvation or ATP depletion

(Fig. 6.8). Incubation with CHX during recovery prevented tolerance acquisition after heat shock treatment,[169] although some delay (for the first 3 hours) of DNA fragmentation was found in heat-shocked cells even after CHX treatment. This may indicate the blocking effect of heating without HSP accumulation (Gabai, unpublished data). It is important that CHX treatment per se neither suppresses nor induces apoptosis, as was observed in some other systems (see ref. 173 for review and chapter 8). Another indication that observed resistance to ischemia-induced apoptosis depends on HSP synthesis is that both heat-pretreated exponential cells as well as stationary cells become tolerant to apoptosis induced by hyperthermia.[169] Such resistance to heat-induced apoptosis associated with HSP70 accumulation was demonstrated earlier by Mosser and Martin in the human leukemic cell line.[174]

Thus, we conclude that EAC cells with elevated content of HSP70/68 become resistant to ischemia-induced apoptotic death. To our knowledge, there are no data on this issue for other tumor cells. In future experiments, it is necessary to test this conclusion with HSP70-transfected cells and to evaluate the role of other HSPs in protection against ischemia-induced apoptosis.

CONCLUDING REMARKS

Summarizing all the facts available, we conclude that HSP70-mediated protection from necrotic death has been clearly established in neuronal and astrocyte cultures. Many data on the preconditioning phenomenon in the brain also show the involvement of this chaperone in rescuing different cerebral cells in vivo. The experiments with transgenic animals overexpressing HSPs are still necessary, since some other proteins (e.g., antioxidative enzymes, Bcl-2) also participate in the anti-stress defense. Recent works clearly show that delayed neuronal death in the brain, where protective function of HSP70 is observed, represents apoptosis. One may predict enhanced resistance of neuronal cell cultures over-

expressing HSP70 to ischemia-induced apoptosis. We believe that HSP-mediated protection from apoptosis is an exciting area of future research.

Besides the brain, HSPs probably alleviate ischemic damage in the liver, some muscles and tumor cells, whereas in the kidney ischemic tolerance is not dependent on HSPs. We consider the lack of HSP-mediated anti-ischemic protection in the kidney as an important issue which may help to establish possible intracellular targets for HSPs in different cells.

Presently, it is quite obvious that anti-ischemic protection of various cells and tissues mediated by HSPs is independent of the conservation of ATP during ischemia but is, apparently, associated with the maintenance of some protein structures in energy-deprived cells, in particular the cytoskeleton, as we established in tumor cells.

As a whole, ischemic tolerance conferred by HSPs is a widely distributed phenomenon but, surprisingly, little is known about the mechanisms of their protective action during ATP depletion. We discuss this problem scrupulously in the next two chapters.

RECENT NEWS

The association between stress protein induction and ischemic tolerance to focal cerebral ischemia in rat cortex was studied by Chen and co-workers.[174] Three 10 minute intervals of transient ischemia separated by 45-minutes of reperfusion was the most effective pretreatment and substantially reduced the volume of infarction after subsequent 100-min MCA occlusion if the challenging ischemia was applied 2-5 days after preconditioning; time course of HSP72 expression rather than GRP75 or GRP78 most closely correlated with tolerance. In a model of permanent MCA occlusion, States et al[176] demonstrated for the first time that cells accumulating HSP70 during ischemic stress did not undergo DNA fragmentation, which supports our suggestion of protective function for HSP70 against ischemia-induced apoptosis (see section 6.1). Several studies also confirmed activation of the proapoptotic Bax gene and suppression of anti-apoptotic Bcl-2 and Bcl-x genes in models of focal and global ischemia;[177-179] moreover, their expression correlated with neuronal vulneability to DNA fragmentation. An increase in ICE-like protease activity (see Fig.6.3) was observed in the CA1 region of gerbil hippocampus after global ischemia.[179,180]

The novel data of Perdrizet et al[181] show that whole body hyperthermia (42.5° C, 15 min) followed by recovery (6-8 hour) successfully protects several tissues and organs in diverse animal models of surgically induced ischemia/reperfusion. The heat pretreatment considerably enhanced post-ischemic survival of rat and pig kidney transplants, and rat pancreatic islets. Likewise, the heat preconditioning prevented paralysis induced in rabbits by ischemia of spinal cord as a result of 20 min of aorta occlusion. All the above cases of protection demonstrated a positive temporal correlation between the functional recovery and increase in expression of inducible HSP70.

Finally, Kume et al[182] described an ischemic preconditioning phenomenon in rat liver. The researchers observed that short-term (15-min) ischemia increased HSP72 expression after 48 h and afforded significant protection of hepatic function and higher rat survival following longer (30 min) ischemia.

REFERENCES

1. Novak TS, Abe H. Post-ischemic stress response in brain. In: Morimoto RI, Tissieres A, Georgopoulos C, eds. The Biology of Heat Shock Proteins and Molecular Chaperones. Cold Spring Harbor, NY: Cold Spring Harbor Laboratory Press, 1994: 553-576.

2. Dwyer BE, Nishimura RN. Heat shock proteins in hypoxic-ischemic brain injury: a perspective. Brain Pathol 1992; 2:245-251.

3. Koroshetz WJ, Bonventre JV. Heat shock response in the central nervous system. Experientia 1994; 50:1085-1091.

4. Kitagawa K, Matsumoto M, Tagaya M et al. Ischemic tolerance phenomenon found in the brain. Brain Res 1990; 528: 21-24.

5. Kirino T, Tsujita Y, Tamura A. Induced tolerance to ischemia in gerbil hippocampal neurons. J Cereb Blood Flow Metab 1991; 11: 299-307.

6. Kitagawa K, Matsumoto M, Kuwabara K et al. Ischemic tolerance phenomenon detected in various brain regions. Brain Res 1991; 561: 203-211.

7. Ohtsuki T, Matsumoto M, Kitagawa K et al. Induced resistance and susceptibility to cerebral ischemia in gerbil hippocampal neurons by prolonged but mild hypoperfusion. Brain Res 1993; 614: 279-284.

8. Aoki M, Abe K, Kawagoe J et al. The preconditioned hippocampus accelerates HSP70 heat shock gene expression following transient ischemia in the gerbil. Neurosci Lett 1993; 155: 7-10.

9. Aoki M, Abe K, Kawagoe J et al. Acceleration of HSP70 and HSC70 heat shock gene expression following transient ischemia in the preconditioned gerbil hippocampus. J Cereb Blood Flow Metab 1993; 13: 781-788.

10. Liu RY, Li X, Li L et al. Expression of human hsp70 in rat fibroblasts enhances cell survival and facilitates recovery from translational and transcriptional inhibition following heat shock. Cancer Res 1992; 52: 3667-3673.

11. Kato H, Liu XH, Nakata N et al. Immunohistochemical visualization of heat shock protein-70 in the gerbil hoppocampus following repeated brief cerebral ischemia. Brain Res 1993; 615: 240-244.

12. Chen T, Kato H, Liu XH et al. Ischemic tolerance can be induced repeatedly in the gerbil hippocampal neurons. Neurosci Lett 1994; 177: 159-161.

13. Liu Y, Kato H, Nakata N et al. Protection of rat hippocampus against ischemic neuronal damage by pretreatment with sublethal ischemia. Brain Res 1992; 586: 121-124.

14. Liu Y, Kato H, Nakata N et al. Temporal profile of heat shock protein 70 synthesis in ischemic tolerance induced by preconditioning ischemia in rat hippocampus. Neuroscience 1993; 56: 921-927.

15. Chopp M, Chan H, Ho K-L et al. Transient hyperthermia protects against subsequent

forebrain ischemic cell damage in the rat. Neurology 1989; 39: 1396-1398.

16. Kitagawa K, Matsumoto M, Tagaya K et al. Hyperthermia-induced neuronal protection against ischemic injury in gerbils. J Cereb Blood Flow Metab 1991; 11: 449-452.

17. Marini AM, Kozuka M, Lipsky RH et al. 70-Kilodalton heat shock protein induction in cerebellar astrocytes and cerebellar granule cells in vitro: comparison with immunocytochemical localization after hyperthermia in vivo. J Neurochem 1990; 54: 1509-1516.

18. Marcuccilli CJ, Mathur SK, Morimoto RI et al. Regulatory differences in the stress response of hippocampal neurons and glial cells after heat shock. J Neurosci 1996; 16: 478-485.

19. Ohtsuki T, Matsumoto M, Kuwabara K et al. Influence of oxidative stress on induced tolerance to ischemia in gerbil hippocampal neurons. Brain Res 1992; 599: 246-252.

20. Simon RP, Cho H, Gwinn R et al. The temporal profile of 72-kDa heat-shock protein expression following global ischemia. J Neurosci 1991; 11: 881-889.

21. Nowak TS. Synthesis of heat shock/stress proteins during cellular injury. Ann N Y Acad Sci 1993; 679: 142-156.

22. Ferrer I, Soriano MA, Vidal A et al. Survival of parvalbumin-immunoreactive neurons in the gerbil hippocampus following transient forebrain ischemia does not depend on HSP-70 protein induction. Brain Res 1995; 692: 41-46.

23. Kawagoe J, Abe K, Kogure K. Reduction of HSP70 and HSC70 heat shock mRNA induction by pentobarbital after transient global ischemia in gerbil brain. J Neurochem 1993; 61: 254-260.

24. Shaver EG, Welsh FA, Sutton EG. Deep hypothermia diminishes the ischemic induction of heat-shock protein-72 mRNA in piglet brain. Stroke 1995; 26: 1273-1277.

25. Kumar K, Wu XL, Evans AT et al. The effect of hypothermia on induction of heat shock protein (HSP)-72 in ischemic brain. Metabolic Brain Disease 1995; 10: 283-291.

26. Yager JY, Asselin J. Effect of mild hypothermia on cerebral energy metabolism during the evolution of hypoxic-ischemic brain damage in the immature rat. Stroke 1996; 27: 919-925.

27. Hearteaux C, Bertaina V, Widmann C et al. K^+ channel openers prevent global ischemia-induced expression of c-fos, c-jun, heat shock protein, and amyloid beta-protein precursor genes and neuronal death in rat hippocampus. Proc Natl Acad Sci USA 1993; 90: 9431-9435.

28. Quraishi H, Brown LR. Expression of heat shock protein 90 (hsp90) in neural and nonneural tissues of the control and hyperthermic rabbit. Exp Cell Res 1995; 219: 358-363.

29. Satoh JI, Kim SU. Differential expression of heat shock protein HSP27 in human neurons and glial cells in culture. J Neurosc Res 1995; 41: 805-818.

30. Kato H, Liu Y, Kogure K et al. Induction of 27-kDa heat shock protein following cerebral ischemia in a rat model of ischemic tolerance. Brain Res 1994; 634: 235-244.

31. Kato H, Araki T, Itoyama Y et al. An immunohistochemical study of heat shock protein-27 in the hippocampus in a gerbil model of cerebral ischemia and ischemic tolerance. Neuroscience 1995; 68: 65-71.

32. Satoh J, Kim SU. Cytokines and growth factors induce HSP27 phosphorylation in human astrocytes. J Neuropathol Exp Neurol 1995; 54: 504-512.

33. Noga M, Hayashi T. Ubiquitin gene expression following transient forebrain ischemia. Mol Brain Res 1996; 36: 261-267.

34. Kato H, Chen T, Liu XH et al. Immunohistochemical localization of ubiquitin in gerbil hippocampus with induced tolerance to ischemia. Brain Res 1993; 619: 339-343

35. Hayashi T, Takada K, Matsuda M. Subcellular distribution of ubiquitin-protein conjugates in the hippocampus following transient ischemia. J Neuroscience Res 1992; 31: 561-564.

36. Hayashi T, Takada K, Matsuda M. Posttransient ischemia increase in ubiquitin conjugates in the early reperfusion. 1992; 3: 519-520.

37. Morimoto T, Ide T, Ihara Y et al. Transient ischemia depletes free ubiquitin in the gerbil hippocampal CA1 neurons. Am J Pathol 1996; 148: 249-257.

38. Kamikubo T, Hayashi T. Changes in proteasome activity following transient ischemia. Neurochem Int 1996; 28: 209-212.

39. Uyama O, Matsuyama T, Michishita H et al. Protective effect of human recombinant superoxide dismutase on transient ischemic injury of CA1 neurons in gerbils. Stroke 1992; 23: 75-81.

40. Liu XH, Kato H, Nagata N et al. An immunohistochemiacal study of copper/zinc superoxide dismutase and manganese superoxide dismutase in rat hippocampus after transient cerebral ischemia. Brain Res 1993; 625: 29-37.

41. Gorgias N, Maidatsi P, Tsolaki M et al. Hypoxic pretreatment protects against neuronal damage of the rat hippocampus induced by severe hypoxia. Brain Res 1996; 714: 215-225.

42. Rosenbaum DM, Kalberg J, Kessler JA. Superoxide dismutase ameliorates neuronal death from hypoxia in culture. Stroke 1994; 25: 857-862.

43. Li Y, Chopp M, Garcia JH et al. Distribution of the 72-kd heat-shock protein as a function of transient focal ischemia in rats. Stroke 1992; 23: 1292-1298.

44. Sharp FR, Kinouchi H, Koistinaho J et al. HSP70 heat shock gene regulation during ischemia. Stroke 1993; 24: 172-175.

45. Li Y, Chopp M, Zhang ZG et al. Expression of glial fibrillary acidic protein in areas of focal cerebral ischemia accompanies neuronal expression of 72-kDa heat shock protein. J Neurol Sci 1995; 128: 134-142.

46. Soriano MA, Ferrer I, Rodriguezfarre E et al. Expression of c-fos and inducible hsp-70 mRNA following a transient episode of focal ischemia that had non-lethal effects on the rat brain. Brain Res 1995; 670: 317-320.

47. Chopp M, Li Y, Zhang ZG et al. p53 expression in brain after middle cerebral artery occlusion in the rat. Biochem Biophys Res Comm 1992; 182: 1201-1207.

48. Kato H, Kogure K, Liu XH et al. Immunohistochemical localization of the low molecular weight stress protein HSP27 following focal cerebral ischemia in the rat. Brain Res 1995; 679: 1-7.

49. Wang S, Longo FM, Chen J et al. Induction of glucose regulated protein (grp78) and inducible heat shock protein (hsp70) mRNAs in rat brain after kainic acid seizures and focal ischemia. Neurochem Int 1993; 23: 575-582.

50. Massa SM, Longo FM, Zuo J et al. Cloning of rat grp75, an hsp70-family member, and its expression in normal and ischemic brain. J Neorosci Res 1995; 40: 807-819.

51. Liu XH, Kato H, Araki T et al. An immunohistochemical study of copper/zinc superoxide dismutase and manganese superoxide dismutase following focal cerebral ischemia in the rat. Brain Res 1994; 644: 257-266.

52. Nimura T, Weinstein PR, Massa SM et al. Hemeoxygenase-1 (HO-1) protein induction in rat brain following focal ischemia. Mol Brain Res 1996; 37: 201-208.

53. Siesjo BK. Calcium and cell death. Magnesium 1989; 8: 223-237.

54. Simon RP, Swan JH, Griffiths T et al. Blockade of N-methyl-D-aspartate receptors may protect against ischemic damage in the brain. Science 1984; 226: 850-852.

55. Wahlestedt C, Golanov E, Yamamoto S et al. Antisense oligodeoxynucleotides to NMDA-R1 receptor channel protect cortical neurons from excitotoxicity and reduce focal ischaemic infarctions. Nature 1993; 363: 260-263.

56. Gwag BJ, Lobner D, Koh JY et al. Blockade of glutamate receptors unmasks neuronal apoptosis after oxygen-glucose deprivation in vitro. Neurosci 1995; 68: 615-619.

57. Varming T, Drejer J, Frandsen A et al. Characterization of a chemical anoxia model in cerebellar granule neurons using sodium azide: Protection by nifedipine and MK-801. J Neurosci Res 1996; 44: 40-46.

58. Santos MS, Moreno AJ, Carvalho AP. Relationships between ATP depletion, membrane potential, and the release of neurotransmitters in rat nerve terminals: An in vitro study under conditions that mimic anoxia, hypoglycemia, and ischemia.Stroke 1996; 27: 941-950.

59. Nakata N, Kato H, Liu Y et al. Effects of pretreatment with sublethal ischemia on the extracellular glutamate concentrations during secondary ischemia in the gerbil hippocampus evaluated with intracerebral microdialysis. Neurosci Lett 1992; 138: 86-88.

60. Kato H, Liu Y, Araki T et al. MK-801, but not anisomycin, inhibits the induction of tolerance to ischemia in the gerbil hippocampus. Neurosci Lett 1992; 139: 118-121.

61. Rordorf G, Koroshetz WJ, Bonventre JV. Heat shock protects cultured neurons from glutamate toxicity. Neuron 1991; 7: 1043-1051.

62. Lowenstein DH, Chan PH, Miles MF. The stress protein response in cultured neurons: characterization and evidence for a protective role in excitotoxicity. Neuron 1991; 7: 1053-1060.

63. Lafon-Cazal M, Pietri S, Culcasi M et al. NMDA-dependent superoxide production and neurotoxicity. Nature 1993; 364: 535-537.

64. Patel M, Day BJ, Crapo JD et al. Requirement for superoxide in excitotoxic cell death. Neuron 1996; 16: 345-355.

65. Jacquier-Sarlin MR, Fuller K, Dinh-Xuan AT et al. Protective effects of hsp70 in inflammation. Experientia 1994; 50: 1031-1038.

66. Zhang J, Dawson VL, Dawson TM et al. Nitric oxide activation of poly(ADP-ribose) synthetase in neurotoxicity. Science 1994; 263: 687-689.

67. Ho DY, Saydam TC, Fink SL et al. Defective herpes simplex virus vectors expressing the rat brain glucose transporter protect cultured neurons from necrotic insults. J Neurochem 1995; 65: 842-850.

68. Furukawa K, Smithswintosky VL, Mattson MP. Evidence that actin depolymerization protects hippocampal neurons against excitotoxicity by stabilizing $[Ca^{2+}]_i$. Exper Neurol 1995; 133: 153-163.

69. Posner A, Raser KJ, Hajimohammadreza I et al. Aurintricarboxylic acid is an inhibitor of μ- and m-calpain. Biochem Mol Biol International 1995; 36:291-299.

70. Amin V, Cumming DV, Coffin RS et al. The degree of protection provided to neuronal cells by a pre-conditioning stress correlates with the amount of heat shock protein 70 it induces and not with the similarity of the subsequent stress. Neurosci Lett 1995; 200: 85-88.

71. Amin V, Cumming DV, Latchman DS. Over-expression of heat shock protein 70 protects neuronal cells against both thermal and ischaemic stress but with different efficiencies. Neurosci Lett 1996; 206: 45-48.

72. Papadopoulos MC, Sun XY, Cao JM et al. Over-expression of HSP-70 protects astrocytes from combined oxygen-glucose deprivation. Neuroreport 1996; 7: 429-432.

73. Bergeron M, Mivechi NF, Giaccia AJ et al. Mechanism of heat shock protein 72 induction in primary cultured astrocytes after oxygen-glucose deprivation. Neurol Res 1996; 18: 64-72.

74. Shreiber SS, Baudry M. Selective neuronal vulnerability in the hippocampus—a role for gene expression. Trends Neurosci 1995; 18: 446-451.

75. Johnson EM, Greenlund LJS, Akins PT et al. Neuronal apoptosis: Current understanding of molecular mechanisms and potential role in ischemic brain injury. J Neurotrauma 1995; 12: 843-852.

76. Dragunow M, Preston K. The role of inducible transcription factors in apoptotic nerve cell death. Brain Res Rev 1995; 21: 1-28.

77. Goto K, Ishige A, Sekiguchi K et al. Effect of cycloheximide on delayed neuronal death in rat hippocampus. Brain Res 1990; 534: 299-302.

78. Okamoto M, Matsumoto M, Ohtsuki T et al. Internucleosomal DNA cleavage involved in ischemia-induced neuronal death. Biochem Biophys Res Comm 1993; 196: 1356-1362.

79. Volpe BT, Wessel TC, Mukherjee B et al. Temporal pattern of internucleosomal DNA fragmentation in the striatum and hippocampus after transient forebrain ischemia. Neurosci Lett 1995; 186: 157-160.

80. Islam N, Aftabuddin M, Moriwaki A et al. Detection of DNA damage induced by apoptosis in the rat brain following incomplete ischemia. Neurosci Lett 1995; 188: 159-162.

81. Macmanus JP, Hill IE, Preston E et al. Differences in DNA fragmentation following transient cerebral or decapitation ischemia in rats. J Cereb Blood Flow Metab 1995; 15: 728-737.

82. Li Y, Chopp M, Jiang N et al. In situ detection of DNA fragmentation after focal

cerebral ischemia in mice. Mol Brain Res 1995; 28: 164-168.

83. Li Y, Sharov VG, Jiang N et al. Ultrastructural and light microscopic evidence of apoptosis after middle cerebral artery occlusion in the rat. Am J Pathol 1995; 146: 1045-1051.

84. Li Y, Chopp M, Jiang N et al. Temporal profile of in situ DNA fragmentation after transient middle cerebral artery occlusion in the rat. J Cereb Blood Flow Metab 1995; 5: 389-397.

85. Li Y, Chopp M, Jiang N et al. Induction of DNA fragmentation after 10 to 20 minutes of focal cerebral ischemia in rats. Stroke 1995; 26: 1252-1257.

86. Du C, Hu R, Csernansky et al. Very delayed infarction after middle focal cerebral ischemia: A role for apoptosis? J Cereb Blood Flow Metab 1996; 16: 195-201.

87. Charriaut-Marlangue C, Benari Y. A cautionary note on the use of the TUNEL stain to determine apoptosis. Neuroreport 1995; 7: 61-64.

88. Steller H. Mechanisms and genes of cellular suicide. Science 1995; 267: 1445-1449.

89. Thompson CB. Apoptosis in the pathogenesis and treatment of disease. Science 1995; 267: 1456-1462.

90. Li Y, Chopp M, Zhang ZG et al. p53-immunoreactive protein and p53 mRNA expression after transient middle cerebral artery occlusion in rats. Stroke 1994; 25: 849-855.

91. Crumrine RC, Thomas AL, Morgan PF. Attenuation of p53 expression protects against focal ischemic damage in transgenic mice. J Cereb Blood Flow Metab 1994; 14: 887-891

92. Tobita M, Nagano I, Nakamura S et al. DNA single-strand breaks in post-ischemic gerbil brain detected by in situ nick translation procedure. Neurosci Lett 1995; 200: 129-132.

93. Sakhi S, Bruce A, Sun N et al. p53 induction is associated with neuronal damage in the central nervous system. Proc Natl Acad Sci USA 1994; 91: 7525-7529.

94. Morrison RS, Wenzel HJ, Kinoshita Y et al. Loss of p53 tumor suppressor gene protects neurons from kainate-induced cell death. J Neurosci 1996; 16: 1337-1345.

95. Bonfoco E, Krainc D, Ankarcrona M et al. Apoptosis and necrosis: Two distinct events induced, respectively, by mild and intense insults with N-methyl-d-aspartate or nitric oxide/superoxide in cortical cell cultures. Proc Natl Acad Sci USA 1995; 7162-7166.

96. Dreyer EB, Zhang DX, Lipton SA. Transcriptional or translational inhibition blocks low dose NMDA-mediated cell death. Neurorep 1995; 6: 942-944.

97. Hara A, Iwai T, Niwa M et al. Immunohistochemical detection of Bax and Bcl-2 proteins in gerbil hippocampus following transient forebrain ischemia. Brain Res 1996; 711: 249-253.

98. Zhan Q, Fan S, Bae I et al. Induction of *bax* by genotoxic stress in human cells correlates with normal p53 status and apoptosis. Oncogene 1994; 9: 3743-3751.

99. Reed JC. Bcl-2 and the regulation of programmed cell death. J Cell Biol 1994; 124: 1-6.

100. Nunez G, Clarke MF. The Bcl-2 family of proteins: regulators of cell death and survival. Trend Cell Biol 1994; 4: 399-403.

101. Shimazaki K, Ishida A, Kawai N. Increase in bcl-2 oncoprotein and the tolerance to ischemia-induced neuronal death in the gerbil hippocampus. Neurosci Res 1994; 20: 95-99.

102. Chen J, Graham SH, Chan PH et al. bcl-2 is expressed in neurons that survive focal ischemia in the rat. Neurorep 1995; 6: 394-398.

103. Myers KM, Fiskum G, Liu YB et al. Bcl-2 protects neural cells from cyanide/aglycemia-induced lipid oxidation, mitochondrial injury, and loss of viability. J Neurochem 1995; 65: 2432-2440.

104. Lawrence MS, Ho DY, Sun GH et al. Overexpression of bcl-2 with herpes simplex virus vectors protects CNS neurons against neurological insults in vitro and in vivo. J Neurosci 1996; 16: 486-496.

105. Linnik MD, Zahos P, Geschwind MD et al. Expression of bcl-2 from a defective herpes simplex virus-1 vector limits neuronal death in focal cerebral ischemia. Stroke 1995; 26: 1670-1674.

106. Martinou JC, Dubois-Dauphin M, Staple JK et al. Overexpression of BCL-2 in transgenic

mice protects neurons from naturally occurring cell death and experimental ischemia. 1994; 13: 1017-1030.

107. Krajewski S, Tanaka S, Takayama S et al. Cancer Res 1993; 53: 4707-4714.

108. Zhong L-T, Sarafian T, Kane DJ et al. bcl-2 inhibits death if central neural cells induced by multiple agents. Proc Natl Acad Sci USA 1993; 90; 4533-4537.

109. Abe K, Araki T, Kawagoe J et al. Phospholipid metabolism and second messenger system after brain ischemia. Adv Exp Med Biol 1992; 318: 183-195.

110. Abe K, Kawagoe J, Aoki M et al. Changes of mitochondrial DNA and heat shock protein gene expressions in gerbil hippocampus after transient forebrain ischemia. J Cereb Blood Flow Metab 1993; 13: 773-780.

111. Hartley A, Stone JM, Heron C et al. Complex I inhibitor induced dose-dependent apoptosis in PC12 cells: relevance to Parkinson's disease. J Neurochem 1994; 63: 1987-1990.

112. Tsubokawa H, Oguro K, Robinson HP et al. Abnormal Ca^{2+} homeostasis before cell death revealed by whole cell recording of ischemic CA1 hippocampal neurons. Neurosci 1992; 49: 807-817.

113. Limbrick DD, Churn SB, Sombati S et al. Inability to restore resting intracellular calcium levels as an early indicator of delayed neuronal cell death. Brain Res 1995; 690: 145-156.

114. Kumar K, Savithiry S, Madhukar BV. Comparison of alpha-tubulin mRNA and heat ahock protein-70 mRNA in gerbil brain following 10 min ischemia. Mol Brain Res 1993; 20: 130-136.

115. Kumar K, Wu XL. Expression of beta-actin and alpha-tubulin mRNA in gerbil brain following transient ischemia and reperfusion up to 1 month. Mol Brain Res 1995; 30: 149-157.

116. Raleysusman KM, Murata J. Time course of protein changes following in vitro ischemia in the rat hippocampal slice. Brain Res 1995; 694: 94-102.

117. Saito N, Kawai K, Nowak TS. Reexpression of developmentally regulated MAP2c mRNA after ischemia: colocalization with hsp72 mRNA in vulnerable neurons. J Cereb Blood Flow Metab 1995; 15: 205-215.

118. Malinak C, Silverstein FS. Hypoxic-ischemic injury acutely disrupts microtubule-associated protein 2 immunostaining in neonatal rat brain. Biology Neonate 1996; 69: 257-267.

119. Deshpande J, Bergstedt K, Linden K et al. Ultrastructural changes in the hippocampal CA1 region following transient cerebral ischemia: evidence against programmed cell death. Exp Brain Res 1992; 88: 91-105.

120. Borkan SC, Emami A, Schwartz JH. Heat stress protein-associated cytoprotection of inner medullary collecting duct cells from rat kidney. Am J Physiol 1993; 265: F333-F341.

121. Perdrizet GA, Kaneko H, Buckley TM et al. Heat shock and recovery protects renal allografts from warm ischemia injury and enhances HSP72 production. Transplant Proc 1993; 25: 1670-1673.

122. Joannidis M, Cantley LG, Spokes K et al. Induction of heat-shock proteins does not prevent renal tubular injury following ischemia. Kidney Int 1995; 47: 1752-1759.

123. Zager RA. Heme protein-induced tubular cytoresistance: expression at the plasma membrane level. Kidney Int 1995; 47: 1336-1345.

124. Zager RA. Obstruction of proximal tubules initiates cytoresistance against hypoxic damage. Kidney Int 1995; 47: 628-637.

125. Zager RA, Burkhart KL, Gmur DJ. Post-ischemic proximal tubular resistance to oxidant stress and Ca^{2+} ionophore-induced attack. Implications for reperfusion injury. Lab Invest 1995; 72: 592-600.

126. Paller MS, Nath KA, Rosenberg ME. Hemeoxygenase is not expressed as a stress protein after renal ischemia. J Lab Clin Med 1993; 122: 341-345.

127. Iwata M, Herrington J, Zager RA. Protein synthesis inhibition induces cytoresistance in cultured human proximal tubular (HK-2) cells. Am J Physiol 1995; 268: F1154-F1163.

128. Iwata M, Herrington J, Zager RA. Sphingosine: a mediator of acute renal tubular injury and subsequent cytoresistance. Proc Natl Acad Sci USA 1995; 92: 8970-8974.

129. Sheridan AM, Schwartz JH, Kroshian VM et al. Renal mouse proximal tubular cells are more susceptible than MDCK cells to chemical anoxia. Am J Physiol 1993; 265: F342-F350.

130. Alhunaizi AM, Yaqoob MM, Edelstein CL et al. Arachidonic acid protects against hypoxic injury in rat proximal tubules. Kidney Int 1996; 49: 620-625.

131. Edelstein CL, Wieder ED, Yaqoob MM et al. The role of cysteine proteases in hypoxia-induced rat renal proximal tubular injury. Proc Natl Acad Sci USA 1995; 92: 7662-7666.

132. Ueda N, Walker PD, Hsu S-M et al. Activation of a 15-kDa endonuclease in hypoxia/reoxygenation injury without morphologic features of apoptosis. Proc Natl Acad Sci USA 1995; 92: 7202-7206.

133. Van Why SK, Hildebrandt F, Ardito T et al. Induction and intracellular localization of HSP-72 after renal ischemia. Am J Physiol 1992; 263: F769-F775.

134. Leiser J, Molitoris BA. Disease processes in epithelia: the role of the actin skeleton and altered surface membrane polarity. Biochim Biophys Acta 1993; 1225: 1-13.

135. Lobner D, Choi DW. Preincubation with protein synthesis inhibitors protects cortical neurons against oxygen-glucose deprivation-induced death. Neurosci 1996; 72: 335-341.

136. Borrelli MJ, Stafford DM, Rausch CM et al. Reduction of levels of nuclear-associated protein in heated cells by cycloheximide, D_2O, and thermotolerance. Radiat Res 1992; 131:204-213.

137. Komatsuda A, Wakui H, Imai H et al. Renal localization of the constitutuve 73 kD heat shock protein in normal and AN rats. Kidney Int 1992; 41: 1204-1212.

138. Matsubara O, Kasuga T, Maruma F et al. Localizaion of 90-kDa heat shock protein in the kidney. Kidney Int 1990; 38: 830-834.

139. Khan W, Mcguirt JP, Sens MA et al. Expression of heat shock protein 27 in developing and adult human kidney. Toxicol Lett 1996; 84: 69-79.

140. Saad S, Kanai M, Awane M et al. Protective effect of heat shock pretreatment with heat shock protein induction before hepatic warm ischemic injury caused by Pringle's maneuver. Surgery 1995; 118: 510-516.

141. Stojadinovic A, Kiang J, Smallridge R et al. Induction of heat-shock protein 72 protects against ischemia reperfusion in rat small intestine. Gastroenterology 1995; 109: 505-515.

142. Musch MW, Ciancio MJ, Sarge K et al. Induction of heat shock protein 70 protects intestinal epithelial IEC-18 cells from oxidant and thermal injury. Am J Physiol (Cell Physiol) 1996; 39: C429-C436.

143. Hotter G, Closa D, Prados M et al. Intestinal preconditioning is mediated by a transient increase in nitric oxide. Biochem Biophys Res Communicat 1996; 222: 27-32.

144. Malyshev IY, Manukhina EB, Mikoyan VD et al. Nitric oxide is involved heat-induced HSP70 accumulation. FEBS Lett 1995; 370: 159-162.

145. Garramone RRJ, Winters RM, Das DK et al. Reduction of skeletal muscle injury through stress conditioning using the heat-shock response. Plastic Reconstructive Surgery 1994; 93: 1242-1247.

146. Liauw SK, Rubin BB, Lindsay TF et al. Sequential ischemia reperfusion results in contralateral skeletal muscle salvage. Am J Phys 1996; 39: H1407-H1413.

147. Ianuzzo CD, Ianuzzo SE, Feild M et al. Cardiomyoplasty: preservation of the latissimus darsi muscle. J Cardiac Surgery 1995; 10: 104-110.

148. Polla BS, Bonventre JV. Heat shock protects cells dependent on oxidative metabolism from inhibition of oxidative phosphorylation. Clin Res 1987; 35: 555A.

149. Gabai VL, Kabakov AE. Rise in heat-shock protein level confers tolerance to energy deprivation. FEBS Lett 1993; 327: 247-250.

150. Kabakov AE, Gabai, VL. Heat shock-induced accumulation of 70-kDa stress protein (HSP70) can protect tumor cells from necrosis. Exp Cell Res 1995; 217: 15-21.

151. Aujame L. The major heat-shock protein hsp68 is not induced by stress in mouse erythroleukemia cell lines. Biochem Cell Biol 1988; 66: 691-701.

152. Gabai VL, Kabakov AE. Induction of heat-shock protein synthesis and thermotolerance in EL-4 ascites tumor cells by transient ATP depletion after ischemic stress. Exp Molec Pathol 1994; 60: 88-99.

153. Gabai VL, Kabakov AE. Tumor cell resistance to energy deprivation and hyperthermia can be determined by the actin skeleton stability. Cancer Lett 1993; 70: 25-31.

154. Kabakov AE, Gabai VL. Protein aggregation as primary and characteristic cell reaction to various stresses. Experientia 1993; 49: 706-710.

155. Olivetto M, Paoletti F. The role of respiration in tumor cell transition from the noncycling to the cycling state. J Cell Physiol 1981; 107: 243-249.

156. Sweet S, Singh G. Accumulation of human promyelocytic leukemic (HL-60) cells at two energetic cell cycle checkpoints. Cancer Res 1995; 55: 5164-5167.

157. Matsunaga T, Kudo J, Takahashi K et al. Rotenone, a mitochondrial NADH dehydrogenase inhibitor, induces cell surface expression of CD13 and CD38 and apoptosis in HL-60 cells. Leukemia Lymphoma 1996; 20: 487-494.

158. Tannock IF, Kopelyan I. Influence of glucose concentration on growth and formation of necrosis in spheroids derived from a human bladder cancer cell line. Cancer Res 1986; 46: 3105-3110.

159. Hlatky L, Sachs RK, Alpen EL. Joint oxygen-glucose deprivation as the cause of necrosis in a tumor analog. J Cell Physiol 1988; 134: 167-178.

160. Kabakov AE, Molotkov AO, Budagova KR et al. Adaptation of Ehrlich ascites carcinoma cells to energy deprivation in vivo can be associated with heat shock protein accumulation. J Cell Physiol 1995; 165: 1-6.

161. Mosin AF, Gabai VL, Makarova YuM et al. Damage and interphase death of the Ehrlich ascite carcinoma tumor cells, being at different growth phases, due to energy deprivation and heat shock. Cytology 1994; 36: 384-391.

162. Gabai VL, Mosina VA, Budagova KR et. Spontaneous overexpression of heat-shock proteins in Ehrlich ascites carcinoma cells during in vivo growth. Biochem Molec Biol Int 1995; 35: 95-102.

163. Arora AS, Degroen PC, Croall DE et al. Hepatocellular carcinoma cells resist necrosis during anoxia by preventing phospholipase-mediated calpain activation. J Cell Physiol 1996; 167: 434-442.

164. Schiaffonati L, Pappalardo C, Tacchini L. Expression of the HSP70 gene family in rat hepatoma cell lines of different growth rate. Exp Cell Res 1991; 196: 330-336.

165. Jaattela M. Overexpression of major heat shock protein hsp70 inhibits tumor necrosis factor-induced activation of phospholipase A_2. J Immunol 1993; 151: 4286-4294.

166. Wolvetang EJ, Johnson KL, Krauer K et al. Mitochondrial respiratory chain inhibitors induce apoptosis. FEBS Lett 1994; 339: 40-44.

167. Yao KS, Clayton M, Odwyer PJ. Apoptosis in human adenocarcinoma HT29 cells induced by exposure to hypoxia. J Natl Cancer Inst 1995; 87: 117-122.

168. Shimizu S, Eguchi Y, Kosaka H et al. Prevention of hypoxia-induced cell death by Bcl-2 and Bcl-xL. Nature 1995; 374: 811-813.

169. Gabai VL, Zamulaeva IV, Mosin AF. Resistance of Ehrlich tumor cells to apoptosis can be due to accumulation of heat shock proteins. FEBS Lett 1995; 375: 21-26.

170. Vaux DL, Strasser A. The molecular biology of apoptosis. Proc Natl Acad Sci USA 1996; 93: 2239-2244.

171. Borner MM, Schneider E, Pirnia F et al. The detergent Triton X-100 induces a death pattern in human carcinoma cell lines that resembles cytotoxic lymphocyte-induced apoptosis. FEBS Lett 1994; 353: 129-132.

172. Gabai VL, Kabakov AE, Makarova YuM et al. DNA fragmentation in Ehrlich ascites carcinoma under exposures causing cytoskeletal protein aggregation. Biochemistry (Moscow) 1994; 59: 399-404.

173. Martin SJ. Apoptosis: suicide, execution or murder? Trends Cell Biol 1993; 3: 141-144.

174. Mosser DD, Martin L. Induced thermotolerance to apoptosis in a human T lymphocyte cell line. J Cell Physiol 1992; 151: 561-570.

175. Chen J, Graham SH, Zhu RL, Simon RP. Stress proteins and tolerance to focal cere-

bral ischemia. J Cereb Blood Flow Metab 1996; 16: 566-577.

176. States BA, Honkaniemi J, Weinstein PR, Sharp FR. DNA fragmentation and HSP70 protein induction in hippocampus and cortex occurs in separate neurons following permanent middle cerebral artery occlusions. J Cereb Blood Flow Metab 1996; 16: 1165-1175.

177. Gillardon F, Lenz C, Waschke KF et al. Altered expression of Bcl-2, Bcl-x, Bax and c-fos colocalizes with DNA fragmentation and ischemic cell damage following middle cerebral artery occlusion in rats. Mol Brain Res 1996; 40: 254-260.

178. Chen J, Zhu RL, Nakayama M et al. Expression of the apoptosis-effector gene, Bax, is up-regulated in vulnerable hippocampal CA1 neurons following global ischemia. J Neurochem 1996; 67: 64-71.

179. Honkaniemi J, Massa SM, Sharp FR. Apoptosis associated genes are induced in gerbil hippocampus following global ischemia. Restorative Neurol Neurosci 1996; 9: 227-230.

180. Bhat RV, Dirocco R, Marcy VR et al. Increased expression of IL-1-beta converting enzyme in hippocampus after ischemia - selective localization in microglia. J Neurosci 1996; 16: 4146-4154.

181. Perdrizet GA, Garcia JC, Lena CJ et al. Surgical stress and the heat shock response. Abstracts of 1996 Meeting on Molecular Chaperones & The Heat Shock Response (May 1-5, 1996, Cold Spring Harbor, NY), p. 240.

182. Kume M, Yamamoto Y, Saad S et al. Ischemic preconditioning of the liver in rats -implication of heat shock protein induction to increased tolerance of ischemia-reperfusion injury. J Lab Clin Med 1996; 128: 251-258.

CHAPTER 7

WHAT ARE THE MECHANISMS OF HEAT SHOCK PROTEIN-MEDIATED CYTOPROTECTION UNDER ATP DEPRIVATION?

As was reviewed in chapters 4-6: first, transient ATP depletion induces heat shock gene expression and HSP synthesis; and, second, the elevated level of HSP(s) is associated with the protection of various mammalian cells from injury and death under metabolic (or ischemic) stress. Both conclusions were well-grounded and in fact we address here a special adaptive reaction which confers tolerance to energy starvation, thus allowing a cell to withstand sustained deprivation of ATP. Nevertheless, it is still poorly understood how excess HSP(s) compensates for ATP deficiency and maintains the viability of ATP-deprived cells. In this chapter, we present some speculations and hypotheses which might, at least in part, clarify this intriguing problem.

7.1. TWO PARADIGMS OF CELL TOLERANCE TO ATP DEPRIVATION: REDUCED INJURY DURING THE STRESS AND THE IMPROVED POST-STRESS RECOVERY

"To be or not to be...?" Hamlet's famous question is greatly true for each cell undergoing stress. This amusing analogy, in our opinion, well reflects the ambiguous fate of stressed cells. Indeed, depending on the degree of injury, any cell involved can either die or remain viable and acquire tolerance. The rate and mechanisms of stress-provoked cell death are also different in various cases. For example, an affected cell can die promptly as a result of the plasma membrane disintegration (necrosis). Another method of cell death is that an affected cell escapes necrotic death but thereafter is eliminated by triggering its suicide mechanism (apoptosis). While stresses kill resting cells via necrosis or apoptosis (two kinds of interphase death), proliferating cells being stressed can die due to a failure in mitosis (reproductive death). Finally, cells within affected tissue can remain viable but their physiological functions and intercellular communications are temporarily interrupted, which may be fatal for some organs or even for whole organism. All these possibilities must be taken into account in defining a beneficial role of HSPs in the protection of mammalian cells and tissues from injury and death under metabolic (ATP-depleting) stress.

As reviewed in chapters 5 and 6, enhanced HSP expression appears to confer resistance in various cells and organs to energetically unfavorable conditions (ischemia, hypoxia, blocking of ATP generation). Briefly, a number of in vitro studies indicate that the elevated HSP content protects cultured mammalian cells of different origins from necrotic death during ischemia-mimicking ATP depletion.[1-8] At the tissue level, the increased level of HSP70 clearly correlates with a reduction of necrotic lesions in ischemic heart, brain and probably other mammalian organs undergoing severe ischemia or hypoxia (see chapters 5 and 6). Thus, there is no doubt that the accumulation of HSP(s) before metabolic (or ischemic) stress rescues an ATP-depleted cell from necrotic death. Obviously, an excess of HSPs is able to prevent other forms of cell death as well. In particular, the elevated colony survival in the HSP-overexpressing rat myogenic cell line experiencing an ischemic-like stress[5] implies cytoprotection against ischemia-induced reproductive death. We have demonstrated an inverse correlation between the level of HSP expression and the intensity of apoptosis provoked in Ehrlich ascites tumor cells by transient ATP depletion.[9] Likewise, so-called delayed neuronal death in post-ischemic brain strongly resembles apoptosis and it may be suppressed by the previous induction of HSPs in the cells (see chapter 6). This allows us to suggest an antiapoptotic activity of HSPs as their specific contribution to cell resistance to ischemia. Summarizing, we conclude that the high level of HSP expression considerably enhances survival of mammalian cells exposed to ATP-depleting conditions.

In the case of ischemic insult or related stress, besides the elevated post-stress

cell survival, the HSP-overexpressing tissues, organs and cell lines display improved recovery of metabolic processes, energy balance and physiological functions. Indeed, intracardiac HSP70 accumulation as a result of either preconditioning treatment or transgene overexpression significantly improves contractile recovery of post-ischemic myocardium (summarized in chapter 5). Acceleration of both the recovery of high energy phosphate stores and the correction of metabolic acidosis following global ischemia was found in murine transgenic hearts overexpressing human HSP70.[10] Similarly, in rat liver, accelerated post-ischemic recovery of energy metabolism as a consequence of HSP70 accumulation has been achieved by heat preconditioning.[11] Certainly one may interpret the observations as a contribution of excess (overexpressed) HSP(s) to the regeneration of tissues damaged during ischemia.

To explain all the above listed findings, we postulate that the tolerance to energy deprivation consists of two components acting in succession one after another: (1) reduction of cell injury *during* ATP depletion (or ischemia, or hypoxia); and (2) improvement of the *post-stress* cell recovery when energy becomes available (i.e., when ATP generation resumes in cells). Although the lesser initial damage of stressed cells a priori implies better repair and faster return to the normal state, an efficiency of reparative processes seems to be higher in the tolerant cells. Importantly, both the former and the latter component of tolerance may be mediated by excess HSP(s) which are accumulated in cells as a result of preconditioning or constitutive transgene overexpression. Below we provide some relevant facts and comments to substantiate our postulate.

In regard to reduced cell injury, many data indicate that under conditions of equal ATP deprivation, thermotolerant (subjected to challenging pretreatment or constitutively overexpressing foreign HSPs) mammalian cells are less damaged during ischemic stress than nontolerant ones (reviewed in chapters 5 and 6). First of all, reduced cell injury is evinced in the better preservation of mitochondrial function in both thermotolerant (preheated) and constitutively overexpressing HSP70 heart-derived H9c2 cells in vitro exposed to ATP-depleting conditions.[5] Earlier the improved mitochondrial function, namely a higher index of respiratory control and a reduced overloading with calcium, was found in rabbit myocardium undergoing ischemia/reperfusion after heat preconditioning.[12] Consistent with the data, protection of mitochondrial ultrastucture in ischemic skeletal muscle was achieved by previous heat shock inducing HSP72.[13]

Such facts of mitochondrial preservation appear to contribute to the phenomenon of tolerance to ischemia, since first, sustained dysfunction of mitochondria per se can kill the energy-deprived cell (see chapter 2) and second, normal operation of mitochondria during the post-stress period should quickly restore the ATP level in the affected cell to provide energy to all reparative processes. In this case, the improved post-ischemic repair of an entire cell may be a direct consequence of the reduced damage or faster recovery of its mitochondria. The possible involvement of HSPs in mitochondrial protection against ischemia/reperfusion injury is considered below.

One of the most dangerous consequences of mitochondrial dysfunction in anoxic cells is a burst in free radical generation (i.e., oxidative stress) following reoxygenation (see chapter 2). In this respect, the level of reactive oxygen species or oxidized products is a very important indicator of cell injury by ischemia/reperfusion. Accumulation and release of oxidized glutathione during ischemia/reperfusion of rabbit myocardium is significantly reduced by heat preconditioning[12] which suggests the suppression of post-ischemic oxidative stress and reduced cell injury correspondingly. Besides the free radical generation inside postanoxic mitochondria, ischemic cells can be damaged by superoxide and NO radicals produced by blood-derived phagocytes (macrophages, neutrophils, eosinophils) which attack affected tissues. It was shown by Polla's group that heat shock inhibits superoxide generation by NADPH oxidase of human neutrophils[14,15] and thus protects thermotolerant cells from oxidative injury, probably owing to HSP accumulation.[15] Hirvonen et al[16] revealed that a two-fold increase in the expression of HSC70 in rat macrophages confers tolerance to the cytotoxic action of NO generated during macrophage activation. We ourselves observed lower cytotoxicity of hydrogen peroxide and vikasol (both are inducers of oxidative stress) toward Ehrlich tumor cells accumulating HSPs.[3] It seems quite possible that the high level of HSPs is able to protect cells and tissues from the post-ischemic oxidative stress evoked by both the internal and external free radicals. Therefore, a capacity of excess HSPs to withstand oxidative stress also contributes to cell tolerance to ischemia/reperfusion injury and it is considered one of the examples of the improved post-stress recovery.

More evidence that mammalian cells with high content of HSP(s) being ATP-deprived undergo reduced injury was obtained by Williams et al.[17] In the study, murine cells constitutively overexpressing a human *HSP70* transgene demonstrated better morphology and tighter attachment to substrates during an ATP-depleting treatment than control (nontransfected) cells of the parental line during the same stress.[17] Since both the morphology of spread cells and the integrity of their contacts with substrate are mainly determined by intact cytoskeletal structures, the cytoprotection observed may be connected with HSP70-mediated preservation of the cytoskeleton from the destructive effects of ATP depletion. Likewise, bleb formation on the cell surface is an early morphological indicator of ATP decrease and cytoskeletal lesions (reviewed in chapter 3). We have revealed significant attenuation of the plasma membrane blebbing in ATP-depleted murine tumor cells with high content of HSPs,[1-4] which also suggests both reduced injury of the tolerant cells and HSP-mediated stabilization of the cytoskeleton during energy deprivation. Discussion of this problem will continue in the next sections.

Finally, the most persuading evidence that the tolerant cells undergo less injury during stress is the low level of protein aggregation present in the ATP-depleted cells with high content of HSPs;[1-4] the rate of ATP decrease is independent of HSP induction (see section 6.4 and ref. 2 this chapter). According to the conclusion made in chapter 3, aggregation is directly associated with the necrotic death of energy-deprived cells. Therefore, the observed attenuation of protein aggregation in ATP-depleted thermotolerant cells is considered a probable cause for their resistance to necrosis under metabolic stress (see sections 6.4 and 7.4.). In addition to the evaluation of aggregation of total cellular protein and certain cytoskeletal proteins,[1-4] we also analyzed the behavior of three reporter cytosolic enzymes during ATP depletion/replenishment in control and tolerant mammalian cells (Fig. 7.1). In chapter 3, we wrote that the interferon-induced 68 kDa dsRNA-dependent protein kinase (p68) and two transfected enzymes, firefly luciferase and β-galactosidase, become rapidly insoluble in response to cellular ATP depletion.[18] Such a feature allows us to use the enzymes as markers of the stress-induced protein aggregation in living cells. One of the authors of this book (AEK) in collaboration with Bensaude and Nguyen compared the dynamics of insolubility/resolubilization of these enzymes during ATP decrease/restoration in control and thermotolerant cells recovered after a challenging heat shock. Significant accumulation of HSP70 in cells made thermotolerant was confirmed by immunoblotting. It is clearly seen in Fig. 7.1 that all three enzymes aggregate less in the thermotolerant than control cells under equivalent ATP depletion. Consistent with the revealed difference in the enzyme aggregation during ATP deprivation, the post-stress

enzyme solubility recovery to the initial level also occurs faster in thermotolerant cells (Fig. 7.1). Although the solubility of the enzymes is not determinative for cell viability under metabolic stress, it appears to be a parameter quite reliably reflecting aggregation of cytosolic proteins in ATP-depleted mammalian cells.

7.2. CHAPERONING UNDER ATP DEPLETION AS A SPECIAL FUNCTION OF HSP70

A number of studies using the constitutive overexpression of *HSP70* transgene in cell lines[5-8,17] and in vivo (see chapter 5) provides direct evidence that HSP70 is indeed involved in the protection of ATP-deprived mammalian cells from injury and death. For all of that, a molecular mechanism of the HSP70-mediated cytoprotection under ATP-depleting metabolic (or ischemic) stress remains unresolved. For instance, the most important function of HSP70 as a molecular chaperone was shown to be dependent on ATP and K^+ (see chapters 1 and 3), whereas the intracellular content of both these cofactors sharply decreases during energy-depleting exposures (reviewed in chapter 2). Moreover, in the case of severe ATP depletion, HSP70 may even enhance protein aggregation in the cytosol (see ref. 18 and chapter 3 for detailed review), which is difficult to consider as beneficial action for the stressed cells. Nevertheless, both protein aggregation and death due to lack of ATP is suppressed in cells accumulating HSP70 (see chapter 6, the previous subsection and ref. 2). As a matter of fact, we see no paradox here and, in our opinion, it is the stress-evoked ATP/ADP exchange in the nucleotide-binding site of HSP70 ATPase that defines the unique cytoprotective properties of this chaperone under ATP depletion.

7.2.1. PREVENTION OF PROTEIN AGGREGATION DURING ATP DEPLETION: THE ROLE OF ADP AND HIP

As we wrote in chapter 3, prolonged energy deprivation in mammalian cells

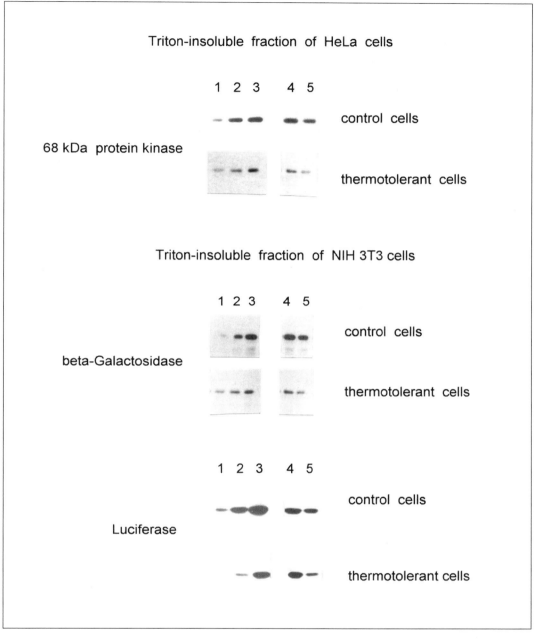

Fig. 7.1. Decreased aggregation of reporter enzymes in ATP-depleted thermotolerant cells. The HSP accumulation and thermotolerance were induced in cultured cells by sublethal heat shock (44°C for 15 minutes) followed by recovery at 37°C overnight. Then HeLa cells expressing the 68 kDa dsRNA-dependent protein kinase and murine NIH 3T3 fibroblasts overexpressing E. coli β-galactosidase and P. pyralis luciferase[18] were incubated in normal growth medium (1) or in glucose-free medium with 20 μM CCCP (an uncoupler) for 30 minutes (2) and 60 minutes (3) to deplete cellular ATP. Restoration of the ATP level occurred during incubation of the treated cells in rich growth medium for 30 minutes (4) and 60 minutes (5). Equal numbers of the cells were extracted with 1% Triton X-100 and the enzymes were detected in aliquots of Triton-soluble and insoluble fractions by immunoblotting as described.[18] Note that the stress-induced enzyme insolubility is markedly attenuated in ATP-depleted thermotolerant cells and accordingly, post-stress recovery of the enzyme solubility occurs faster in thermotolerant cells restoring the ATP level. Obtained by Kabakov, Nguyen and Bensaude (previously unpublished).

leads to a sharp decrease in the ATP/ADP ratio in the cytosol that, in turn, results in the depletion of ATP-HSP70 and nucleotide-free HSP70 pools and the accumulation of ADP-HSP70 complexes. This conversion of HSP70 into its "ADP form" is accompanied by the dimerization and acquisition of a high affinity for unfolded proteins. Recent data suggest that the "ADP state" of HSP70 is maintained by Hip, a novel cochaperone together with HSP40 involved in the eukaryotic HSP70 reaction cycle.[19] Hip oligomers bind at least two HSP70 molecules (possibly a dimer),[19] thus stabilizing the "ADP form" of HSP70 in a certain conformational state which has a maximal binding capacity toward protein substrates (Fig. 7.2). In cells, the molar concentration of ADP multifold exceeds that of HSP70; moreover the latter binds ADP more strongly than ATP. Therefore, a dramatic drop in the cytosolic ATP/ADP ratio during metabolic stress evokes total complexing of all available HSP70 with ADP. Apparently, such hyperactivation of HSP70 by increased ADP should inevitably result in rapid complex formation between the chaperone and its targets, since the activated HSP70 begins to greedily bind any unfolded proteins.

Further, we believe there are two alternative possibilities:

1. HSP70 is a limiting component and the amount of its targets is much higher. Such a situation takes place in the ATP-depleted nontolerant cells which constitutively express HSP70 (reviewed in detail in chapter 3). Briefly, the activated HSP70 (i.e., its dimer complexed with ADP and stabilized by Hip) quickly associates with unfolded proteins recognizing certain binding sites. The latter may be unmasked hydrophobic domains[20] and/or malfolded polypeptide loops[21] of intracellular proteins, damaged as a result of ATP depletion or unfolded spontaneously. Being in a deficit, the constitutive HSP70 cannot saturate the binding sites of unfolded proteins

that could markedly minimize their aggregation; this chaperone/substrate disproportion results in the rapid sequestration of HSP70 and its inclusion in Triton-insoluble aggregates (Fig. 7.2). Taking into account Bensaude's assumption,[18] we believe that limiting ADP-HSP70 might even promote aggregation of intracellular thermolabile proteins by trapping the unfolded conformations of their molecules and thus fixing them in the "aggregable" state. Importantly, both the preexisting and the newly formed chaperone-protein complexes become very stable within ATP-depleted cells, since their dissociation requires ATP, K^+ and is inhibited by ADP (see chapter 3). Hence, sustained energy deprivation in nontolerant mammalian cells leads to co-aggregation of the constitutive HSP70 with its protein substrates; obviously, this process per se does not attenuate the stress-induced protein aggregation but triggers heat shock gene expression followed by development of tolerance (specially reviewed in chapter 4).

2. Another situation can also occur. Tolerant ATP-depleted cells contain a lot of HSP70 accumulated as a result of previous stress or the transgene overexpression. In this case, an excess of the high affinity (bound to ADP) HSP70 may be quite sufficient to minimize the protein aggregation by chaperoning the sites responsible for the formation of undesirable intermolecular bonds (Fig. 7.2). Of course, even being overexpressed the constitutive and inducible HSP70s do not shelter all sites potentially involved in the protein aggregation; however, excess HSP70 might neutralize the most reactive sites exposed by unfolded and misfolded proteins and thus prevent or at least slow down aggregation. Therefore, blocking

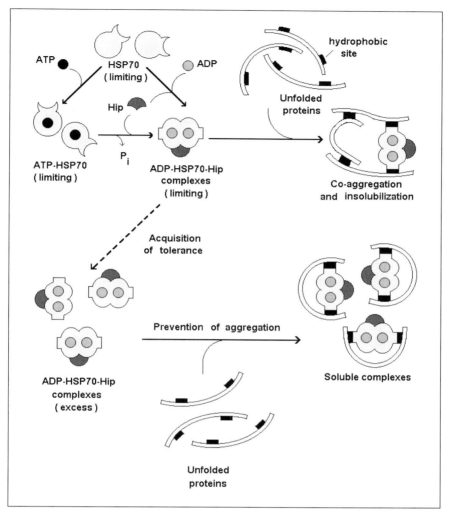

Fig. 7.2. Hypothetical scheme showing the possible mechanism of HSP70-mediated protection from protein aggregation in ATP-depleted tolerant cells. See the text for detailed analysis.

some critical number of reactive sites in cellular proteins by excess HSP70 should in fact inhibit protein aggregation in tolerant cells lacking ATP (see Fig. 7.2). The hypothesis is supported by our finding that there is significantly reduced protein aggregation in ATP-depleted murine tumor cells which possess an approximately 2.5 fold excess of HSP70.[2] Possibly, a HSP70 dimer simultaneously targets two binding sites within one substrate molecule, thus retaining some de-

natured proteins in a tighter (semi-folded) state with low conformational mobility (Fig. 7.2).

7.2.2. PROTEIN DISAGGREGATION AND REFOLDING: REQUIREMENT OF ATP AND HSP40

In further developing the above concept on the two paradigms of the cell tolerance, we suggest that overexpressed HSP70 is able not only to reduce protein aggregation during ATP depletion (lesser damage) but also to accelerate the post-stress disaggregation and refolding of denatured proteins when ATP is restored (better repair). Support for this suggestion came from our experiments on cell fractionation with Triton X-100 (see chapter 6). On the one hand, in thermotolerant cells considerably more protein matter remains in the Triton-soluble fraction during ATP depletion.[1-4] Importantly, the level of soluble HSP70 in thermotolerant cells markedly exceeds that in nontolerant ones under equal ATP decrease (Fig. 7.3). Perhaps the excess of the cytosolic chaperone may be responsible for the maintenance of affected proteins in a soluble state during ATP depletion (i.e., better protection against aggregation). On the other hand, the content of HSP70 in Triton-insoluble fraction is also higher in thermotolerant cells (Fig. 7.3), which in turn may preserve cytosolic proteins from association with the insoluble cytostructures and/or facilitate the post-stress disaggregation (improved repair).

Indeed, the increase in insoluble HSP70 may indicate the enhanced chaperoning of aggregated proteins and saturation of "adhesive sites" exposed as a result of ATP depletion by the preexisting insoluble cytostructures, e.g., cyto- and nucleoskeleton. The latter event should really reduce the stress-provoked insoluble state of many cytosolic proteins (see chapter 3). In regard to the enhanced insolubility of HSP70 (Fig. 7.3), it may also contribute to cell tolerance as a special device for faster disaggregation during the post-stress recovery. The chaperone-mediated process of disaggregation (or resolubilization) of damaged cellular proteins remains enigmatic. In Figure 7.4, we present a hypothetical scheme that shows how the increased HSP70 inclusion in the protein aggregates upon ATP depletion might in turn accelerate disaggregation and renaturation of involved proteins. Evidently, both restoration of the ATP/ADP ratio and availability of HSP40 are obligatory conditions to carry out disaggregation. When Hip and HSP40 catalyze the

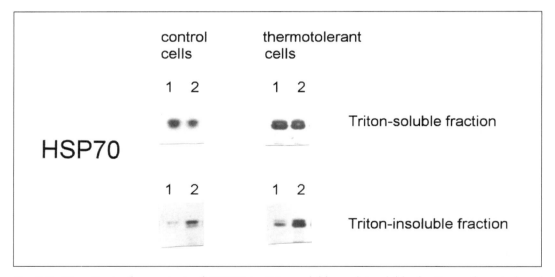

Fig. 7.3. Diversity in the content of HSP70 in Triton-soluble and insoluble fractions of control and thermotolerant HeLa cells undergoing ATP depletion. The HSP70 accumulation and thermotolerance were induced in the cells by sublethal heat shock (44°C for 15 minutes) followed by recovery at 37°C overnight. The control and thermotolerant cells were incubated in normal growth medium (1) or in glucose-free medium with 20 μM CCCP (an uncoupler) for 1 hour (2) to deplete cellular ATP. Equal numbers of the cells were extracted with 1% Triton X-100 and HSP70 was detected in aliquots of Triton-soluble and insoluble fractions by immunoblotting as described.[18] Here it is clearly seen that the ATP-depleted thermotolerant cells contain considerably more HSP70 in both Triton-soluble and insoluble fractions. Obtained by Kabakov and Bensaude, previously unpublished.

ATP/ADP exchange in the nucleotide-binding domain of HSP70, it is probably accompanied by fast and dramatic conformational changes in the chaperone molecule tightly bound to aggregated proteins. These changes include the dissociation of dimeric HSP70 to monomers and the loss of affinity to protein substrate that immediately causes the release of the ATP-carrying chaperone from its targets (Fig. 7.4). It seems likely that the ATP-induced conformational perturbations in molecules of HSP70, preceding its release, transmit a powerful stretch to molecules of aggregated proteins. The resultant tension in a network from aggregated proteins may break down hydrophobic and other intermolecular bonds, thus carrying out the disaggregation. It is easily to imagine here that the disaggregation efficiency directly depends on the amount of co-aggregated HSP70, and the higher the HSP70 content, the more effective the disaggregation (Fig. 7.4). The liberated (soluble) proteins appear to undergo refolding in the HSP70 chaperone machine, which also requires ATP and

participation of Hip and HSP40. The functions of the two latter cofactors probably consist of the regulation of HSP70-protein substrate interaction and stimulation of HSP70 ATPase to accelerate hydrolysis of ATP, the rate-limiting step in the chaperone cycle.[19] Thereafter, as a result of the hydrolysis, the active ADP-HSP70 complex is again formed and binds to Hip and unfolded proteins (see Fig. 7.2). It seems obvious that the refolding of disaggregated proteins occurs better and faster in tolerant cells with elevated levels of HSPs and improved energy metabolism.

7.2.3. Is the Nucleotide-Binding Domain of HSP70 Necessary for Protection of ATP-Deprived Cells?

One can ask this question after reading the articles in which HSP70 lacking the ATP-binding domain exhibited, nevertheless, some cytoprotective and chaperoning capacities during heat shock.[22-24] In particular, Li and colleagues showed that overexpression of the mutant form of human

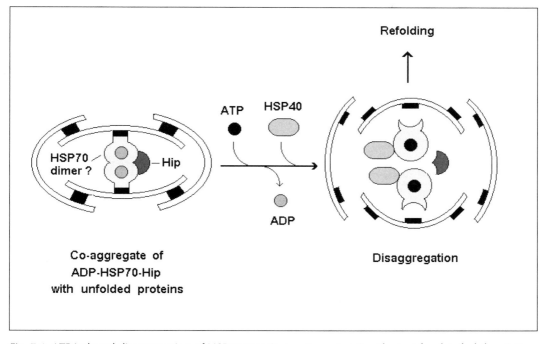

Fig. 7.4. ATP-induced disaggregation of HSP70-protein co-aggregates. See the text for detailed description.

HSP70 with the deletion of the ATP-binding domain reliably elevates survival of the transfected rat fibroblast line Rat-1 after thermal stress.[22] To explain this surprising result the researchers hypothesized that HSP70 lacking the ATP-binding domain is still able to bind to thermolabile cellular proteins, thus preventing their aggregation at extremely high temperature; the ADP/ATP exchange and ATP hydrolysis in HSP70 ATPase are rather needed for the dissociation of the chaperone from its substrates and the dissolution of stress-induced protein aggregates.[22] This supposition was confirmed by the next studies of Stege et al[23,24] who revealed on the same transfectants (Rat-1) that overexpression of the deletion mutant of human HSP70 without the ATP-binding domain not only confers thermoresistance but also clearly reduces intranuclear aggregation of cellular proteins during heat shock. The loss of ATP-binding and hydrolyzing activities of HSP70 does not abolish its protective effects on survival and protein aggregation in heat-shocked cells; hence the authors made a conclusion about the minor importance of the ATP-binding domain for HSP70-mediated thermoresistance. It should be noted, however, that the overexpressed chaperone lacking its ATP-binding domain exerted a lesser inhibitory effect on heat shock-induced intranuclear protein aggregation than the overexpressed intact HSP70.[24] Therefore, the missing domain plays some role in the protection against protein aggregation under thermal stress.

From this it follows that in ATP-depleted mammalian cells, HSP70 might prevent the stress-induced protein aggregation even without any nucleotide exchange in the chaperone molecule. Although the ADP-HSP70 complex possesses a higher affinity for protein substrates than free HSP70, the latter can associate with unfolded polypeptides.[25-28] Hence even after deletion of the nucleotide-binding domain, the residual fragments of the chaperone molecule might form complexes with damaged proteins to protect them from

aggregation upon ATP depletion as it probably occurs in the case of heat shock.[22-24] The experiments on ATP depletion in cells transfected with the deletion mutant HSP70 lacking the nucleotide domain have not yet been performed but we assume that even such defective chaperone molecules, being in excess, may partly protect against protein aggregation and cell death during ATP deprivation. At the same time, we are sure that the expected protective effects will be less than those of intact HSP70 overexpression. Arguably, we would like to underline the evident importance of the ADP-HSP70 binding for suppression of stress-provoked protein aggregation by excess HSP70. As discussed above, HSP70 coupled with ADP acquires the highest affinity for unfolded polypeptides[25-28] that defines rapid recognition and sequestration of damaged proteins by the chaperone. On the one hand, this ADP-mediated activation of HSP70 results in saturating chaperone-target complex formation, thus conferring maximal protection against protein aggregation in ATP-depleted cells (see previous subsection and Fig. 7.2). On the other hand, the activation of HSP70 by the elevated ADP is a direct consequence of cellular ATP depletion when the ATP/ADP ratio sharply drops in the cytosol (see chapters 2 and 3). All these facts and reasonings made us revise the conclusion regarding the minor importance of the nucleotide-binding domain for HSP70-mediated cytoprotection.[24] Contrary to that, we consider the enhancement of a protein-binding capacity of HSP70 through its coupling with ADP as not only one of the steps in regulation of chaperone machinery but also as a special adaptive device preventing the protein aggregation in ADP-repleted cells devoid of ATP. Perhaps our hypothesis could be examined on the same system with Rat-1 transfectants;[23,24] however, in order to predict and to interpret any results, additional information is required about the deletion mutant properties. The important questions are: what is the affinity to unfolded proteins of the HSP70 mutant form lacking the nucleotide-binding domain; and it is able to form oligomers? Meanwhile, a role of the nucleotide-binding domain of HSP70 as well as of HSP70 itself in post-stress protein disaggregation remains undefined. Earlier importance of the ATP-binding and hydrolyzing activities of HSP70 in post-heat shock protein disaggregation in cell nuclei and nucleoli was suggested;[20,29,30] however, Stege and colleagues found that the overexpression of neither HSP70 lacking the nucleotide-binding domain nor even intact HSP70 affects the rate of intranuclear protein disaggregation in heat-shocked cells.[23,24] Since in the latter studies only nuclear surplus protein was assessed, the data do not exclude the possible acceleration of post-heat shock protein disaggregation in the cytoplasm; moreover, Stege's results say nothing about disaggregation following metabolic stress. In regard to the cells recovering the ATP level, we expect that in contrast to the chaperone fragments with deletion of the nucleotide-binding domain, an excess of intact HSP70 molecules will accelerate the post-stress protein disaggregation (see subsection 7.2.2. and Fig. 7.4). In future, we plan to examine this suggestion thoroughly in cells transfected with intact HSP70 and its deletion mutant forms.

7.2.4. WHAT ARE THE CRITICAL TARGETS?

Despite the detailed analysis of this problem in chapters 3 and 4, here we again would like to list the main types of protein targets of HSP70 during ATP deprivation and to consider the cytoprotective aspects of such targeting. Consistent with the above expressed reasons, in the case of ATP depletion, all preexisting HSP70-substrate complexes remain and are stabilized. Indeed, before the stress, some part of cytosolic HSP70 is associated with immature proteins, clathrin and other intranuclear proteins (see chapters 3, 4 and ref. 28 this chapter). Among all of the preexisting chaperone targets, the nascent polypeptides are the most abundant material for aggregation under cellular stresses. Because of

this, the saturation of nascent polypeptide chains by HSP70 during ATP depletion[31] might thus protect the energy-deprived cells from misfolding and aggregation of immature proteins.

Some amount of cellular HSP70 appears to bind to the cytoskeleton. HSP70 associates with actin microfilaments,[32] and the interaction of HSP70 with actin and/or actin-regulating proteins is affected by ATP.[33,34] Moreover, binding of HSP70 to keratin intermediate filaments also occurs in an ATP-dependent manner.[35] It seems likely that HSP70 bound to the preexisting cytoskeletal structures stabilizes them during ATP depletion. In mammalian cells with inhibited ATP generation, the cytoskeleton is altered and disrupted, thus providing new targets for HSP70 (see chapter 3). The latter probably blocks some reactive sites exposed by the affected cytoskeletal networks and their debris to minimize the protein aggregation in energy-deprived cells. Although direct binding of HSP70 to any cytoskeletal proteins upon cellular ATP decrease has not been yet established, such indirect data as resistance of cell-substrate contacts[16] and suppression of both blebbing and actin aggregation[1,2] revealed in ATP-depleted cells with high content of HSP70 suggest HSP70-mediated protection of the cytoskeleton during metabolic stress. Excess HSP70 may also be able to facilitate the post-stress restoration of the cytoskeletal framework when cellular ATP becomes sufficient for this. In any event, the proposed protection of the cytoskeleton by HSP70 might really enhance the viability of mammalian cells undergoing energy starvation.

The question of what the targets of HSP70 inside the nucleus are remains unresolved. The stress-induced translocation of HSP70 to nuclei of cardiomyocytes in vitro experiencing ischemic-like conditions was demonstrated;[36] however, neither mechanism nor a functional significance of this phenomenon has been explained. Of course, some intracellular proteins interacting with HSP70 may have nuclear localization, for instance, topoisomerase I, HSF and p53. Furthermore, HSP70 binds to the nuclear localization sequence (NLS) of cellular proteins before their import into the nucleus.[37] At the same time, expected damage and dysfunction of the chromatin, nucleoli, nuclear matrix proteins, ribonucleoproteins and nuclear enzymes in ATP-deprived cells may cause, if not necrosis, delayed cell death (i.e., apoptosis or reproductive death). It seems likely that by protecting nuclear proteins upon ATP deficiency and/or repairing them during cell recovery, HSP70 could prevent apoptosis and reproductive death following the transient decrease in ATP. Nevertheless, concrete HSP70 targets within the nucleus and their role in cell survival under energy starvation remain to be established.

In our opinion, intracellular enzymes represent the most intriguing class of HSP70 targets. As we described above, according to Bensaude's hypothesis[18] any cytosolic proteins whose molecules possess high conformational mobility may become targets of HSP70 in cells lacking ATP (see also chapter 3). It seems important to emphasize that not only proteins denatured due to ATP fall, but also many enzymes, being in a transiently unfolded state, are trapped by the active ADP-HSP70 complexes formed as a result of energy deprivation (see Fig. 7.2). In regard to the transiently unfolded state of protein molecules, this happens without any stress because some thermolabile proteins undergo spontaneous unfolding, whereas other enzymes can be in partially unfolded unstable conformations during usual catalysis. If HSP70 indeed catches transiently unfolded enzymes, how does this capture affect their catalytic activity? It seems most likely that these enzymes, being bound to HSP70, are inactive, since they are fixed by the chaperone in a nonnative conformation and, additionally, their catalytic domain may be blocked. Therefore, in the case of ATP depletion, HSP70 could form stable complexes with some intracellular enzymes and thereby inhibit their catalytic activity during the

stress. In terms of tolerance to energy deprivation, the HSP70-mediated transient inhibition of certain degradative enzymes (proteases, lipases, nucleases) might prevent or at least postpone cell death. Besides the reversible inactivation of lytic enzymes, inhibitory chaperoning by HSP70 of apoptotic effectors, some ATPases, ion pumps, key enzymes of biosynthesis, etc. may also enhance the viability of mammalian cells under ATP-depleting exposures. Possibly, excess HSP70 in ATP-deprived tolerant cells can "freeze" those enzymes whose activity is harmful for cell survival under these conditions. Although direct HSP70-phospholipase interaction has not been demonstrated, inhibition of the tumor necrosis factor (TNF)-provoked activation of phospholypase A_2 in WEHI-S fibrosarcoma cells overexpressing HSP70 is suggested as an actual mechanism of the HSP70-mediated cytoprotection against cytokine-induced cytotoxicity.[38] In that study, cellular ATP content was not examined but it is very likely that it decreased in the TNF-treated cells, since this cytokine is known to induce powerful oxidative stress within target cells, thus damaging cellular systems of ATP generation. We hypothesize that excess HSP70 is also able to prevent the stress-induced activation of phospholipase A_2 in ATP-depleted cells, thus preventing cell death. This seems possible since the phospholipase A_2 activation is one of the critical events in the mechanism of necrotic death under energy deprivation (see chapters 2, 3 and ref. 39 this chapter).

We do not exclude, though, the alternative that the HSP70-enzyme binding provoked by ATP depletion may maintain some cytosolic enzymes in a soluble and catalytically active state or even increase their activity during the stress. Moreover, the activation of some enzymatic reactions within the ATP-deprived cell is probably mediated by HSP70 at the level of substrate when the chaperone fixes a bound protein in unfolded conformation, thus demasking certain amino acid residues for the attack of site-specific enzymes (protein kinases, phosphatases, proteases, etc.). Of course, the activation of some enzymatic reaction by excess HSP70 might exert beneficial effects on the cell viability under energy starvation. In a cell-free system, HSP70 activates phosphoprotein phosphatases;[40] however, it is unclear whether this phenomenon occurs in vivo and, if so, how it affects the viability of cells lacking ATP. Intriguingly, the often described elevation of catalase activity in the myocardium tolerant to ischemia seems to be associated with a rise in the amount of cardiac HSP70 (see chapter 5), which allowed researchers to speculate about the activation of catalase by induced HSP70.[41] However, later the examination of transgenic mice overexpressing HSP70 failed to reveal any increase in catalase activity.[42] In any event, we consider the proposed modulation of enzyme activities in energy-deprived cells by HSP70 as the special stress-specific cytoprotective device, based on the formation of the stable (long-living) complexes of the chaperone with molecules of enzymes or enzyme's protein substrates. Both the formation of such complexes and their stabilization appear to be due to a decrease in the ATP/ADP ratio in the cytosol (see subsection 7.2.1. above).

In addition, as oxidative stress follows ischemia, oxidized intracellular proteins may also be critical targets for chaperoning by HSP70 in post-ischemic tissues. Possible mechanisms of cytoprotective actions of HSP70 upon oxidative stress are reviewed in ref. 43.

Finally, enhanced targeting of excess HSP70 to irreversibly damaged cellular proteins might maintain the proteins in a proteolytically susceptible (unfolded) conformation (discussed in ref. 44). This hypothetical HSP70 function may provide for faster proteolysis of the aberrant proteins intended for degradation, thus facilitating post-stress cell repair. Contrary to that, excess HSP70 might serve in masking some cellular proteins to prevent their cleavage by proteases activated as a result of ATP depletion or ischemia/reperfusion.

If such a device is indeed realized in vivo, this may be the molecular basis for resistance of tolerant cells to post-ischemic apoptosis (see chapter 8 for further discussion).

7.3. HOW OTHER STRESS PROTEINS MIGHT MAINTAIN THE VIABILITY OF ATP-DEPRIVED CELLS

While the experiments with constitutive *HSP70* transgene overexpression[5-8,17] have proved the cytoprotective function of HSP70 under ATP-depleting metabolic stress, the role of other HSPs in energy-deprived cells remains to be established. At present, no direct data have been obtained that any HSP(s) besides HSP70 really contributes to cell resistance to energy deprivation. It should be noted that the cells preconditioned to simulated ischemia as a result of challenging heat shock (i.e., cells overexpressing the whole spectrum of HSPs) exhibit better survival after ATP-depleting exposure than the same cells constitutively overexpressing HSP70 alone.[5] This suggests the cytoprotective action of some other HSPs as well. In principle, all HSPs, being molecular chaperones, might be involved in either protection of cellular proteins during ATP depletion or their repair afterward. Here we wish to discuss the possible contribution of different stress proteins to resistance of energy-deprived cells.

7.3.1. HSP100s

This class of stress proteins was mainly studied in prokaryotes or yeasts (see chapter 1) and its role in mammalian cells is not clear. Expression and functions of HSP100s in energy-deprived cells or ischemic tissues have not yet been examined. Like HSP70, HSP100s are ATPases and therefore they might sense ATP depletion and respond to it in the same manner as HSP70.

Intriguing facts were revealed in Lindquist's group: in yeast cells HSP104 did not preserve transfected luciferase from heat-induced inclusion in protein aggregates but mediated the enzyme recovery from aggregates during cell recovery.[45] The researchers suggested that HSP70 and HSP104 act one after another—the former prevents cellular proteins from aggregation during heat shock, whereas the latter carries out the post-stress dissolution of protein aggregates.[45] Extrapolating this suggestion to metabolic stress, one may imagine an analogous mechanism: inhibition of protein aggregation by HSP70 during ATP depletion followed by disaggregating actions of HSP104 during cell recovery. If it is true, here we again encounter two paradigms of cell tolerance, each due to a different HSP. Nevertheless, we do not disregard the above proposed role of HSP70 in post-stress protein disaggregation (see Fig. 7.4); it seems likely that HSP70 and HSP104 perform this important function in concert.

Moreover, Parsell and Lindquist actively discussed in their review a role of HSP104 in proteolysis that may be relevant to stress tolerance.[44] According to their hypothesis, HSP104 might promote the degradation of damaged cellular proteins which are not salvageable by HSP70. It remains undefined whether HSP100s perform protective functions in energy-deprived mammalian cells.

7.3.2. HSP90s

Expression of these chaperones increases in mammalian tissues and cells undergoing ischemic episodes or energy starvation (see chapters 4-6); however, their role in cytoprotection from ischemic-like stress is unresolved. In contrast to overexpressed HSP70, HSP90 stably transfected in cultures of myocytes[6] and neurons[8] does not protect the cells from necrotic death under ATP-depleting conditions.

At the same time, some properties of HSP90 allow us to speculate about its possible contribution to cell resistance to energy deprivation. For instance, HSP90 was shown to interact with actin microfilaments in a Ca^{2+}-dependent manner[46,47] and the high level of its expression appears to stimulate cell motility.[48] From this it follows that HSP90 is involved in the regulation of actin microfilament dynamics and such activity might be applied towards the stabilization of the actin skeleton during ATP depletion and/or reassembly of the microfilament network when ATP is restored. We already discussed the role of HSP90 as a potential protector of the actin skeleton in ATP-deprived cells.[4,49] In a cell-free system, HSP90 interacts with F-actin via an ATP-dependent mechanism, thus interfering with myosin.[50] If this phenomenon really occurs in vivo, it might affect the formation of rigor actin-myosin complexes in ATP-depleted cells (see chapter 3). Likewise, nothing is known about the role of HSP90 in post-stress apoptosis.

HSP90 is able in vitro to enhance binding of histones to DNA, thus condensing the chromatin structure,[51] condensation of the chromatin in situ being one of the pre-apoptotic symptoms (see chapters 2, 6 and 8). It seems possible that this activity of HSP90 allows an excess of the chaperone to modulate the process of apoptosis in tolerant post-ischemic tissues. Of course, it is still only a speculation.

7.3.3. GRPs

The expression of glucose-regulated proteins, GRP75, GRP78 and GRP94, is strongly activated by ischemia or hypoxia (reviewed in chapter 4), which permits us to suppose their beneficial function against the inducing exposures. However, no data were obtained which might directly confirm the cytoprotective function of any GRPs under ATP-depleting metabolic stress.

In unstressed mammalian cells, GRP78 and GRP94 act as molecular chaperones regulating maturation and assembly of proteins inside the endoplasmic reticulum. GRP75 (or mitochondrial HSP70) participates in protein import into mitochondria (see chapter 1). One may imagine that GRPs accumulated after primary ischemia would locally attenuate protein aggregation in the endoplasmic reticulum and mitochondria during subsequent ischemic episodes. Moreover, excess GRPs might preserve their compartments from severe injury, thus facilitating the post-stress recovery of cellular metabolism and functions.

The fact that the endoplasmic reticulum and mitochondria are known to be regulators of intracellular $[Ca^{2+}]$ in turn provides some material for speculation. Indeed, damage of these compartments as a result of metabolic stress may be accompanied by dysfunction of ATP-dependent calcium pumps, Ca^{2+} efflux, stimulation of Ca^{2+}-activated degradative enzymes and eventually cell death (reviewed in chapter 2 and subsection 7.4.). Moreover, at least two enzymes, an endonuclease (DNase I) and a protease (probably ICE) involved in apoptosis are localized to the endoplasmic reticulum and mitochondria, respectively, and may exit the damaged compartments to promote apoptotic death (see chapter 8). We hypothesize that excess GRP(s) in tolerant cells serves for better protection and stabilization of the compartments during ATP-depleting ischemia to delay fatal efflux of Ca^{2+} and/or the apoptotic enzymes. Experiments with constitutive overexpression of GRPs could test our hypothesis.

7.3.4. HSP60

This HSP is a major prokaryotic chaperonin. In eukaryotes, HSP60 is localized to the mitochondrial matrix, where together with GRP75 it participates in polypeptide folding and assembly of protein oligomers (see chapter 1). Both HSP70 and HSP60 are K^+-dependent ATPases and perform chaperone functions in unstressed cells; however, only the former was shown to be directly involved in cytoprotection from ischemia or ATP-depleting stress.[5-8,10,17] In contrast to HSP70, induction of HSP60 does not correlate with cardioresistance to ischemic insult (reviewed in chapter 5) and overexpression of a *HSP60* transgene does not confer tolerance to simulated ischemia in cultured myocytes,[6] which enables us to eliminate HSP60 from the potential anti-ischemic effectors.

Based on general chaperone properties of HSPs one may expect that excess HSP60

will protect mitochondria, especially since these organelles are primary targets for injurious action of ischemia or hypoxia (chapter 2), they being less damaged in tolerant cells (see above subsection 7.1.). However, experimental data does not confirm this. Perhaps, HSP60 functions as chaperonin rather than chaperone, i.e., it carries out folding and assembly of intramitochondrial proteins only under normal conditions when ATP is available but it cannot preserve protein molecules during deenergization of mitochondria and ATP depletion. Moreover, a partner of HSP60, HSP10, may be a limiting factor that excludes the involvement of surplus HSP60 in the beneficial chaperone cycle. We think that although overexpressed HSP60 does not save ATP-depleted mammalian cells from death, excess HSP60 might contribute to the improved post-stress repair of mitochondria and recovery of cellular energy and ionic homeostasis.

7.3.5. THE SMALL HSPS

Up to now there are no data directly indicating the cytoprotective capacity of HSP27 or α-crystallins under energy deprivation. Nevertheless, these stress proteins clearly respond to cellular ATP decrease and their expression is enhanced after ischemia (see chapters 3-6), which allows us to discuss the possible role of the small HSPs in the mechanism of cell adaptation to energy deprivation. Comparing HSP27 distribution in tolerant and nontolerant gerbil brain, Kato et al[52] concluded that the accumulation of the small HSP is not associated with post-ischemic survival of neurons but it may be related to the protection of astroglial cells (see chapter 6). Likewise, an increase in cardiac HSP27 seems to play a role in myocardial preconditioning by tumor necrosis factor.[53] The expected protection of the actin filaments by excess HSP27 is discussed as one of the factors determining resistance of thermotolerant cells to energy deprivation.[4,49]

We are sure that the small HSPs are able to protect mammalian cells during

ATP depletion or ischemia/reperfusion, despite the fact that the results confirming this are presently absent. Our confidence is based on a number of data indicating the beneficial action of HSP27 at the level of the actin skeleton under various stresses. On the one hand, the overexpressed HSP27 was shown to maintain the integrity of actin microfilaments under heat shock, treatment with cytochalasin $D^{54,55}$ or oxidative stress.[56,57] On the other hand, the stress-induced disruption of the actin framework causes necrotic death of ATP-depleted mammalian cells (reviewed in chapter 3). Perhaps excess HSP27 protects cortical actin microfilaments from fragmentation and bundling during ATP decrease, thus postponing cell swelling, blebbing and death.

However, the mechanism of HSP27-mediated protection of the actin skeleton remains to be established. The known properties of HSP27 do not characterize it as a factor stabilizing filamentous actin. On the contrary, HSP27, possessing actin barbed-end capping activity, inhibits actin polymerization and depolymerizes F-actin.[58,59] It seems enigmatic how such a protein can preserve actin microfilaments during stress. At the same time, in unstressed cells HSP27 is involved in signal transduction from cell-activating external stimuli to actin microfilaments, this function being regulated via HSP27 phosphorylation.[60,61] Briefly, HSP27 is phosphorylated in vivo in response to the action of serum, growth factors, hormones, cytokines, etc. that modulate its activity toward actin. Lavoie et al[60] demonstrated that it is the phosphorylation of overexpressed HSP27 that defines the rapid polymerization of submembranous actin filaments under mitogenic stimulation of fibroblasts. In contrast, the cells overexpressing the nonphosphorylatable mutant form of HSP27 had reduced cortical F-actin concentration.[60]

According to Arrigo and Landry, phosphorylation of HSP27 during stimulation by growth factors results in the dissociation of HSP27 from actin that frees barbed ends for addition of monomers and thus stimulates actin polymerization instead of its inhibition.[61] Perhaps the same suggestion can be applied towards the explanation of how HSP27 acts under stresses. Since HSP27 does undergo phosphorylation by stress-activated protein kinase(s) during heat shock [54,55] and treatment with oxidants,[56,57] this may mediate the restoration of microfilaments in stressed cells by promoting elongation of F-actin fragments. Actually, in the cells overexpressing HSP27, its phosphorylation was shown to be necessary for protection of the actin skeleton from destructive actions of heating, cytochalasin D and oxidants.[54-57] In support of this, Benndorf et al found in a cell free system that phosphorylation of murine small HSP25 abolishes its capacity to inhibit actin polymerization.[62]

A question arises whether analogous mechanism takes place under ATP depletion or ischemia/reperfusion. As we wrote in chapter 3, in ATP-depleted human endothelial cells, HSP27 is dephosphorylated and concentrated in nuclei in large granules.[63] Therefore, in contrast to heat shock and oxidative stress which increase in HSP27 phosphorylation,[54-57] ATP deprivation exerts the opposite effect and apparently the device proposed by Arrigo and Landry[61] cannot be realized here. Nevertheless, the dephosphorylation, nuclear compartmentalization and granulation of HSP27 occurring in ATP-depleted cells may be considered as special adaptive reactions that prevent actin aggregation and F-actin bundling or disassembly during energy starvation.

To explain how HSP27 could affect the actin skeleton during ATP depletion we propose the following scenario. Before ATP loss, HSP27 is preferentially localized in the cytoplasm (see chapter 3). Being involved in regulation of microfilament dynamics, some part of HSP27 performs the capping of actin barbed-ends (see Fig. 7.5). According to Benndorf et al,[62] only monomers of unphosphorylated HSP27 possess the capping activity and inhibit actin polymerization. Phosphorylated HSP27 appears to be nonbound to actin, the equilibrium between the pools of phosphorylated and unphosphorylated HSP27 being retained by protein kinases/phosphatase machinery (Fig. 7.5). What happens in cells lacking ATP? During ATP depletion, the equilibrium shifts toward dephosphorylation of HSP27[63] likely owing to the dysfunction of protein kinase(s) and/or activation of a protein phosphatase. Both suggested causes seem quite real since, first, the activity of protein kinases involved in the HSP27 phosphorylation decreases during ATP depletion[64] and second, calcium/calmodulin-

dependent phosphatase (calcineurin) dephosphorylating HSP27[65] may be activated in ATP-deprived cells through the stress-induced rise in cytosolic Ca^{2+} (see chapter 2). Monomers of HSP27 lacking phosphates should cap free barbed-ends of F-actin fragments (Fig. 7.5); furthermore, in vivo an excess of dephosphorylated HSP27 or rather its monomers reduces F-actin.[60,62] Perhaps to escape this, a special mechanism exists which removes HSP27 lacking phosphates from the cytoplasm (nuclear compartmentalization)

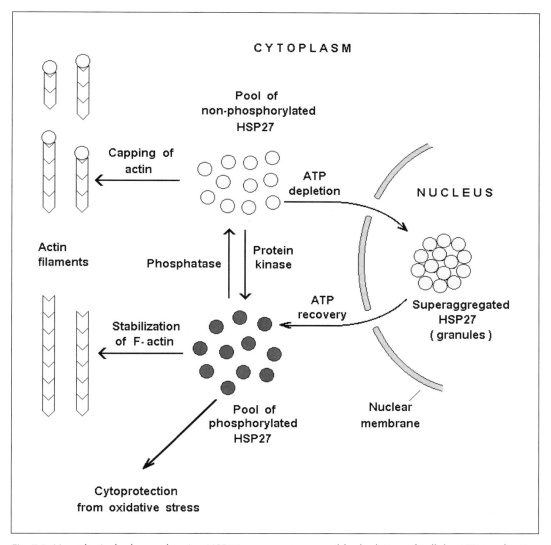

Fig. 7.5. Hypothetical scheme showing HSP27 response to reversible depletion of cellular ATP. Implication of the proposed mechanism for cytoprotection is considered in the text.

and/or assembles the dephosphorylated monomers into granules, since unphosphorylated HSP27 integrated in large particles does not affect actin polymeryzation.[62] Thus, all the observed responses of HSP27 to ATP deprivation, namely its dephosphorylation, translocation to the nucleus and intranuclear granulation (Fig. 7.5) may be aimed at achieving better preservation of actin filamens in energy-deprived cells. Certainly our model is purely hypothetical and we agree that additional information on the HSP27-actin interaction under conditions of ATP deficiency is required.

It seems likely that excess HSP27 also contributes to the improved post-stress repair of tolerant cells. Restoration of the ATP level in endothelial cells was accompanied by the disappearance of the HSP27-containing granules from nuclei and recovery of the soluble HSP27 pool in the cytosol.[63] Obviously, the protein kinases reactivated through ATP replenishment[64] promote this process by phosphorylating the granulated HSP27 when the risk of actin aggregation is absent. Such degranulation allows HSP27 to again be involved in the regulation of microfilament dynamics, which is beneficial for the post-stress restoration of the actin framework (Fig. 7.5). Moreover, overexpressed HSP27 protects actin microfilaments[56,57] and enhances cell survival[57,66,67] upon oxidative stress. This suggests an HSP27-mediated antioxidant defense mechanism that may function during postanoxic reoxygenation (reperfusion).

At the same time, the cytoprotective capacity of HSP27 upon ATP deprivation may not be restricted only by stabilization (or repair) of actin filaments. As molecular chaperone, HSP27 might help in renaturing various cellular proteins damaged during ATP depletion or ischemia/reperfusion, the chaperoning function of HSP27 being independent of ATP[68] or phosphorylation status.[69] Interestingly, Kampinga et al showed that the overexpression of HSP27 accelerates cell recovery from heat-induced nuclear protein aggregation.[70] By analogy, one may speculate that the reversible

intranuclear granulation of HSP27 during ATP depletion/replenishment also serves in the chaperoning of intranuclear proteins to improve their post-stress disaggregation. This assumption could be easily examined on the system employed by Kampinga et al.[70]

In regard to α-crystallins, these small stress proteins have a sequence homology with HSP27 (see chapter 1) and exhibit a number of similar properties. In particular, like HSP27, the overexpression of αB-crystallin protects cells from oxidative stress.[66,67] In vitro, both HSP27 and α-crystallin function as molecular chaperones.[68,71,72] Cardiac α-crystallin forms large amorphous clumps in ischemic rat hearts[73] that resemble the granulation of endothelial HSP27 in response to ATP depletion.[63]

At present, there are not direct data confirming the cytoprotective role of α-crystallin under ATP depletion or ischemia. However, the beneficial actions of α-crystallin during the post-stress cell recovery (or reperfusion) seem quite likely on account of its chaperone activity[68,71,72] and its apparent involvement in the antioxidant defense.[66,67] The most intriguing results were obtained in studies of cardiac αB-crystallin (reviewed in chapters 3 and 5). Several research groups reported the association of cardiac αB-crystallin with cytoskeletal proteins[73-75] and connected this effect with low pH. Since acidification of the cytosol is known to be an obligatory attribute of ischemic insult (see chapter 2), an increase in binding affinity of αB-crystallin to actin and desmin filaments at a slightly acidic pH may direct the small chaperone to cytoskeletal and contractile proteins for their stabilization during ischemia. It is difficult to imagine that clumping of αB-crystallin (probably its co-aggregation with cytosolic actin) in response to myocardial ischemia[73] gives cytoprotection. In contrast, in vitro ability of αB-crystallin to effectively prevent the aggregation of actin filaments (i.e., formation of F-actin paracrystals)[74] may have a direct relation to cytoprotection, especially since actin aggregation and microfilament bundling under energy deprivation may result in rapid necrotic death of mammalian cells (reviewed in chapter 3 and ref. 49 this chapter). Moreover, the association of α-crystallin with myofibrils in ischemic rat hearts[75] may affect contractile activity of energy-depleted cardiomyocytes that in its turn may adapt the myocardium to ischemia (see for review chapter 2).

Taken together, all the above facts and reasonings are in support of the suggestion that the small HSPs protect mammalian cells during ATP depletion or ischemia/reperfusion. Perhaps the cardioprotective function of α-crystallin will be evaluated in the near future, since transgene mice with the heart-specific overexpression of this chaperone are available.[76]

7.3.6. UBIQUITIN

No one has yet demonstrated that overexpressed ubiquitin improves the viability of ATP-deprived mammalian cells. As we described in chapter 3, ubiquitin-dependent proteolysis requires ATP and therefore this pathway is blocked under energy starvation. Meanwhile, the beneficial function of the ubiquitin system under cell recovery from ATP-depleting metabolic stress or ischemia seems more than likely. Powerful ubiquitination of proteins and depletion of the free ubiquitin pool in ischemic tissues (see chapter 3) suggest an active involvement of ubiquitin in post-ischemic protein degradation. We observed that some intracellular proteins insoluble during ATP depletion rapidly degrade when ATP is restored (Kabakov, Bensaude and Gabai, unpublished data), which may mean their ubiquitin-mediated proteolysis. If it is true, an excess of ubiquitin might better rescue the post-ischemic cells from irreversibly damaged proteins, thus facilitating cell repair.

Studying the distribution of ubiquitin in ischemic gerbil hippocampus Kato et al[77] observed the irreversible disappearance of ubiquitin immunoreactivity in nontolerant CA1 neurons which undergo delayed neuronal death following ischemia, while recovery of the immunoreactivity took place in the tolerant neurons (see chapter 6). There-

fore, recovery of ubiquitin may be a prerequisite to neuronal resistance to ischemia.

Since deficiency of ubiquitin or block in ubiquitin-mediated proteolysis may have fatal consequences for a cell,[78] depletion of the free ubiquitin pool in gerbil hippocampal neurons during transient ischemia[79] was suggested as one of the possible causes of post-ischemic derangement of metabolism and delayed neuronal death. Since the latter occurs as apoptosis, a special interest represents expected failure in ubiquitin-dependent degradation of p53, a protein which may mediate post-ischemic apoptosis (reviewed in chapter 8). Perhaps an excess of induced ubiquitin in tolerant cells might help cells escape the total depletion of free ubiquitin and harmful consequences of ubiquitin deficiency including apoptosis. However, to check this possibility additional experiments are required on models of ischemia with the cells constitutively overexpressing ubiquitin.

7.3.7. HEMEOXYGENASE (HSP32)

A function of this stress protein in the case of ATP loss or ischemia has not yet been studied but its positive role in cellular antioxidant defense seems evident. The expression of hemeoxygenase 1 in fibroblasts is associated with their resistance to oxidative stress.[80,81] This suggests the potential contribution of hemeoxygenase 1 to cytoprotection from post-ischemic burst of reactive oxygen species, especially as cardiac hemeoxygenase expression is augmented in response to ischemia.[82,83]

Moreover, hemeoxygenase may attenuate ischemic injury at the tissue level. One of the most dangerous complications of ischemic lesions is hemorrhage. Besides the ischemic deprivation of nutrient and oxygen, hemorrhage results in the appearance of additional injurious factors such as cytotoxic products of hemolysis. In this connection, the in vitro protection of microvessel endothelial cells transfected with the hemeoxygenase gene from heme and hemoglobin toxicity[84] may be relevant to the reparative function of this stress protein during post-hemorrhagic tissue regeneration.

7.3.8. OTHER HSPs

A role of peptidyl prolyl cis-trans isomerases (PPIases), HSP47, HSP40 and HSP10 under ATP depletion or ischemia remains undefined. HSP40 and HSP10 are auxiliary components of HSP70 and HSP60 chaperone machines and at least HSP40 may really assist HSP70 in its protective actions in ATP-depleted cells (see above subsection 7.2.2. and Fig. 7.4). Both HSP40 and HSP10 are weakly inducible by stresses and the intriguing question is whether these HSPs are limiting factors in cells. If so, their deficiency may attenuate the beneficial effects of the excess of major chaperones, HSP70 and HSP60, in case of their constitutive overexpression.

In regard to HSP47, this collagen-binding protein appears to be involved in collagen maturation and assembly of the triple-helical collagen structures.[85] Nothing is known about its role in ATP-deprived cells or ischemic tissues. We can only speculate that at the tissue level, HSP47 might help to regenerate ischemic lesions by carrying out the collagen matrix formation at areas of necrosis.

Finally, it is absolutely unknown how PPIases (immuno- and cyclophilins) act during ATP depletion. However, abundance of these ubiquitous enzymes and their characteristic properties of ATP-independent chaperones (see chapter 1) allow us to propose the involvement of PPIase in the refolding of cellular proteins denatured as a result of ATP depletion or post-ischemic oxidative stress.

7.4. HOW DO HSPs PROTECT CELLS AGAINST NECROSIS DURING ATP DEPLETION?

Excess HSP(s) in mammalian cells is directly associated with their resistance to necrotic death during energy deprivation. This resistance is not due to a slower decrease in cellular ATP but only to a milder injury of cells during metabolic stress (see for review chapters 5, 6 and refs. 1-8 this chapter). At least the involvement of HSP70 in the maintenance of cell viability under ATP-depleting conditions is beyond any doubt.[5-8,17] However, molecular mechanisms of the HSP-mediated protection from cell injury and death during energy starvation remain a mystery.

Apparently, the delay of lethal protein aggregation by the presence of excess HSP(s) is one of the main points in this cytoprotection. In a series of articles reviewed in chapter 3, we demonstrated a direct link between the level of protein aggregation, blebbing and necrotic death in ATP-depleted murine tumor cells. Hereafter, we found that HSP accumulation preceding the loss of ATP in these cells significantly reduces necrosis resulting from protein aggregation and blebbing during metabolic stress,[1-4] approximately 2- to 2.5-fold increase in HSP70 alone being sufficient for marked suppression of both protein aggregation and necrotic death.[2]

In chapter 3, we suggested that the cytoskeleton destruction and ionic imbalance resulting from ATP depletion lead to swelling and blebbing of the energy-deprived cell that ends in the rupture of the plasma membrane, i.e., necrotic death. Likewise, a similar mechanism of necrosis was proposed for anoxic cardiomyocytes by Steenbergen, Hill and Jenings.[86] Consistent with our observations, the stress-induced actin aggregation is the key event that promotes rapid necrotic death in ATP-depleted mammalian cells (see chapter 3). We hypothesize that an excess of HSP70 or other HSPs is able to inhibit the ATP depletion-induced side-to-side aggregation of cortical actin, thus preventing blebbing and rapid necrosis.[49] The findings of reduced insolubility of actin in ATP-depleted tumor cells with high content of HSPs[1-4] are in favor of our idea. Since both the acquisition of resistance to necrosis and the appearance of "actin tolerance" to ATP deprivation coincide with the heat-induced HSP70 accumulation that in turn precedes the accumulation of HSP90 and HSP27,[2] we suggested that HSP70 plays the major role in protecting the actin skeleton and therefore in preventing necrosis in energy-deprived cells. Actually, it has been shown in myocytes and neurons that differentially overexpress various HSPs that HSP70, rather than HSP90 or HSP60, protects cells from necrosis under ATP-depleting conditions.[6,8] We believe, though, that the small HSPs may also contribute to the stabilization of actin filaments during metabolic stress (see above subsection 7.3.5. and Figs. 7.5 and 7.6).

What is the mechanism of HSP-mediated protection from actin aggregation owing to ATP depletion? In contrast to HSP27 whose major function seems to be protection of microfilaments[54-57,61] and αB-crystallin whose antiaggregative effect toward F-actin has been directly demon-

strated,[73] the beneficial influence of excess HSP70 on the actin skeleton is poorly understood. Although a number of publications characterize HSP70 as interacting with actin and/or actin-associated proteins,[32-34,87-90] a role of these interactions in the HSP70-mediated cytoprotection under ATP depletion is not clear. Margulis and Welsh[33] and Haus et al[34] found that HSP70 may interact with actin in an ATP-dependent fashion. In the latter study (ref. 34), HSP70 inhibits actin polymerization in *Dictyostelium* by complexing with an F-actin capping protein cap32/34 that sharply increases its capping activity. Since

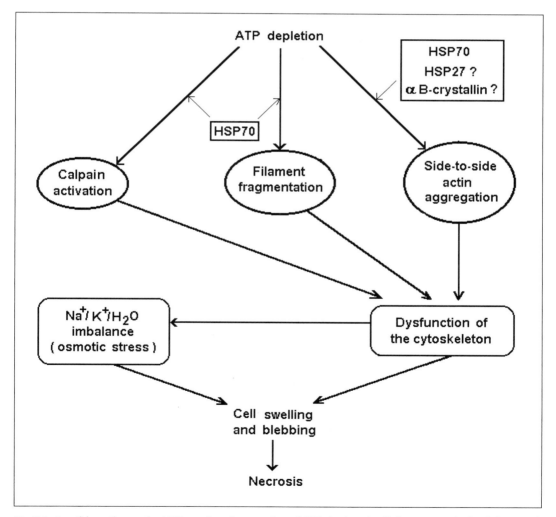

Fig. 7.6. Possible pathways for HSP-mediated protection of ATP-depleted cells from necrotic death (see text).

ATP dissociates the HSP70-cap32/34 complex,[34] one may expect enhanced capping of F-actin in case of ATP depletion. If an analogous regulatory mechanism takes place in mammalian cells, this resembles the above suggested situation with HSP27 in which the protein undergoing dephosphorylation during ATP deprivation initiates the capping of actin barbed ends (subsection 7.3.5. and Fig. 7.5). It is unclear how such enhanced capping of actin could protect microfilaments during the depletion of ATP; we only assume that this device minimizes a risk of the stress-induced side-to-side actin aggregation resulting from ATP deficiency (Fig. 7.7).[49] Contrary to the finding of Haus et al,[34] Condeelis' group reported that *Dictyostellium* HSP70 (or HSC70) per se acquires the F-actin capping activity in response to cAMP stimulation.[89] Intriguingly, before the discovery of the inhibitory effects of HSP70 on actin polymerization,[34,89] Macejak and Luftig revealed that HSP70 promotes the formation of F-actin from monomers in vitro.[88] The polymerizing activity better suits HSP70 as a potential protector of the actin skeleton; however, it is unknown whether this function is intrinsic to HSP70 in vivo. In any case, the proposed modulation of the actin skeleton state by excess HSP70 should be taken into consideration to unravel the real mechanism of HSP70-mediated cytoprotection under energy deprivation.

Such prominent features of HSP70 as the structural similarity of its ATPase domain with G-actin[91] gives occasion to speculate that HSP70 is able to bind directly to actin or even to be copolymerized with G-actin into a filamentous strand.[49,92] If this is true, molecules of HSP70 included into F-actin fibers might perform the function of anchors or clamping cross-linkers which stabilize microfilaments to prevent their fragmentation and-or side-to-side bundling under ATP depletion (Fig. 7.7).[49] However, no data are available which confirm this hypothesis.

Likewise, the maintenance of the monomer pool of cytoskeletal proteins in the stable state to avoid their misassembly (or aggregation) during stress seems to be an important function of cytosolic chaperones. We suggested previously[49] that the depletion of the G-actin pool on account of the ATP loss-provoked side-to-side monomer aggregation (see Fig. 7.7) should promote dissolution of the F-actin filament network by affecting the equilibrium in G ↔ F-actin transitions towards depolymerization. Therefore, chaperoning of G-actin during ATP depletion might serve to protect the preexisting actin framework and the effective restoration of the disrupted microfilaments. In the cytosol, HSP70 associates with chemically-denatured actin monomers during their refolding[90] and it seems likely that HSP70 stabilizes G-actin in ATP-deprived cells as well. In turn, the HSP-mediated preservation of both F-actin and G-actin pools may be relevant to cell resistance to post-stress apoptosis (reviewed in chapter 8).

It should be noted that excess HSP70 may protect the cytoskeleton indirectly. Such protection may be achieved by inhibiting the stress-provoked proteolysis of the actin skeleton constituents by calpain or other proteases. For instance, the overexpression of HSP70 inhibits phospholipase A_2,[38] while the phospholipase-mediated activation of calpain during ATP depletion or anoxia may result in necrosis.[39] Although it is unclear how excess HSP70 can inhibit stress-activated phospholipase A_2, this inhibitory effect may define better protection of the cytoskeleton in tolerant cells and the anti-necrotic capacity of HSP70 on the whole (Fig. 7.6).

Besides the cytoskeleton dysfunction, increase in intracellular Ca^{2+} and injury of mitochondria may result in necrotic death during ischemia/reperfusion (see for review chapter 2). In fact, it is difficult to explain how excess HSP70 in the cytoplasm of tolerant cells protects the mitochondrion and/or decreases the calcium level in case of energy deprivation. Probably, a surplus HSP70 does not interfere with the elevation of {Ca^{2+}} in the cytosol but is able to suppress the fatal reactivity of Ca^{2+}-activated lytic enzymes (e.g., phospholipase

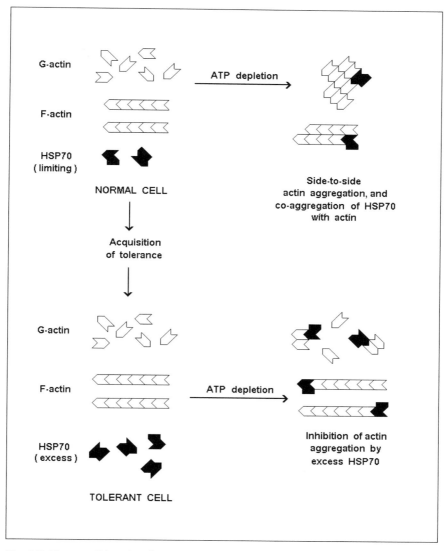

Fig. 7.7. The possible role of HSP70-actin interactions in preventing side-to-side actin aggregation as a result of energy deprivation. Reprinted with permission from Kabakov AE, Gabai VL. Trends Cell Biol 1994; 4: 193-196.

and calpain, see above) thus delaying necrotic death. In regard to the proposed mitochondrial protection by HSPs, this may also occur indirectly, in particular via stabilization of the cytoskeletal anchorage for mitochondria during the stress or via better provision for protein import into mitochondria during recovery. Apparently, this problem requires further serious investigation.

RECENT NEWS

Possible mechanisms of the HSP70-mediated cardioprotection against ischemia are briefly discussed in a recent review by Plumier and Currie.[93] As one of the possibilities the authors suggest that excess HSP70 may facilitate delivery of newly synthesized proteins into mitochondria, thus contributing to recovery of mitochondrial functions after ischemia/reperfusion injury.[93] Such an assumption looks quite reasonable, but it is poorly understood why HSP60, being essential for refolding and oligomeric assembly of intramitochondrial proteins, in its turn does not increase cell survival upon ischemia[6] (see also chapter 5). Other suggestions regarding the potential capacity of HSP70 to prevent the ischemia-provoked protein aggregation and collapse of the cytoskeleton[93] resemble our own speculations adduced in this chapter.

A recent report of Martin et al[94] confirms our prediction that the small HSPs protect ATP-depleted (or ischemic) cells from lethal injury. As we discussed above, the mechanism of the small HSP-mediated cytoprotection may be due to (I) chaperoning of contractile and cytoskeletal proteins during energy deprivation and (II) attenuation of the post-stress oxidative burst; however both suggestions need experimental examination.

REFERENCES

1. Gabai VL, Kabakov AE. Rise in heat-shock protein level confers tolerance to energy deprivation. FEBS Lett 1993; 327: 247-250.
2. Kabakov AE, Gabai VL. Heat shock-induced accumulation of 70-kDa stress protein (HSP70) can protect ATP-depleted tumor cells from necrosis. Exp Cell Res 1995; 217: 15-21.
3. Gabai VL, Mosina VA, Budagova KR, Kabakov AE. Spontaneous over-expression of heat-shock proteins in Ehrlich ascites carcinoma cells during in vivo growth. Biochem Mol Biol Int 1995; 35: 95-102.
4. Kabakov AE, Molotkov AO, Budagova KR et al. Adaptation of Ehrlich ascites carcinoma cells to energy deprivation in vivo can be associated with heat shock protein accumulation. J Cell Physiol 1995; 165: 1-6.
5. Mestril R, Chi S-H, Sayen R et al. Expression of inducible stress protein 70 in rat heart myogenic cells confers protection against simulated ischemia-induced injury. J Clin Invest 1994; 93: 759-767.
6. Heads RJ, Yellon DM, Latchman DS. Differential cytoprotection against heat stress or hypoxia following expression of specific stress protein genes in myogenic cells. J Mol Cell Cardiol 1995; 27: 1669-1678.
7. Papadopoulos MC, Sun XY, Cao JM et al. Over-expression of HSP-70 protects astrocytes from combined oxygen-glucose deprivation. Neuroreport 1996; 7: 429-432.
8. Amin V, Cumming DV, Latchman DS. Over-expression of heat shock protein 70 protects neuronal cells against both thermal and ischaemic stress but with different efficiences. Neurosci Lett 1996; 206: 45-48.
9. Gabai VL, Zamulaeva IV, Mosin AF et al. Resistance of Ehrlich tumor cells to apoptosis can be due to accumulation of heat shock proteins. FEBS Lett 1995; 375: 21-26.
10. Radford NB, Fina M, Benjamin IJ et al. Cardioprotective effects of 70-kDa heat shock protein in transgenic mice. Proc Natl Acad Sci USA 1996; 93: 2339-2342.
11. Saad S, Kanai M, Awane M et al. Protective effect of heat shock pretreatment with heat shock protein induction before hepatic warm ischemic injury caused by Pringle's maneuver. Surgery 1995; 118: 510-516.
12. Yellon DM, Pastini E, Cargoni A et al. The protective role of heat stress in the ischaemic and reperfused rabbit myocardium. J Mol Cell Cardiol 1992; 24: 895-907.
13. Garramone RR Jr, Winters RM, Das DK et al. Reduction of skeletal muscle injury through stress conditioning using the heat-shock response. Plastic Reconst Surg 1994; 93: 1242-1247.
14. Maridonneau-Parini I, Clerc J, Polla BS. Heat shock inhibits NADPH oxidase in human neutrophils. Biochem Biophys Res Commun 1988; 154: 179-186.
15. Maridonneau-Parini I, Malawista SE, Stubbe H et al. Heat shock in human neutrophils: superoxide generation is inhibited by a mechanism distinct from heat-denaturation of NADPH oxidase and is protected by heat shock proteins in thermotolerant cells. J Cell Physiol 1993; 156: 204-211.
16. Hirvonen MR, Brune B, Lapetina EG. Heat shock proteins and macrophage resistance to the nitric oxide. Biochem J 1996; 315: 845-849.
17. Williams RS, Thomas JA, Fina M et al. Human heat shock protein 70 (hsp70) pro-

tects murine cells from injury during metabolic stress. J Clin Invest 1993; 92: 503-508.

18. Nguyen VT, Bensaude O. Increased thermal aggregation of proteins in ATP-depleted mammalian cells. Eur J Biochem 1994; 220: 239-246.

19. Hohfeld J, Minami Y, Hartl FU. Hip, a novel cochaperone involved in the eukaryotic Hsc70/Hsp40 reaction cycle. Cell 1995; 83: 589-598.

20. Pelham HR. Speculations on the functions of the major heat shock and glucose-regulated proteins. Cell 1986; 46: 959-961.

21. Rothman JE. Polypeptide chain binding proteins: catalysts of protein folding and related processes in cells. Cell 1989; 59: 591-601.

22. Li GC, Li L, Liu RY et al. Heat shock protein hsp70 protects cells from thermal stress even after deletion of its ATP-binding domain. Proc Natl Acad Sci USA 1992; 89: 2036-2040.

23. Stege GJJ, Li GC, Li L et al. On the role of hsp72 in heat-induced intranuclear protein aggregation. Int J Hyperthermia 1994; 10: 659-674.

24. Stege GJJ, Li L, Kampinga HH et al. Importance of the ATP-binding domain and nucleolar localization domain of HSP72 in the protection of nuclear proteins against heat-induced aggregation. Exp Cell Res 1994; 214: 279-284.

25. Palleros DR, Welch WJ, Fink AL. Interaction of hsp70 with unfolded proteins: Effects of temperature and nucleotides on the kinetics of binding. Proc Natl Acad Sci USA 1991; 88: 5719-5723.

26. Palleros DR, Reid KL, Shi L et al. ATP-induced protein-Hsp70 complex dissociation requires K^+ but not ATP hydrolysis. Nature 1993; 365: 664-666.

27. Palleros DR, Shi L, Reid KL, Fink AL. Hsp70-protein complexes; complex stability and conformation of bound substrate protein. J Biol Chem 1994; 269: 13107-13114.

28. Hightower LE, Sadis SE. Interaction of vertebrate hsc70 and hsp70 with unfolded proteins and peptides. In: Morimoto RI, Tissieres A, Georgopoulos C, eds. The Biology of Heat Shock Proteins and Molecular Chaperones. Cold Spring Harbor, NY:

Cold Spring Harbor Laboratory Press, 1994: 179-207.

29. Lewis MJ, Pelham HRB. Involvement of ATP in the nuclear and nucleolar functions of the 70 kd heat shock protein. EMBO J 1985; 4: 3137-3143.

30. Ohtsuka K, Utsumi KR, Kaneda T, Hattori H. Effect of ATP on the release of hsp 70 and hsp 40 from the nucleus in heat-shocked HeLa cells. Exp Cell Res 1993; 209: 357-366.

31. Beckmann RP, Lovett M, Welch WJ. Examining the function and regulation of hsp70 in cells subjected to metabolic stress. J Cell Biol 1992; 117: 1137-1150.

32. La Thangue NB. A major heat shock protein defined by a monoclonal antibody. EMBO J 1984; 3: 1871-1879.

33. Margulis BA, Welsh M. Analysis of protein binding to heat shock protein 70 in pancreatic islet cells exposed to elevated temperature or interleukin 1β. J Biol Chem 1991; 266: 9295-9298.

34. Haus U, Trommler P, Fisher PR et al. The heat shock cognate protein from *Dictyostelium* affects actin polymerization through interaction with the actin-binding protein cap32/34. EMBO J 1993; 12: 3763-3771.

35. Liao J, Lowthert LA, Ghori N, Omary MB. The 70-kDa heat shock proteins associate with glandular intermediate filaments in an ATP-dependent manner. J Biol Chem 1995; 270: 915-922.

36. Tuijl MJM, van Bergen en Henegouwen PMP, van Wijk R et al. The isolated neonatal rat-cardiomyocyte used in an in vitro model for 'ischemia'. II. Induction of the 68 kDa heat shock protein. Biochim Biophys Acta 1991; 1091: 278-284.

37. Imamoto N, Matsuoka Y, Kurihara T et al. Antibodies against 70-kD heat shock cognate protein inhibit mediated nuclear import of karyophilic proteins. J Cell Biol 1992; 119: 1047-1061.

38. Jaattela M. Overexpression of major heat shock protein hsp70 inhibits tumor necrosis factor-induced activation of phospholipase A_2. J Immunol 1993; 151: 4286-4294.

39. Arora AS, Degroen PC, Croall DE et al. Hepatocellular carcinoma cells resist necrosis during anoxia by preventing phospholi-

pase-mediated calpain activation. J Cell Physiol 1996; 167: 434-442.

40. Mivechi NF, Trainor LD, Hahn GM. Purified mammalian HSP-70 kDa activates phosphoprotein phosphatases in vitro. Biochem Biophys Res Commun 1993; 192: 954-963.

41. Kukreja RC, Hess ML. The oxygen free radical system: from equations through membrane protein interactions to cardiovascular injury and protection. Cardiovasc Res 1992; 26: 641-655.

42. Marber MS, Mestril R, Chi SH et al. Overexpression of the rat inducible 70-kD heat stress protein in a transgenic mouse increases the resistance of the heart to ischemic injury. J Clin Invest 1995; 95: 1446-1456.

43. Jacquier-Sarlin MR, Fuller K, Dinh-Xuan AT, Richard M-J, Polla BS. Protective effects of hsp70 in inflamation. Experientia 1994; 50: 1031-1038.

44. Parsell DA, Lindquist S. Heat shock proteins and stress tolerance. In: Morimoto RI, Tissieres A, Georgopoulos C, eds. The Biology of Heat Shock Proteins and Molecular Chaperones. Cold Spring Harbor, NY: Cold Spring Harbor Laboratory Press, 1994: 457-494.

45. Parsell DA, Kowal AS, Singer MA, Lindquist S. Protein disaggregation mediated by heat-shock protein Hsp104. Nature 1994; 372: 475-478.

46. Koyasu S, Nishida E. Kadowaki T et al. Two mammalian heat shock proteins, HSP90 and HSP100, are actin-binding proteins. Proc Natl Acad Sci USA 1986; 83: 8054-8058.

47. Nishida E, Koyasu S, Sakai H, Yahara H. Calmodulin-regulated binding of the 90 kDa heat shock protein to actin filaments. J Biol Chem 1986; 261: 16033-16036.

48. Yahara I, Iida H, Koyasu S. A heat shock-resistant variant of Chinese hamster cell line contitutively expressing heat shock protein of Mr. 90,000 at high level. Cell Struct Funct 1989; 11: 65-73.

49. Kabakov AE, Gabai VL. Heat shock proteins maintain the viability of ATP-deprived cells: what is the mechanism? Trends Cell Biol 1994; 4: 193-196.

50. Kellermayer MSZ, Csermely P. ATP induces dissociation of the 90 kDa heat shock protein (hsp90) from F-actin: Interference with the binding of heavy meromyosin. Biochem Biophys Res Commun 1995; 211: 166-174.

51. Csermely P, Kajtar J, Hollosi M et al. The 90 kDa heat shock protein (hsp90) induces the condensation of the chromatin structure. Biochem Biophys Res Commun 1994; 202: 1557-1663.

52. Kato H, Araki T, Itoyama Y et al. An immunohistochemical study of heat shock protein-27 in the hippocampus in a gerbil model of cerebral ischemia and ischemic tolerance. Neuroscience 1995; 68: 65-71.

53. Maulik N, Das DK. Cardioprotective effect of TNFα is mediated through the myocardial adaptation to oxidative stress. J Mol Cell Cardiol 1993; 25 [suppl.III]: S43.

54. Lavoie JN, Gingras-Breton G, Tanguay RM, Landry J. Induction of Chinese hamster HSP27 gene expression in mouse cells confers resistance to heat shock. J Biol Chem 1993; 268: 3420-3429.

55. Lavoie JN, Lambert H, Hickey E et al. Modulation of cellular thermoresistance and actin filament stability accompanies phosphorylation-induced changes in the oligomeric structure of heat shock protein 27. Mol Cell Biol 1995; 15: 505-516.

56. Huot J, Lambert H, Lavoie JN et al. Characterization of 45-kDa/54-kDa HSP27 kinase, a stress-sensitive kinase which may activate the phosphorylation-dependent protective function of mammalian 27-kDa heat-shock protein HSP27. Eur J Biochem 1995; 227: 416-427.

57. Huot J, Houle F, Spitz DR, Landry J. HSP27 phosphorylation-mediated resistance against actin fragmentation and cell death induced by oxidative stress. Cancer Res 1996; 56: 273-279.

58. Miron T, Wilchek M, Geiger B. Characterization of an inhibitor of actin polymerization in vinculin-rich fraction of turkey gizzard smooth muscle. Eur J Biochem 1988; 178: 543-553.

59. Miron T, Vancompernolle K, Vanderkerckhove J et al. A 25-kD inhibitor of actin polymerization is a low molecu-

lar mass heat shock protein. J Cell Biol 1991; 114: 255-261.

60. Lavoie JN, Hickey E, Weber LA, Landry J. Modulation of actin microfilament dynamics and fluid phase pinocytosis by phosphorylation of heat shock protein 27. J Biol Chem 1993; 268: 24210-24214.

61. Arrigo A-P, Landry J. Expression and function of the low-molecular weight heat shock proteins. In: Morimoto RI, Tissieres A and Georgopoulos C, eds. The Biology of Heat Shock Proteins and Molecular Chaperones. Cold Spring Harbor, NY: Cold Spring Harbor Laboratory Press, 1994: 335-373.

62. Benndorf R, Hayeb K, Ryazantsev S et al. Phosphorylation and supra-molecular organization of murine small heat shock protein HSP25 abolish its actin polymerization-inhibiting activity. J Biol Chem 1994; 269: 20780-20784.

63. Loktionova SA, Ilyinskaya OP, Gabai VL, Kabakov AE. Distinct effects of heat shock and ATP depletion on distribution and isoform patterns of human Hsp27 in endothelial cells. FEBS Lett 1996; 392:100-104.

64. Pombo CM, Bonventre JV, Avruch J et al. Stress-activated protein kinases are major c-jun N-terminal kinases activated by ischemia and reperfusion. J Biol Chem 1994; 269: 26546-26551.

65. Gaestel M, Benndorf R, Hayess K et al. Dephosphorylation of the small heat shock protein hsp25 by calcium/calmodulin-dependent (Type 2B) protein phosphatase. J Biol Chem 1992; 267: 21607-21611.

66. Mehlen P, Preville X, Chareyron P et al. Constitutive expression of human hsp27, or human αB-crystallin confers resistance to TNF- and oxidative stress-induced cytotoxicity in stably transfected murine L929 fibroblasts. J Immunol 1995; 154: 363-374.

67. Mehlen P, Kretzrremy C, Preville X, Arrigo AP. Human hsp27, Drosophila hsp27 and human alpha B-crystallin expression-mediated increase in glutathione is essential for the protective activity of these proteins against TNF alpha-induced cell death. EMBO J 1996; 15: 2695-2706.

68. Jacob U, Gaestel M, Engel K, Buchner J. Small heat shock proteins are molecular chaperones. J Biol Chem 1993; 268: 1517-1520.

69. Knauf U, Jakob U, Engel K et al. Stress- and mitogen-induced phosphoryl.ation of the small heat shock protein Hsp25 by MAPKAP kinase 2 is not essential for chaperone properties and cellular thermoresistance. EMBO J 1993; 13: 54-60.

70. Kampinga HH, Brunsting JF, Stege GJJ et al. Cells overexpressing Hsp27 show accelerated recovery from heat-induced nuclear protein aggregation. Biochem Biophys Res Commun 1994; 204: 1170-1177.

71. Horwitz J. Alpha-crystallin can function as a molecular chaperone. Proc Natl Acad Sci USA 1992; 89: 10449-10453.

72. Das KP, Petrash JM, Surewicz WK. Conformational properties of substrate proteins bound to a molecular chaperone alpha-crystallin. J Biol Chem 1996; 271: 10449-10452.

73. Chiesi M, Longoni S, Limbruno U. Cardiac alpha-crystallin. III. Involvement during heart ischemia. Mol Cell Biochem 1990; 97: 129-136.

74. Bennardini F, Wrzosek A, Chiesi M. αB-Crystallin in cardiac tissue: association with actin and desmin filaments. Circ Res 1992; 71: 288-294.

75. Barbato R, Menabo R, Dainese P et al. Binding of cytosolic proteins to myofibrils in ischemic rat hearts. Circ Res 1996; 78: 821-828.

76. Gopalsrivastava R, Haynes JI, Piatigorsky J. Regulation of the murine alpha B-crystallin/small heat shock protein gene in cardiac muscle. Mol Cell Biol 1995; 15: 7081-7090.

77. Kato H, Chen T, Liu XH et al. Immunohistochemical localization of ubiquitin in gerbil hippocampus with tolerance to ischemia. Brain Res 1993; 619: 339-343.

78. Jentsch S. Ubiquitin-dependent protein degradation: a cellular perspective. Trends Cell Biol 1992; 2: 98-103.

79. Morimoto T, Ide T, Ihara Y et al. Transient ischemia depletes free ubiquitin in the gerbil hippocampal CA1 neurons. Am J Pathol 1996; 148: 249-257.

80. Vile GF, Basu-Modak S, Waltner C, Tyrrell RM. Hemeoxygenase 1 mediates an adaptive response to oxidative stress in human skin fibroblasts. Proc Natl Acad Sci USA 1994; 91: 2607-2610.

81. Guyton KZ, Spitz DR, Holbrook NJ. Expression of stress response genes GADD153, c-jun, and hemeoxygenase in H_2O_2- and O_2-resistant fibroblasts. Free Rad Biol Med 1996; 20: 735-741.

82. Sharma HS, Maulik N, Gho BCG et al. Coordinated expression of hemeoxygenase-1 and ubiquitin in the porcine heart subgected to ischemia and reperfusion. Mol Cell Biochem 1996; 157: 111-116.

83. Maulik N, Sharma HS, Das DK. Induction of the haem oxygenase gene expression during the reperfusion of ischemic rat myocardium. J Mol Cell Cardiol 1996; 28: 1261-1270.

84. Abraham NG, Lavrovsky Y, Schwartzman ML et al. Transfection of the hemeoxygenase gene into rabbit coronary microvessel endothelial cells: protective effect against heme and hemoglobin toxicity. Proc Natl Acad Sci USA 1995; 92: 6798- 6802.

85. Nakai A, Satoh M, Hirayoshi K, Nagata K. Involvement of the stress protein HSP47 in procollagen processing in the endoplasmic reticulum. J Cell Biol 1992; 117: 903-914.

86. Steenbergen C, Hill ML, Jennings RB. Volume regulation and plasma membrane injury in aerobic, anaerobic, and ischemic myocardium in vitro. Effects of osmotic cell swelling on plasma membrane integrity. Circ Res 1985; 57: 864-875.

87. Margulis BA, Welsh M. Isolation of hsp70-binding proteins from bovine muscle. Biochem Biophys Res Commun 1991; 178: 1-7.

88. Macejak DG, Luftig RB. Stabilization of actin filaments at early times after adenovirus infection and in heat-shocked cells. Virus Res 1991; 19: 31-45.

89. Eddy RJ, Sauterer RA, Condeelis JS. Aginactin, an agonist-regulated F-actin capping activity is associated with an Hsc70 in *Dictyostelium*. J Biol Chem 1993; 268: 23267-23274.

90. Frydman J, Hartl FU. Principles of chaperone-assisted protein folding: differences between in vitro and in vivo mechanisms. Science 1996; 272: 1497-1502.

91. Flaherty KM, DeLuca-Flaherty C, McKay DB. Three dimensional structure of the ATPase fragment of a 70K heat-shock cognate protein. Nature 1990; 346: 623-628.

92. Tsang TC. New model for 70 kDa heat-shock proteins' potential mechanisms of function. FEBS Lett 1993; 323: 1-3.

93. Plumier J-CL, Currie RW. Heat shock-induced myocardial protection against ischemic injury: a role for Hsp70? Cell Stress & Chaperones 1996; 1: 13-17.

94. Martin JL, Mestril R, Hilal-Dandan R, Dillmann W. Small heat shock proteins and protection against ischemic injury in cardiomyocytes. Abstracts of 1996 Meeting on Molecular Chaperones & The Heat Shock Response (May 1-5, 1996, Cold Spring Harbor, NY), p. 342.

CHAPTER 8

MECHANISMS OF HSP-MEDIATED PROTECTION FROM ISCHEMIA-INDUCED APOPTOSIS

Apoptosis can be divided into the three following phases: induction (or triggering), transduction of signal, and execution (see refs. 1-3 for review). Theoretically, HSPs may modulate any of these apoptotic phases with the same final result, namely rescue of the cells. Below we consider what is presently known about ischemia-induced apoptosis and speculate how HSPs may affect its distinct phases.

8.1 HSPs AND TRIGGERING OF APOPTOSIS IN ISCHEMIC CELLS

In 1992, Price and Calderwood[4] found that hypoxic exposure of NIH-3T3 cells, besides induction of GRP78 (glucose-regulated protein), also resulted in the marked stimulation of the expression of *gadd* (growth arrest and DNA damage) genes, *gadd45* and *gadd153*. These genes are usually induced after DNA damage through a p53-dependent mechanism. In 1994, Graeber and co-workers[5] from Giaccia's lab demonstrated in a number of tumor cell lines that hypoxia resulted in the accumulation of p53 within the nucleus and the stimulation of *gadd45* gene transcription. Although in these two works apoptosis was not examined, in the subsequent study Graeber et al[6] demonstrated the suppression of hypoxia-induced apoptosis in tumor cells lacking p53; furthermore, in transplanted tumors expressing wild-type p53, highly apoptotic regions strongly correlated with hypoxic regions whereas limited apoptosis occurred in hypoxic regions of p53-deficient tumors. Therefore, similar to apoptosis in ischemic neurons (see section 6.1), hypoxia-induced apoptosis in tumor cells apparently depends on p53 expression. Recently, increased expression of p53 under mild ischemia was found also in liver[7] and adrenal cortex;[8] in the latter case, enhanced transcription of a p53-dependent gene, *p21(WAF1/CIP1)*, an inhibitor of cyclin-dependent kinase, was also detected.

Usually considered a "guardian" of the genome, p53 performs an adaptive cellular response to genotoxic stress leading to G_1 arrest (due to *p21* expression) and DNA repair.[9,10] However, if DNA damage is extensive and cannot be completely repaired, a cell suicide mechanism (apoptosis) is triggered (see ref. 11 for review). Almost any exposure leading to DNA breaks can result in p53 accumulation, the latter being associated mainly with greatly increased stability of this protein.[12,13] As was shown in brain, reperfusion-induced oxidative stress can damage DNA and this may be one of the mechanisms of p53 induction (see section 6.1). However, as Graeber et al[4] found, p53 is accumulated during hypoxia without any subsequent oxygenation and DNA damage; in addition to hypoxia, p53 elevation was also observed in heat-shocked cells. Recently Sugano and co-workers[14] found that various cellular stresses which evoke the heat shock response (heat shock, heavy metals, amino acid analogs, H_2O_2) also caused the nuclear accumulation of p53 in human fibroblasts. More importantly, such an effect is detected after an exposure of cells to a respiratory inhibitor, NaN_3.[14]

On the other hand, it is well-known that various proteotoxic exposures are powerful inducers of apoptosis. As early as 1990, Lennon and co-workers[15] found in HL-60 cells that hyperthermia, ethanol or H_2O_2 evokes apoptosis at low dosage and necrosis at higher dosages. Later hyperthermia- or oxidant-induced apoptosis was observed on a number of other cell lines (see ref. 16-22). In our studies on EAC cells, we also observed that besides ATP depletion, hyperthermia and H_2O_2 also greatly activate DNA fragmentation.[23] Thus, protein-damaging exposures can principally provoke three diverse cellular responses: activation of stress genes, apoptosis or necrosis; apparently, these responses are determined mainly by the intensity of the stress. It is possible that diverse proteotoxic stresses, including ATP depletion, turn on quite different apoptotic pathways, but it seems more likely that some common mechanism exists for all these stresses. Below we consider a hypothetical scheme based on the apoptosis-eliciting properties of p53, which was shown

to be directly involved in hypoxia-induced apoptosis (see above).

As we already noted, an increase in p53 level is mainly determined by its increased stability (i.e., delayed degradation). Under normal conditions, p53 rapidly degrades through the ubiquitin-proteasome system,[24,25] but the appearance of strand breaks in DNA increases p53-DNA binding and makes p53 undigestable by this system.[26] Comparing p53 accumulation in tumor cells under genotoxic stress (gamma-irradiation) and hypoxia or hyperthermia, Graeber et al[5] concluded that the mechanism for hypoxic and heat induction of p53 is different from that for DNA damage and consists in resistance of p53 to the ubiquitin degradation machinery and/or inhibition of this machinery involved in "normal" turnover of p53. The conclusion is based on results with cells overexpressing human papillomavirus E6 gene, which show increased degradation of p53 and a failure to accumulate p53 in response to DNA-damaging agents; however, the cells do accumulate p53 following heating or hypoxia. One can easily imagine that ATP depletion, heat shock and other proteotoxic stresses either directly unfold p53 to make it resistant to proteolysis or, more likely, redirect ubiquitination and proteolysis to new targets, namely to the abundant unfolded proteins in the cytosol of stressed cells (see chapter 3). Following heat shock, Parag et al[27] observed in hepatoma cells a burst of ubiquitin conjugation and degradation of the long-lived proteins accompanied by a fall in free ubiquitin and ubiquitinated histone 2A, a "normal" target of ubiquitination. Carlson et al[28] found in HeLa cells about a two-fold decrease in the pool of free ubiquitin and histone-ubiquitin conjugates and an increase in high molecular weight (HMW) conjugates following even mild heat shock (45°C, 5 minutes). Likewise, after transient ischemia, a decrease in the free ubiquitin pool was demonstrated in neurons, and stimulation of HMW protein ubiquitination was detected in the brain and the heart (see chapter 3, sections 4.2, 6.1). Thus, ubiquitination of ischemia-damaged proteins may dramatically alleviate the ubiquitination and degradation of p53, thus augmenting its intracellular content.

As a consequence, p53-mediated apoptosis is triggered (see Fig. 8.1).

Although energy deprivation may increase p53 stability by directly turning off

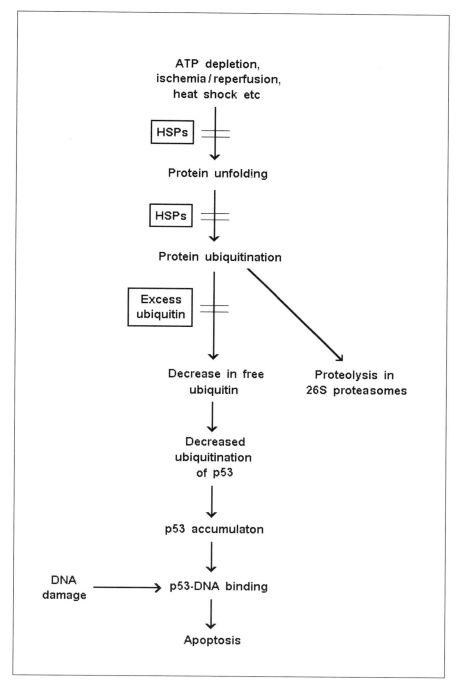

Fig. 8.1. Triggering of ischemic apoptosis through p53 pathway and protective function of HSPs (see text for explanation).

the ubiquitin-proteasome pathway, since this proteolytic system requires ATP (see ref. 29 for review), we do not think this is the case. First of all, apoptosis proceeds only when ample ATP is available, otherwise cells rapidly die via necrosis.[30,31] Thus, either chronic mild energy deficiency or transient acute ATP depletion are necessary for apoptosis induction (see sections 6.1, 6.4). Furthermore, the observed stimulation of ubiquitination in post-ischemic tissues also indicates that this system is activated rather than inhibited. However, it remains to be assessed whether a similar effect takes place under chronic mild energy deficiency.

What can we expect to occur in the HSP-overexpressing cells? In such cells, proteotoxic exposures (ATP-depletion, heat shock, etc.) will lead to decreased intracellular protein damage owing to the action of excess chaperones (see chapters 1, 6, 7 for discussion) and thereby, to reduced ubiquitination and proteolysis of these proteins. Because of this, degradation of p53 under stressful conditions in HSP-overexpressing cells will be increased compared to the cells with normal HSP content; in turn a decreased p53 level will lead to the attenuation of apoptosis in such cells (Fig. 8.1). Thus, both ubiquitin and HSPs, in particular HSP70, may confer tolerance to p53-mediated apoptosis after proteotoxic exposures.

As we described above, heat-induced p53 accumulation is a well-known phenomenon (see also ref. 32); activation of p53-mediated transcription of p21 was also demonstrated in heat-treated glioblastoma cells.[33] On the other hand, there are several studies where HSP-overexpressing cells exhibit reduced apoptosis under heat shock treatment. Initially this phenomenon was described in 1992 by Mosser and Martin[34] for a human leukemia cell line. They observed that in the thermotolerant cells with elevated HSP70 content the percentage of apoptotic cells and DNA fragmentation following hyperthermia was decreased more than five-fold compared to nontolerant cells. Later, Martin and co-workers[35] observed that

the increased sensitivity of some thymocyte subpopulations to heat shock-induced apoptosis is associated with their restricted capacity to synthesize HSPs. In our studies on rat thymocytes, we also failed to observe any appreciable HSP accumulation following the usual preconditioning heat shock treatment (42°C, 30 minutes, or 44°C, 10 minutes) and this was accompanied by rapid thymocyte apoptosis (Gabai et al, unpublished data). Such impaired stress response due to increased proteolysis of heat shock factor was recently described in lymphoma cells by Blake et al.[36] In contrast, EAC cells rapidly accumulate HSP68 and become resistant to both ischemia and hyperthermia-induced apoptosis (see section 6.4 and ref. 37). Furthermore, recently Li and co-workers[38] using rat fibroblasts transfected with the human HSP70 gene directly demonstrated the protective effect of HSP70 against hyperthermia-induced apoptosis.

Unfortunately, in most of the above works the mechanism by which hyperthermia induces apoptosis has not been evaluated. Besides p53 expression, heat shock induces *c-fos* and *c-jun* genes[21,39] and these genes are also activated following ischemia-reperfusion (see section 4.5). Similarly to p53, c-fos protein is rapidly degraded in vivo and can elicit apoptosis when overexpressed.[40] Interestingly, c-fos protein breakdown is also mediated by the ubiquitin-proteasome system,[29,41] and its accumulation after proteotoxic exposures may be governed by a similar mechanism to p53.

Besides heat shock and ischemia, an analogous mechanism of apoptosis triggering may exist for other proteotoxic exposures such as oxidative stress, heavy metals, amino acid analogs, etc. Recently, a rather unexpected finding was made by Tishler et al. Microtubule-active agents, taxol (an inhibitor of microtubule depolymerization), and vinblastine or nocodazole (inhibitors of polymerization) stimulated p53 binding to DNA in a dose-dependent manner and activated transcription of p21.[42] Accumulation of c-fos was also detected in cerebella granule cells treated with colchicine, another microtubule-disrupting agent.[43] Microtubule-active drugs (colcemide, colchicine, taxol, vinblastine), on the one hand, are able to induce apoptosis (see for instance, refs. 43-45); on the other hand, energy deprivation can affect microtubules (see chapter 3 and section 6.1), and a destructive effect such as ATP depletion in the tubulin skeleton may be conducive to apoptosis. In this case, the protective effect of chaperones can be expected as well. Indeed, in EAC cells we observed mitigation of vinblastin-induced apoptosis in thermotolerant cells with elevated HSP68 content;[37] likewise, Lee et al[46] found that the induction of HSP70 in brain tumor cells was associated with a resistance to vincristine (a vinblastine analog), although in the latter case cell survival was determined by colony-forming ability. We suggest the possibility that HSP70 directly maintains microtubules, since in thermotolerant cells both hyperthermia- and vinblastine-induced disruption of microtubules were alleviated;[46,47] moreover, Lee et al demonstrated co-precipitation of HSP70 and tubulin.[46]

Two important works have recently been published by Uchida's lab. In the first study Wei and co-workers[48] observed that treating various tumor cell lines with quercetin, which inhibits HSP70 synthesis and diminishes its content, causes apoptosis of the cells. More importantly, in the second study these researchers found that treating MOLT-4 tumor cell culture with HSP70 anti-sense oligonucleotides not only enhances the hyperthermia-induced apoptosis about three-fold, but it evokes by itself almost the same apoptotic reaction as hyperthermia alone.[49] The data suggest, in our opinion, that the accumulation of unfolded proteins (either spontaneous or heat-induced) in the cells with abrogated HSP70 expression may trigger apoptosis.

Excitingly, the deterioration of protein folding not only in the cytosol but also within other cellular organelles, in particular, the endoplasmic reticulum (ER), can also induce apoptosis. Perezsala and Mollinedo[50] recently discovered that specific inhibition of protein glycosylation by tunicamycin induces apoptosis in HL-60 cells; the ability of tunicamycin to induce DNA fragmentation in several other myeloid cell lines correlates with its potency as a glycosylation inhibitor. Hypoxia can also suppress protein glycosylation within the ER (see sections 4.1, 4.5), thus activating apoptosis similarly to tunicamycin or Ca^{2+}-ionophores. These compounds as well as hypoxia can also elicit the activation of *GRP* genes, and it is tempting to speculate that GRPs, similar to HSPs, can save cells from apoptosis provoked by hypoxia.

One may ask why mammalian cells respond to unfolding (damage) of intracellular proteins by activation of a suicide mechanism. We suppose that such apoptosis, being fatal for a cell, is beneficial for a whole organism. Indeed, as we considered in section 2.2, apoptosis, in contrast to necrosis, is a harmless device to eliminate unnecessary cells during embryogenesis, tissue homeostasis, etc. (refs. 1-3). In addition, it can also kill cells which have received irreversible injuries to DNA after environmental stresses (see above). When a cell encounters proteotoxic stress, it first tries to repair protein damage and activates stress genes to combat this damage. However, if protein damage is severe, the chaperone machinery (both the constitutive and inducible) will not be able completely repair it and this, ultimately, may lead to necrotic death. To prevent necrosis which is accompanied by release of intracellular constituents and harmful inflammation, such irreversibly damaged cells "decide" not to wait until necrosis develops and dies via apoptosis. However, when the same proteotoxic stress is encountered by a cell with elevated HSP content, it will overcome this stress and save its own life. Such "altruistic" cell behavior apparently intrinsic to most normal cells may be impaired in various tumor cells (i.e., possessing mutant p53 or lacking p53 at all; see also ref. 6).

Certainly, the above described model is quite hypothetical. There is a paradox, however, in why such great numbers of various cytotoxic agents (at the present time more than 100, ref. 51) elicit such stereotypic cell reactions as apoptosis. Vaux and Hacker[51] recently suggested that apoptosis is used as an organism's anti-viral strategy: cells interpret any metabolic disturbance as evidence of infection by a virus and thereby kill themselves in response to these toxins before they are killed by the toxin itself. Whatever the model, it is necessary to estimate in future research how a cell senses the level of intracellular damage and decides "to be or not to be."

8.2. HSPs AND TRANSDUCTION OF THE APOPTOTIC SIGNAL

Apoptotic signal transduction is a highly obscure process which may include protein kinase/phosphatase cascades, second messengers such as cAMP, Ca^{2+}, ceramide, phosholipases, oxygen radicals, etc. (see refs. 2, 3 for review). Although little is known about signal transduction during ischemia-induced apoptosis, some data for other apoptotic inducers allow us to propose a few possible mechanisms.

First, we must emphasize that only certain types of apoptotic inducers act via the p53-dependent pathway since p53 knockout mice, being resistant to DNA-damaging agents, are as sensitive as their littermates to other apoptotic stimuli (see ref. 11 for review). Thereby, it will be necessary to assess whether a p53-independent mechanism also exists in ischemia-induced apoptosis.

Accumulation of p53 during ischemia may activate the transcription of some genes and repress the transcription of other genes. Several of the activated genes such as *p21* and *gadd* are implicated in an adaptive response (see above) and irrelevant to apoptosis. Accordingly, Brugarolas et al failed to observe the abrogation of p53-dependent apoptosis following the irradiation of p21 (-/-) cells. Conversely, the expression of *bax*, an antagonist of antiapoptotic bcl-2 (see below), may promote apoptosis.[52] In fact, this has been already described in post-ischemic neurons (see section 6.1).

Another protein involved in the induction of p53-dependent apoptosis is APO-1/FAS receptor,[53] whose increased transcription was detected in hypoxic cardiomyocytes.[54] In some models of apoptosis, however, apoptosis is promoted by the repression rather than the activation of p53-mediated transcription,[55-57] and the inhibitors of macromolecular synthesis do not block (but rather activate) such apoptosis. Indeed, in post-ischemic neurons these inhibitors suppress apoptosis (section 6.1) while in cardiomyocytes they activate it.[58] A possible target of p53 repression may be the Bcl-2 protein,[52] but its expression in post-ischemic cells remains to be assessed. In our study on EAC cells, we detected neither an effect of cycloheximide on apoptosis evoked by energy deprivation,[37] nor a cycloheximide effect on hyperthermia-induced apoptosis in EAC or thymocytes.[59] These data and the data from Raff's lab confirm his suggestion that all mammalian cells constitutively express all of the proteins of the apoptotic machinery though its activation often depends on new RNA and protein synthesis.[60,61] The most impressive finding of this lab is the demonstration of apoptosis in the cytoplasts, i.e., completely enucleated cells.[62]

As Raff proposed, most mammalian cells excluding blastomers are programmed to kill themselves, but the death program is normally suppressed by external survival signals, such as growth factors and hormones.[60-63] Further advancing this idea, one can imagine that an interruption of the survival signal transduction by some toxic factors may turn on an apoptotic cascade.[59] Surely, all proteotoxic exposures including ischemia harshly damage various proteins involved in signal transduction, such as receptors, receptor kinases/phosphatases, etc. One may expect that an increase in exogenous concentration of growth factors may partially protect cells from apoptosis through the stimulation of the "survival" cascade. Indeed, insulin-like growth factor I (IGF-1) suppresses cardiomyocyte apoptosis following ischemia-reperfusion;[58,64] likewise, the basic fibroblast growth factor protects cortical neurons from prolonged hypoxia.[65] We suggest that by protecting proteins involved in signal transduction from inactivation (see section 7.2.4), HSPs can also block apoptotic cell death. Previously, Rutherford and Zuker discussed the participation of chaperones in signaling pathways (see ref. 66 for review). In particular, chaperones are needed for the normal operation of some receptors, receptor-kinases, transcriptional factors, etc. It is quite possible that an excess of HSPs may protect these proteins from inactivation during ischemic insult.

On the other hand, clustering (or aggregation) of some receptors such as APO-1/FAS or TNF (tumor necrosis factor) receptors may also trigger apoptosis.[2,3] Therefore, the prevention of such aggregation under proteotoxic exposures (e.g., ischemia) may also protect cells from activation of the death cascade rather than from inactivation of survival signal transduction.

Whatever the initial phases of apoptosis, the important upstream modulators are proteins from the bcl-2 family; some of the proteins promote apoptosis such as bax, while others block it such as bcl-2 and bcl-x. Bcl-2 and bcl-x protect cells from both p53-dependent and p53-independent pathways of apoptosis[67-70] and recently two research groups established the protective action of bcl-2 and bcl-x against hypoxia-induced apoptosis of tumor cells[6,71-73] similar to that of neurons (see section 6.1). Two main activities were initially established for bcl-2 protein, namely anti-oxidating and modulating intracellular Ca^{2+}-fluxes.[67] Since bcl-2 has protective properties even in completely anaerobic conditions,[74] the former activity is apparently dispensable for its guarding capacity. The data obtained show that bcl-2 prevents Ca^{2+}-efflux from the ER,[75-77] though whether this is the case under hypoxia-induced apoptosis remains to be established.

Interestingly, bcl-2 protects cells not only from ischemic apoptosis but also rescues them from many other proteotoxic exposures such as ethanol, oxidative stress (H_2O_2, menadione), TNF (tumor necrosis

factor) and hyperthermia (see refs. 67, 75, 78-80), i.e., from the same stimuli where the protective effect of chaperones was established in models of necrotic and reproductive death (see chapter 1, 7). Surprisingly, bcl-2, similar to HSPs, delays not only the apoptosis but the necrosis following hypoxia or azide treatment.[71-73,81]

Strasser and Anderson[79] studied the effect of simultaneous expression of HSPs and bcl-2 on heat-induced apoptosis of lymphoid cells. They found that bcl-2 co-operates with HSPs in the survival of heated cells (assessed by apoptotic morphology and clonogenicity), although it did not modulate the expression of HSPs nor is bcl-2 itself a stress-inducible protein. These results suggest that bcl-2 and HSPs act via two separate pathways of cell rescue from apoptosis. Considering ischemia-induced apoptosis in neurons (see section 6.1), we also suggest that bcl-2 cooperates with HSP in promoting their survival since bcl-2 is localized to intracellular membranes while HSPs are mainly cytosolic proteins.

At the same time, Jaattela and co-workers[82] recently established that the protection of breast carcinoma cells from TNF-provoked apoptosis by bcl-2 and bcl-x was associated with the inhibition of phospholipase A_2 (PLA_2) activity; the effect was also observed in HSP70-overexpressing cells.[83] PLA_2, being involved in apoptosis,[84] is often activated during necrosis as well (see chapters 6, 7). The enzyme is localized to the membrane and is apparently modulated by some cytosolic proteins and may be one of possible targets of HSP70 in apoptosis (see also section 7.2.4).

Recent works from Kroemer's lab testify to the impairment of mitochondrial functions as a cardinal event of programmed cell death, the main target of bcl-2 being these organelles (see section 2.4 and ref. 85 for review). Accordingly, apoptosis in anucleated cells can be prevented by bcl-2.[74] Recently Vayssiere et al[86] established that p53 overexpression in fibroblasts causes severe mitochondrial damage, possibly through the expression of bax, as was demonstrated in neurons (see section 6.1). As we have already considered in section 7.1, the HSP70-mediated protection of cells from ischemia-reperfusion is apparently associated with the preservation of mitochondrial functions, although it is not clear how the chaperone exerts this role. It seems likely that by maintaining cytoskeleton structures (see chapter 7), HSP70 may indirectly rescue mitochondria since we often observe the association of these organelles with the cytoskeleton, particularly with intermediate filaments and microtubules, which may be of functional importance (see ref. 87 for review). Thus, since ischemic and post-ischemic damage of mitochondria is a well-established phenomenon (see section 2.4, 6.1), such wreckage may promote apoptosis and by impeding it, HSPs may rescue the whole cell (Fig. 8.2).

8.3. HSPs AND THE EXECUTION OF APOPTOSIS

Now it is well-established that the execution phase of apoptosis under almost all studied apoptotic stimuli is governed by specific proteases referred to as the ICE (interleukin-1β converting enzyme) family of cysteine proteases, the common feature among known ICE-family members being cleaving at aspartate residues.[88,89] In the apoptotic cascade, ICE proteases are located downstream of bcl-2 and bcl-x. The upstream proteins can block ICE protease activity although no direct physical interaction between proteases and bcl-2 is detected.[90-92] Shimuzu et al[72,73] recently ascertained that ICE proteases are encompassed in hypoxia-induced apoptosis of tumor cells. The researchers found activation of such proteases under hypoxia, and specific inhibitors of the proteases as well as bcl-2/bcl-x prevented cell death. Likewise, Gottlieb and co-workers demonstrated the prevention of apoptosis by ICE-protease blockers in cardiomyocytes following metabolic inhibition.[93]

Proteolysis is apparently the first irreversible phase of apoptosis, and the addition of ICE protease to isolated nuclei fully simulates apoptotic changes within them, namely chromatin condensation, DNA fragmentation, etc. Among the established substrates of the "death" protease in the nucleus are poly (ADP-ribose) polymerase, topoisomerase I, lamin B1, 70 kDa U1-RNP and histone H1.[88,89] Since cytoplasmic apoptotic changes can be demonstrated in enucleated cells, ICE proteases are most likely localized to the cytosol[89] Although it is presently unknown how ICE proteases are activated, it is tempting to speculate that HSP70 may somehow block these enzymes or mask their targets (Fig. 8.2, see also section 7.2).

One of the drastic extranuclear changes of apoptotic cells is the collapse of the cytoskeleton accompanied by a loss of the cell-substrate contact and membrane blebbing. These changes mainly ensued from proteolytic cleavage of actin[94,95] and other constituents of the actin skeleton, namely, Gas2 protein[96] and fodrin;[97] in the latter case, calpain protease is activated by the digestion of a calpain inhibitor, calpastatin.[88,98] Whether calpain activation is the obligatory step in ischemia-induced apoptosis remains to be established, but Umansky et al[57] detected the suppression of cardiomyocyte apoptosis following ischemia-reperfusion by calpain inhibitors.

As we have mentioned elsewhere (chapter 3, sections 6.4, 7.1-7.3), blebbing is also a prerequisite of necrotic death following ATP depletion, and in tolerant cells with an elevated HSP content we observed marked suppression of blebbing (section 6.4). Although bleb formation during apoptosis is associated with the disruption of the cortical actin-plasma membrane interaction,[99] like that observed during necrosis (see, for instance, ref. 100), some differences between these two processes exists. During apoptosis, the plasma membrane remains impermeable to dyes such as Trypan blue, propidium iodide, etc. and cells shrink; whereas during necrosis, cells become swollen and the membrane is rapidly disrupted (see section 2.2). It is possible that these distinctions may result from the intensity of the cytoskeleton cleavage rather than diverse enzymes or targets involved, since both ICE proteases as well as calpain are involved in necrotic death (see refs. 72, 73 and chapter 6). On the other hand, the apoptotic death of various cells in vitro

(i.e., in the absence of surrounding digesting cells) culminates in necrosis. In any case, HSP-mediated preservation of the cytoskeleton during necrosis may also be important for their protective role in apoptosis.

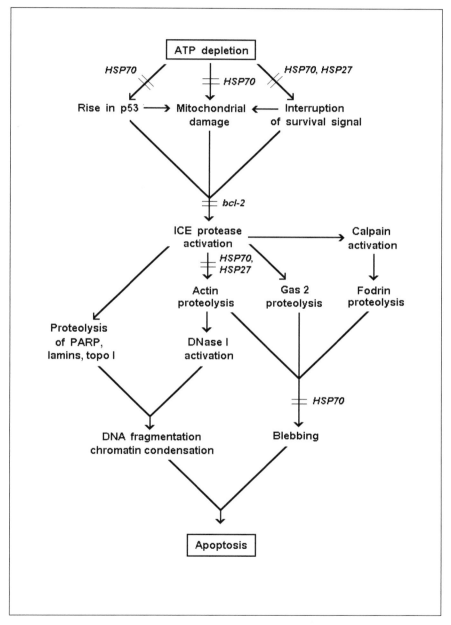

Fig. 8.2. Possible targets of HSPs in apoptotic signal transduction and execution following ATP depletion. PARP, poly (ADP-ribose) polymerase; topo I, topoisomerase I. See text for further explanation.

The question arises as to whether cleavage of cytoskeletal proteins has any relation to apoptotic nuclear changes, in particular DNA fragmentation. An interesting hypothesis was put forward by Ucker and co-workers[101] and further supported by studies of Peitsch and collegues.[102,103] They suggested that the endonuclease involved in DNA fragmentation is DNase I whose characteristic feature is its powerful inhibition by G-actin.[101-103] Such an unusual feature of DNase I is of great physiological importance. G-actin is a natural modulator of this enzyme and respectively, DNA fragmentation. Therefore, one may expect that a decrease in G-actin content (e.g., its polymerization, aggregation or cleavage) would cause an elevation of DNase I activity and DNA fragmentation. Recently, Kayalar et al[104] observed in PC12 tumor cells that apoptosis-induced cleavage of actin by ICE-protease reversed DNase I inhibition; furthermore, such digested actin was unable to polymerize, which may account for the morphology of apoptotic cells (in particular, blebbing). Similar cleavage of actin into 30 kDa and 15 kDa fragments was found during apoptosis of U937 human leukemia cells.[94] In addition, the treatment of cells with cytochalasins, well-known inhibitors of actin polymerization, induces apoptosis in various cells;[104,105] this effect is not strange since cytochalasins interfere with G-actin binding to DNase I.[105] One may imagine that HSP70, interacting with G-actin (see section 7.4), may protect it from ICE protease attack but does not block its DNase I inhibitory activity (Fig. 8.2). Likewise, HSP27, protecting cells from cytochalasin-induced fragmentation of actin filaments (see subsection 7.3.5) may also block apoptosis (Fig. 8.2).

Besides G-actin digestion, uncontrolled G-actin polymerization and/or aggregation may have a similar effect on DNA fragmentation. In 1994, we proposed that a decrease in the G-actin pool in ATP-depleted cells may activate DNase I and apoptosis.[23] Recently, we observed a decrease in G-actin (assayed by DNase I binding) in uncoupler-treated EAC cells (Gabai, Kabakov, unpublished data). Since earlier prevention of actin aggregation in ATP-depleted cells with elevated HSP70 content was established (see section 6.4), such a protective mechanism of HSP70 may also be important for preventing DNA fragmentation during apoptosis (Fig. 8.2).

Discussing the possible protective functions of HSP in the execution phases of apoptosis, it may be expected that HSPs exert their properties not only under proteotoxic exposures such as ATP depletion, heat shock and oxidative stress but also during a number of other pathological and physiological apoptotic stimuli. To date, there are almost no works on this issue, though recently Samali and Cotter[106] observed the suppression of apoptosis induced by actinomycin D (an inhibitor of transcription), camptotecin and etoposide (both topoisomerase II inhibitors) in U937 and Wehi-s cells with elevated HSP content after heat shock treatment; moreover, transfection of Wehi-s cells with HSP70 or HSP27 confer tolerance to these exposures.[106] Of interest is that in the same U937 cells, topoisomerase II inhibitors induce G-actin cleavage[94] and a decrease in F-actin content;[107] therefore, it is quite possible that the cleavage of G-actin not only blocks its polymerization (see above) but F-actin depolymerization and unfolding in stressed cells makes actin digestible by ICE protease. It is tempting to speculate that the cell which will express mutant forms of actin resistant to ICE protease should be also resistant to apoptosis, at least under proteotoxic exposures.

It is important to emphasize that DNase I is not the only nuclease which performs endonucleotic digestion of DNA; among such nucleases, there are at least four other enzymes including DNase II, NUC-18, nuc-1 and Ca^{2+}-independent nuclease (see ref. 103 for review). At the present time, little is known about the role of the nucleases in various models of apoptosis and how these nucleases are activated. However, Gootlieb et al observed the attenuation of acidification and apoptotic DNA fragmentation in preconditioned rabbit cardiomyocytes; the latter may be associated with the suppression of DNase II activity. This nuclease has an acidic pH optimum and may be activated in ATP-depleted cells (see section 2.3). Interestingly, in the transgenic murine heart overexpressing HSP70, ischemic acidification recovered faster (see chapter 5 and ref. 108).

In the future, we greatly hope that rapid progress in apoptosis research will unravel the mechanisms of chaperone-mediated protection from ischemic insult more precisely.

RECENT NEWS

Recently Shinohara and co-workers[109] have demonstrated induction of apoptosis in MOLT-4 leukemia cells by proteasome inhibitors, and this effect was associated with p53 accumulation (see Fig. 8.1).

In infarcted rat myocardium, myocyte apoptosis was accompanied by increase in bax expression and decrease in Bcl-2 expression.[110] In contrast, salvaged myocytes in human myocardium after infarction demonsrated induction of Bcl-2.[111] Thus, ischemia-induced myocyte apoptosis closely resembles that of neurons (see section 6.1). Regarding protective function of Bcl-2 in apoptosis, Petit et al[112] show that this protein blocks mitochondrial permeability transition (see section 2.4) thus preventing release of ICE-like protease from mitochondria.[113]

Similar to hypoxic myocytes, 2-4-fold increase in Apo-1/FAS mRNA accumulation was observed in cell cultures treated with mitochondrial inhibitors.[114] Of great importance is the study of Mehlen et al[115] who directly demontrated protective function of HSP27 and αB-crystallin against Apo-1/FAS-induced apoptosis, although the mechanism of this phenomenon remains to be evaluated.

REFERENCES

1. Martin SJ. Apoptosis: suicide, execution or murder? Trends Cell Biology 1993; 3: 141-144.

2. Martin SJ, Green DR, Cotter TG. Dicing with death: dissecting the components of the apoptosis machinery. Trends Biochem Sci 1994; 19: 26-30.

3. McConkey DJ, Orrenius S. Signal transduction pathways to apoptosis. Trends Cell Biol 1994; 4: 370-375.

4. Price BD, Calderwood SK. Gadd45 and Gadd 153 messenger RNA levels are increased during hypoxia and after exposure of cells to agents which elevate the levels of the glucose-regulated proteins. Cancer Res 1992; 52: 3814-3817.

5. Graeber TG, Peterson JF, Tsai M et al. Hypoxia induces accumulation of p53 protein, but activation of a G_1-phase checkpoint by low-oxygen conditions is independent of p53 status. Mol Cell Biol 1994; 14: 6264-6277.

6. Graeber TG, Osmanian C, Jacks T et al. Hypoxia-mediated selection of cells with diminished apoptotic potential in solid tumours. Nature 1996; 379: 88-91.

7. Cummings MC. Increased p53 mRNA expression in liver and kidney apoptosis. Biochim Biophys Acta 1996; 1315:100-104.

8. Didenko VV, Wang XD, Yang LQ et al. Expression of p21 (WAF1/CIP1/SDI1) and p53 in apoptotic cells in the adrenal cortex and induction by ischemia reperfusion injury. J Clin Invest 1996; 97: 1723-1731.

9. Kuerbitz SJ, Plunkett BS, Walsh WV et al. Wild-type p53 is a cell cycle checkpoint determinant following irradiation. Proc Natl Acad Sci USA 1992; 89: 7491-7495.

10. Brugarolas J, Chandrasekaran C, Gordon JI et al. Radiation-induced cell cycle arrest compromised by p21 deficiency. Nature 1995; 377: 552-557.

11. Hooper ML. The role of the p53 and Rb-1 genes in cancer, development and apoptosis. J Cell Sci 1994; 18: 13-17.

12. Donehower LA, Bradly A. The tumor suppressor p53. Biochim Biophys Acta 1993; 1155: 181-205.

13. Montenarh M. Biochemical properties of the growth suppressor/oncoprotein p53. Oncogene 1992; 7: 1673-1680.

14. Sugano T, Nitta M, Ohmori H et al. Nuclear accumulation of p53 in normal human fibroblasts is induced by various

cellular stresses which evoke the heat shock response, independently of the cell cycle. Jap J Cancer Res 1995; 86: 415-418.

15. Lennon SV, Martin SJ, Cotter TG. Induction of apoptosis (programmed cell death) in tumor cell lines by widely diverging stimuli. Biochem Soc Trans 1990; 18: 343-345.

16. Ghibelli L, Nosseri C, Oliverio S et al. Cycloheximide can rescue heat-shocked L cells from death by blocking stress-induced apoptosis. Exp Cell Res 1992; 201: 436-443.

17. Migliorati G, Nicoletti I, Crocicchio F et al. Heat shock induces apoptosis in mouse thymocytes and protects them from glucocorticoid-induced cell death. Cell Immunol 1992; 143: 348-356.

18. Lee YJ, Kim JH, Ryu S et al. Mechanisms of mild hyperthermia-induced cytotoxicity in human prostatic carcinoma cells: Perturbation of cell cycle progression and DNA fragmentation. J Therm Biol 1994; 19: 305-313.

19. Dypbukt JM, Ankarcrona M, Burkitt M et al. Different prooxidant levels stimulate growth, trigger apoptosis, or produce necrosis of insulin-secreting RINm5F cells. J Biol Chem 1994; 269: 30553-30560.

20. Sakaguchi Y, Stephens LC, Makino M et al. Apoptosis in tumor and normal tissues induced by whole body hyperthermia in rats. Cancer Res 1995; 55: 5459-5464.

21. Cummings M. Increased c-fos expression associated with hyperthermia-induced apoptosis of a Burkitt lymphoma cell line. Int J Radiat Biol 1995; 68: 687-692.

22. Troiano L, Monti D, Cossarizza A et al. Involvement of CD45 in dexamethasone- and heat shock-induced apoptosis of rat thymocytes. Biochem Biophys Res Communic 1995; 214: 941-948.

23. Gabai VL, Kabakov AE, Makarova YuM et al. DNA fragmentation in Ehrlich ascites carcinoma under exposures causing cytoskeletal protein aggregation. Biochemistry (Moscow) 1994; 59: 399-404.

24. Chowdary DR, Dermody JJ, Jha KK et al. Accumulation of p53 in a mutant cell line defective in the ubiquitin pathway. Mol Cell Biol 1994; 14: 1997-2003.

25. Maki CG, Huibregtse JM, Howley PM. In vivo ubiquitination and proteasome-mediated degradation of p53. Cancer Res 1996; 56: 2649-2654.

26. Molinari M, Milner J. p53 in complex with DNA is resistant to ubiquitin-dependent proteolysis in the presence of HPV-16 E6. Oncogene 1995; 10: 1849-1854.

27. Parag HA, Raboy B, Kulka RG. Effect of heat shock on protein degradation in mammalian cells: involvement of the ubiquitin system. EMBO J 1987; 6: 55-61.

28. Carlson N, Rogers S, Rechsteiner M. Microinjection of ubiquitin: changes in protein degradation in HeLa cells subjected to heat-shock. J Cell Biol 1987; 104: 547-555.

29. Ciechanover A. The ubiquitin-proteasome proteolytic pathway. Cell 1994; 79: 13-21.

30. Chou CC, Lam CY, Yung BYM. Intracellular ATP is required for actinomycin D-induced apoptotic cell death in HeLa cells. Cancer Lett 1995; 96: 181-187.

31. Richter C, Schweizer M, Cossarizza A et al. Control of apoptosis by the cellular ATP level. FEBS Lett 1996; 378: 107-110.

32. Matsumoto H, Shimura M, Omatsu T et al. p53 proteins accumulated by heat stress associate with heat shock proteins HSP72/HSC73 in human glioblastoma cell lines. Cancer Lett 1994; 87: 39-46.

33. Ohnishi T, Wang XJ, Ohnishi K et al. p53-dependent induction of WAF1 by heat treatment in human glioblastoma cells. J Biol Chem 1996; 271: 14510-14513.

34. Mosser DD, Martin LH. Induced thermotolerance to apoptosis in a human T lymphocyte cell line. J Cell Physiol 1992; 151: 561-570.

35. Mosser DD, Duchaine J, Bourget L et al. Changes in heat shock protein synthesis and heat sensitivity during mouse thymocyte development. Developmental Genetics 1993; 14: 148-158.

36. Blake MJ, Buckley AR, Zhang MY et al. A novel heat shock response in prolactin-dependent Nb2 node lymphoma cells. J Biol Chem 1995; 270: 29614-29620.

37. Gabai VL, Zamulaeva IV, Mosin AF et al. Resistance of Ehrlich tumor cells to apoptosis can be due to accumulation of heat shock proteins. FEBS Lett 1995; 375: 21-26.

38. Li WX, Chen CH, Ling CC et al. Apoptosis in heat-induced cell killing: The protective

role of hsp-70 and the sensitization effect of the c-myc gene. Radiat Res 1996; 145: 324-330.

39. Bukh A, Martinez-Valdez H, Freedman SJ et al. The expression of c-fos, c-jun, and c-myc genes is regulated by heat shock in human lymphoid cells. J Immunol 1990; 144: 4835-4840.

40. Smeyne RJ, Vendrell M, Hayward M et al. Continuous c-fos expression precedes programmed cell death in vivo. Nature 1993; 363: 166-169.

41. Tsurumi C, Ishida N, Tamura T et al. Degradation of c-fos by the 26S proteasome is accelerated by c-jun and multiple protein kinases. Mol Cell Biol 1995; 15: 5682-5687.

42. Tishler RB, Lamppu DM, Park S et al. Microtubule-active drugs taxol, vinblastine, and nocodazole increase the levels of transcriptionally active p53. Cancer Res 1995; 55: 6021-6025.

43. Bonfoco E, Ceccatelli S, Manzo L et al. Colchicine induces apoptosis in cerebellar granule cells. Exp Cell Res 1995; 218: 189-200.

44. Sherwood SW, Sheridan JP, Schimke RT. Induction of apoptosis by the anti-tubulin drug colcemid: Relationship of mitotic checkpoint control to the induction of apoptosis in HeLa S3 cells. Exp Cell Res 1994; 215: 373-379.

45. Ireland CM, Pittman SM. Tubulin alterations in taxol-induced apoptosis parallel those observed with other drugs. Biochem Pharmacol 1995; 49: 1491-1499.

46. Lee WC, Lin KY, Chen KD et al. Induction of HSP70 is associated with vincristine resistance in heat-shocked 9L rat brain tumor cells. Br J Cancer 1992; 66: 653-659.

47. Wiegant FAC, van Bergen en Henegouwen PMP, van Dongen G al. Stress-induced thermotolerance of the cytoskeleton of mouse neuroblastoma N2A cells and rat reuber H35 hepatoma cells. Cancer Res 1987; 47: 1674-1680.

48. Wei Y, Zhao X, Kariya Y et al. Induction of apoptosis by quercetin: Involvement of heat shock protein. Cancer Res 1994; 54: 4952-4957.

49. Wei Y, Zhao X, Kariya Y et al. Inhibition of proliferation and induction of apoptosis by abrogation of heat-shock protein (HSP)

70 expression in tumor cells. Cancer Immunol Immunother 1995; 40: 73-78.

50. Perezsala D, Mollinedo F. Inhibition of N-linked glycosylation induces early apoptosis in human promyelocytic HL-60 cells. J Cell Physiol 1995; 163: 523-531.

51. Vaux DL, Hacker G. Hypothesis: Apoptosis caused by cytotoxins represents a defensive response that evolved to combat intracellular pathogens. Clin Exp Pharm Physiol 1995; 22: 861-863.

52. Zhan O, Fan S, Bae I et al. Induction of bax by genotoxic stress in human cells correlates with normal p53 status and apoptosis. Oncogene 1994; 9: 3743-3751.

53. Owen-Schaub LB, Zhang W, Cusack JC et al. Wild-type human p53 and a temperature-sensitive mutant induce Fas/APO1 expression. Mol Cell Biol 1995; 15: 3032-3040.

54. Tanaka M, Ito H, Adachi S et al. Hypoxia induces apoptosis with enhanced expression of Fas antigen messenger RNA in cultured neonatal rat cardiomyocytes. Circ Res 1994; 75: 426-433.

55. Caelles C, Helmberg A, Karin M. p53-dependent apoptosis in the absense of transcriptional activation of p53-target genes. Nature 1994; 370: 220-223.

56. Rowan S, Ludwig RL, Haupt Y et al. Specific loss of apoptotic but not cell-cycle arrest function in a human tumor derived p53 mutant. EMBO J 1996; 15: 827-838.

57. Gottlieb E, Lindner S, Oren M. Relationship of sequence-specific transactivation and p53-regulated apoptosis in interleukin 3-dependent hematopoietic cells. Cell Growth Diff 1996; 7: 301-310.

58. Umansky SR, Cuenco GM, Khutzian SS et al. Post-ischemic apoptotic death of rat neonatal cardiomyocytes. Cell Death Diff 1995; 2: 235-241.

59. Gabai VL, Mosina VA, Makarova, YuM et al. Serum withdrawal induces thymocytes apoptosis not requiring protein synthesis or ATP generation. Biochemistry (Moscow) 1995; 60: 907-912.

60. Ishizaki Y, Cheng L, Mudge AW et al. Programmed cell death by default in embryonic cells, fibroblasts, and cancer cells. Mol Biol Cell 1995; 6: 1443-1458.

61. Weil M, Jacobson MD, Coles HSR et al. Constitutive expression of the machinery for programmed cell death. J Cell Biol 1996; 133: 1053-1059.

62. Jacobson MD, Burne JF, Raff MC. Programmed cell death and Bcl-2 protection in the absence of a nucleus. EMBO J 1994; 13: 1899-1910.

63. Raff MC. Social controls on cell survival and cell death. Nature 1992; 356: 397-400.

64. Buerke M, Murohara T, Skurk C et al. Cardioprotective effect of insulin-like growth factor I in myocardial ischemia followed by reperfusion. Proc Natl Acad Sci USA 1995; 92: 8031-8035.

65. Sakaki T, Yamada K, Otsuki H et al. Brief exposure to hypoxia induces bFGF mRNA and protein and protects rat cortical neurons from prolonged hypoxic stress. Neurosci Res 1995; 23: 289-296.

66. Rutherford SL, Zuker CS. Protein folding and the regulation of signaling pathways. Cell 1994; 79: 1129-1132.

67. Reed JC. Bcl-2 and the regulation of programmed cell death. J Cell Biol 1994; 124: 1-6.

68. Hockenbery DM. bcl-2 in cancer, development and apoptosis. J Cell Sci 1994; 18: 51-55.

69. Chiou S-K, Rao L, White E. Bcl-2 blocks p53-dependent apoptosis. Mol Cell Biol 1994; 14: 2556-2563.

70. Schott AF, Apel IJ, Nunez G et al. Bcl-x(L) protects cancer cells from p53-mediated apoptosis. Oncogene 1995; 11: 1389-1394.

71. Shimizu S, Eguchi Y, Kamiike W et al. Induction of apoptosis as well as necrosis by hypoxia and predominant prevention of apoptosis by Bcl-2 and Bcl-X(L). Cancer Res 1996; 56: 2161-2166.

72. Shimizu S, Eguchi Y, Kamiike W et al. Retardation of chemical hypoxia-induced necrotic cell death by Bcl-2 and ICE inhibitors: Possible involvement of common mediators in apoptotic and necrotic signal transductions. Oncogene 1996; 12: 2045-2050.

73. Shimizu S, Eguchi Y, Kamiike W et al. Bcl-2 expression prevents activation of the ICE protease cascade. Oncogene 1996; 12: 2251-2257.

74. Jacobson MD, Raff MC. Programmed cell death and Bcl-2 protection in very low oxygen. Nature 1995; 374: 814-816.

75. Zornig M, Busch G, Beneke R et al. Survival and death of prelymphomatous B-cells from N-myc/bcl-2 double transgenic mice correlates with the regulation of intracellular Ca^{2+} fluxes. Oncogene 1995; 11: 2165-2174.

76. Distelhorst CW, Lam M, Mccormick TS. Bcl-2 inhibits hydrogen peroxide-induced ER Ca^{2+} pool depletion. Oncogene 1996; 12: 2051-2055.

77. Marin MC, Fernandez A, Bick RJ et al. Apoptosis suppression by bcl-2 is correlated with the regulation of nuclear and cytosolic Ca^{2+}. Oncogene 1996; 12: 2259-2266.

78. Tsujimoto Y. Stress-resistance conferred by high level of bcl-2 alpha protein in human B lymphoblastoid cell. Oncogene 1989; 4: 1331-1336.

79. Strasser A, Anderson RL. Bcl-2 and thermotolerance cooperate in cell survival. Cell Growth Diff 1995; 6: 799-805.

80. Cuende E, Ales-Martinez JE, Ding L et al. Programmed cell death by bcl-2-dependent and independent mechanisms in B lymphoma cells. EMBO J 1993; 12: 1555-1560.

81. Vaux DL, Whitney D, Weissman IL. Activation of physiological cell death mechanisms by a necrosis-causing agent. Microscopy Res Technique 1996; 34: 259-266.

82. Jaattela M, Benedict M, Tewari M et al. Bcl-x and bcl-2 inhibit TNF and Fas-induced apoptosis and activation of phospholipase A_2 in breast carcinoma cells. Oncogene 1995; 10: 2297-2305.

83. Jaattela M. Overexpression of major heat shock protein hsp70 inhibits tumor necrosis factor-induced activation of phospholipase A_2. J Immunol 1993; 151: 4286-4294.

84. Macewan DJ. Elevated $cPLA_2$ levels as a mechanism by which the p70 TNF and p75 NGF receptors enhance apoptosis. FEBS Lett 1996; 379: 77-81.

85. Kroemer G, Petit P, Zamzami N et al. The biochemistry of programmed cell death. FASEB J 1995; 9: 1277-1287.

86. Vayssiere JL, Petite PX, Risler Y et al. Commitment to apoptosis is associated with changes in mitochondrial biogenesis and

activity in cell lines conditionally immortalized with simian virus 40. Proc Natl Acad Sci USA 1994; 91: 11752-11756.

87. Saks VA, Kuznetsov AV, Khuchua ZA et al. Control of cellular respiration in vivo by mitochondrial outer membrane and by creatine kinase. A new speculative hypothesis: possible involvement of mitochondrial-cytoskeleton interactions. J Mol Cell Cardiol 1995 27: 625-645.

88. Kumar S, Harvey NL. Role of multiple cellular proteases in the execution of programmed cell death. FEBS Lell 1995; 375: 169-173.

89. Martin SJ, Green DR. Protease activation during apoptosis: Death by a thousand cuts? Cell 1995; 82: 349-352.

90. Boulakia CA, Chen G, Ng FWH et al. Bcl-2 and adenovirus E1B 19 kDa protein prevent E1A-induced processing of CPP32 and cleavage of poly(ADP-ribose) polymerase. Oncogene 1996; 12: 529-535.

91. Chinnaiyan AM, Orth K, Orourke K et al. Molecular ordering of the cell death pathway—Bcl-2 and Bcl-x (L) function upstream of the CED-3-like apoptotic proteases. J Biol Chem 1996; 271: 4573-4576.

92. Monney L, Otter I, Olivier R et al. Bcl-2 overexpression blocks activation of the death protease CPP32/Yama/apopain. Biochem Biophys Res Comm 1996; 221: 340-345.

93. Gottlieb RA, Gruol DL, Zhu JY et al. Preconditioning in rabbit cardiomyocytes—Role of pH, vacuolar proton ATPase, and apoptosis. J Clin Invest 1996; 97: 2391-2398.

94. Mashima T, Naito M, Fujita N et al. Identification of actin as a substrate of ICE and an ICE-like protease and involvement of an ICE-like protease but not ICE in VP-16-induced U937 apoptosis. Biochem Biophys Res Comm 1995; 217: 1185-1192.

95. Kayalar C, Ord T, Testa MP et al. Cleavage of actin by interleukin 1 beta-converting enzyme to reverse DNase I inhibition. Proc Natl Acad Sci USA 1996; 93: 2234-2238.

96. Brancolini C, Benedetti M, Schneider C. Microfilament reorganization during apoptosis: The role of Gas2, a possible substrate for ICE-like proteases. EMBO J 1995; 14: 5179-5190.

97. Martin SJ, Obrien, Nishioka WK et al. Proteolysis of fodrin (non-erythroid spectrin) during apoptosis. J Biol Chem 1995; 270: 6425-6428.

98. Squier MKT, Miller ACK, Malkinson AM et al. Calpain activation in apoptosis. J Cell Phys 1994; 159: 229-237.

99. Laster SM, Mackenzie JM. Bleb formation and F-actin distribution during mitosis and tumor necrosis factor-induced apoptosis. Microscopy Res Technique 1996; 34: 272-280.

100. Miyoshi H, Umeshita K, Sakon M et al. Calpain activation in plasma membrane bleb formation during tertbutyl hydroperoxide-induced rat hepatocyte injury. Gastroenterology 1996; 110: 1897-1904.

101. Ucker DS, Obermiller PS, Eckhart W et al. Genome digestion is a dispensable consequence of physiological cell death mediated by cytotoxic T lymphocytes. Mol Cell Biol 1992; 12: 3060-3069.

102. Peitsch MC, Polzar B, Stephan H et al. Characterization of the endogenous deoxyribonuclease involved in nuclear DNA degradation during apoptosis (programmed cell death). EMBO J 1993; 12: 371-377.

103. Peitsch MC, Mannherz HG, Tschopp J. The apoptosis endonucleases: cleaning up after cell death? Trends Cell Biol 1994; 4: 37-41.

104. Kolber MA, Broschat KO, Landa-Gonzalez B. Cytochalasin B induces cellular DNA fragmentation. FASEB J 1990; 4: 3021-3027.

105. Sauman I, Berry SJ. Cytochalasin-D treatment triggers premature apoptosis of insect ovarian follicle and nurse cells. Int J Dev Biol 1993; 37: 441-450

106. Samali A, Cotter TG. Heat shock proteins increase resistance to apoptosis. Exp Cell Res 1996; 223: 163-170.

107. Endresen PC, Prytz PS, Aarbakke J. A new flow cytometric method for discrimination of apoptotic cells and detection of their cell cycle specificity through staining of F-actin and DNA. Cytometry 1995; 20: 162-171.

108. Radford NB, Fina M, Benjamin IJ et al. Cardioprotective effects of 70-kDa heat shock protein in transgenic mice. Proc Natl Acad Sci USA 1996; 93: 2339-2342.

109. Shinohara K, Tomioka M, Nakano H et al. Apoptosis induction resulting from

proteasome inhibition. Biochem J 1996; 317: 385-388.

110. Cheng W, Kajstura J, Nitahara J et al. Programmed myocyte cell death affects the viable myocardium after infarction in rats. Exp Cell Res 1996; 226: 316-327.

111. Misao J, Hayakawa Y, Ohno M et al. Expression of Bcl-2 protein, an inhibitor of apoptosis, and bax, an accelerator of apoptosis, in ventricular myocytes of human hearts with myocardial infarction. Circulation 1996; 94: 1502-1512.

112. Petit PX, Susin S-A, Zamzani N et al. Mitochondria and programmed cell death:

back to the future. FEBS Lett 1996; 396: 7-13.

113. Susin SA, Zamzani N, Castedo M et al. Bcl-2 inhibits the mitochondrial release of an apoptogenic protease. J Exp Med 1996; 184: 1331-1341.

114. Asoh S, Mori T, Hayashi J-I, Ohta S. Expression of the apoptosis mediator Fas is enhanced by dysfunctional mitochondria. J Biochem 1996; 120: 600-607.

115. Mehlen P, Shulze-Osthoff K, Arrigo A-P. Small stress proteins as novel regulators of apoptosis. J Biol Chem 1996; 271: 16510-16514.

CHAPTER 9

CONCLUSIONS AND PERSPECTIVES

Finishing our book we would like to sum up and to discuss further trends. Despite obvious progress in understanding the molecular mechanisms of cellular adaptation to energy deprivation, some questions remain unresolved and we wish to list them here. Likewise, in this concluding chapter, we intend again to emphasize the clinical relevance of the reviewed problems.

9.1. ADAPTATION TO ENERGY STARVATION ON THE MOLECULAR LEVEL

One can note two major theses of our book: (1) Transient energy-depleting exposures (ischemia, anoxia, metabolic block in ATP generation) induce the synthesis and accumulation of HSPs in the recovering cells; and (2) high content of HSP(s) in mammalian cells confers on them tolerance during sustained ATP deprivation (or ischemia, or anoxia). Undoubtedly, these are two manifestations of the same phenomenon, namely cellular adaptation to alternate episodes of energy starvation. The adaptation is based, first, on a universal mechanism of HSP induction in response to damage of cellular proteins and, second, on a beneficial capacity of the induced HSP(s) toward damaged cellular proteins. To our knowledge, such an adaptive reaction seems especially real for humans whose organs suffer from ischemic attacks. Below we summarize the main facts of each thesis and accent those problems which remain to be clarified.

9.1.1. PROTEIN AGGREGATION IN CELLS LACKING ATP PROMOTES ACTIVATION OF HEAT SHOCK GENES

Mechanisms regulating the level of HSP expression in eukaryotic cells are reviewed in chapter 1. It is generally accepted that the transcriptional response is the major regulatory step in HSP expression. Briefly, synthesis of HSP mRNAs is triggered by the binding of the activated heat shock transcription factor (HSF) to heat shock element (HSE), a specific nucleotide sequence at the promoters of heat shock genes. The activation of HSF1 appears to be achieved by its oligomerization upon heating or some other environmental stress. Many data indicate that the activity of HSF is negatively regulated by HSPs, HSP70 playing an unique role as a regulatory suppressor of the HSF-mediated transcriptional response (see chapter 1 and refs. 1-4 this chapter). According to the modern hypotheses, an excess of free (nonbound to protein substrates) HSP70 molecules inhibits the DNA-binding activity of HSF1, thus interrupting HSF-HSE interactions and preventing the heat shock gene transcription. This gives us an understanding of why heating and many other harmful exposures evoke the same transcriptional response. Indeed, such inducers of HSP expression as high temperature, acidosis, treatment with oxidants, heavy metals, amino acid analogs, etc. are proteotoxic, i.e., all of them damage cellular proteins. In turn, stress-denatured (unfolded) proteins being substrates for chaperoning sequester free HSP70 and thus save HSF from the inhibitor. As a consequence, the activated HSF binds to HSE and initiates the transcription of heat shock genes. Experiments with microinjection of exogenous denatured proteins into an unstressed cell[5,6] suggest that it is the intracellular accumulation of aggregated proteins that triggers the heat shock response.

Is the same scheme realized under ATP depletion? Evidently yes, because on the one hand, ischemia/reperfusion and related metabolic stresses followed by recovery increase the synthesis of HSP mRNAs and HSPs themselves (reviewed in chapter 4) and on the other hand, ATP loss as well as other proteotoxic exposures cause protein aggregation depleting the cytosolic pool of HSP70 (see chapters 3, 4 and refs. 7-10 this chapter). In regard to ischemia, its proteotoxicity can be due in part to ionic imbalance (first of all, acidification) and postanoxic burst in reactive oxygen species (oxidative stress) during reperfusion (see Fig. 9.1). However, ATP depletion per se is shown to activate HSF, this activation occurring before the stress-provoked acidification of the cytosol.[11] Such an "early" transcriptional response appears to be induced by rapid sequestration of free rise of HSP70, which is manifested in sharp HSP70 insolubility and happens immediately after the severe decrease in cellular ATP.[7-10]

The reasons for the HSP70 insolubility in ATP-depleted cells have been thoroughly considered in chapters 3 and 4 and the relevance of this phenomenon to the stress response induction is beyond any doubt. In contrast to heat shock, a drop in the ATP/ADP ratio may result in both stabilization of the preexisting HSP70-protein complexes and generation of new targets for the chaperone. The most abundant preexisting substrates are nascent polypeptides,[12] whereas various intracellular proteins sensitive to ATP decrease become the newly formed HSP70 targets. In particular, constituents of the cytoskeleton[13] and thermolabile enzymes[9] have been suggested as co-aggregating with HSP70 under ATP depletion. Thus, in addition to the decrease in ATP, the rise in ADP seems to play an important role in that it transforms HSP70 into an active "ADP state" with the highest binding affinity for unfolded proteins.[14,15] Perhaps the ADP-mediated hyperactivation enables HSP70 to catch not only proteins denatured due to lack of ATP but also transiently unfolded native proteins (see chapters 3, 7 and ref. 9 this chapter). Moreover, HSP70-protein substrate complexes cannot dissociate in energy-deprived cells that excludes the replenishment of the free chaperone pool. In all other points, the proposed scheme of the stress response induction during ATP depletion or ischemia is indistinguishable from that during any proteotoxic treatments (see Fig. 9.1).

What is yet unclear in this field? Major unresolved issues include:

1. What is the role of protein kinases/phosphatases in the regulation of heat shock gene expression in mammalian cells?

Besides the change in oligomeric structure of HSF1, the HSF-mediated transcriptional response of mammals may be regulated by altering protein kinase/phosphatase activity in a stressed cell. Actually, HSF1 is phosphorylated itself and RNA polymerase II undergoes hyperphosphorylation

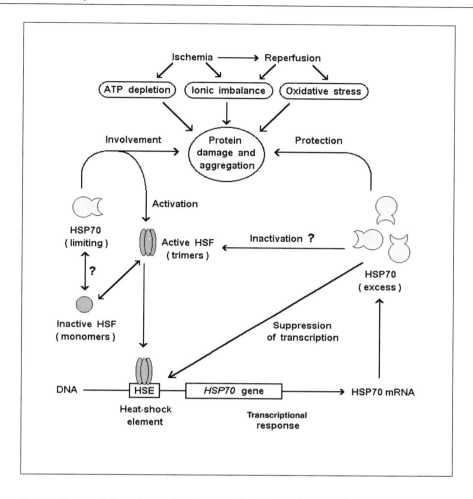

Fig. 9.1. Summarizing scheme that shows molecular mechanism of cellular adaptation to energy deprivation (thoroughly described in the text).

at the CTD domain, which may perform a regulatory function toward heat shock gene transcription (see chapter 1 and 4). This seems all the more intriguing, since stress-activated protein kinases really exist. Furthermore, in vivo activities of some protein kinases (e.g., MAPK kinases) increase just following ATP depletion/replenishment. However, the involvement of protein kinases and phosphatases in the heat shock response of mammalian cells remains undefined.

2. How do excess HSPs turn off the transcription of heat shock genes?

There are several possibilities. For example, an excess of free HSP70 may bind

the activated HSF1 to block or to interrupt its association with HSE, or to facilitate its inactivation (dissociation of trimer to monomers). On the other hand, free HSP70 may form a complex with inactive HSF thus inhibiting its activation (trimerization). This question is unanswered for both ATP depletion and heat shock.

3. Does the HSF-HSE binding in vivo occur during ATP depletion or when ATP has recovered?

It is unknown because, on the one hand, the specific DNA-binding activity is induced in HSF1 by ATP depletion immediately;[11] on the other hand, ATP is required for the GAGA factor-mediated unfolding of nucleosomal DNA that renders HSE at a heat shock promoter available for binding to the activated HSF.[16] On the whole, effects of ATP depletion on the chromatin structure need more attentive investigation.

4. Does the regulation of HSP mRNA translation or of HSP turnover play a role in the intracellular accumulation of HSPs following transient ATP decrease?

In contrast to heat shock, the effects of energy deprivation on the polysome integrity, mRNA stability and rate of HSP degradation are poorly studied, although changes in these parameters might influence the post-stress HSP expression. An intriguing phenomenon is the recently discovered exposition of HSP70 on the cell surface; however, it is unclear whether this device has a significance for the regulation of HSP expression.

9.1.2. EXCESS HSPs PROTECT AGAINST PROTEIN AGGREGATION DURING ATP DEPLETION

One can read in chapter 3 that protein aggregation in mammalian cells lacking ATP correlates with stress-induced cell death. Apparently, the aggregation of actin (side-to-side bundling of filaments and misassembly of monomers) occurring in ATP-depleted cells results in dysfunction of the cortical actin skeleton followed by blebbing and perforation of the plasma membrane (necrosis).[7,13,17] Besides G- and F-actin, many other intracellular proteins aggregate due to lack of ATP, which may also contribute to cell injury and death under ischemic-like metabolic stress (thoroughly reviewed in chapter 3). Importantly, the tolerant cells overexpressing HSP(s) exhibit reduced "aggregability" of actin and cytosolic proteins and elevated survival during ATP depletion when compared with control cells; this diversity is not due to differences in the residual ATP levels (see chapters 6, 7 and refs. 13,18-21 this chapter). From this it follows that excess HSPs are able to protect cellular proteins from the ATP loss-induced aggregation by retaining proteins in a soluble state.

Which of the HSPs are involved in protecting proteins from aggregation in ATP-deprived cells? Based on own observations,[19] we suggest that the major role in the prevention of lethal protein aggregation in ATP-depleted cells belongs to HSP70. Because a direct link exists between protein aggregation and necrotic death, our suggestion is supported by the experiments that demonstrate that constitutive HSP70 overexpression does confer cell resistance to necrosis under ATP-depleting conditions (see for review chapters 5 and 6). At first sight, the HSP70-mediated protection from protein aggregation *during* ATP depletion seems to be surprising, since HSP70 itself needs ATP for the performance of its chaperoning function. However, a slight decrease in the ATP/ADP ratio in the cytosol appears to define the "activation" of HSP70 by ADP and saturating targeting of the activated chaperone to unfolded proteins that should preserve them from aggregation or misfolding upon ATP deprivation (see chapter 7).

Believing in the capacity of HSP70 to protect against intracellular protein aggregation during ATP depletion, we have included this component in Figure 9.1 as determinative of cell tolerance to ATP loss or ischemia/reperfusion. The proposed scheme seems quite logical to explain the molecular mechanism adapting mammalian cells to periods of energy starvation. Obviously, a crucial point here is the stress-induced protein aggregation: being sublethal it recruits limiting HSP70 in nontolerant cells, which is conducive to HSF activation, heat shock gene transcription and HSP synthesis. In turn, the accumulation of HSP70 renders cells tolerant, since excess HSP70 is able to prevent even lethal protein aggregation in case of repeated ATP depletion (see Fig. 9.1 and chapter 7).

Perhaps besides the suppression of protein aggregation provoked by ATP loss, excess HSP70 contributes to disaggregation, refolding or degradation of cellular proteins damaged during ATP depletion or ischemia/reperfusion. Hereafter, being free from performance of its protective and reparative functions, a surplus of HSP70 turns off the transcription of heat shock genes that accomplishes the autoregulatory loop in the mechanism of stress response (see Fig. 9.1). Our concluding scheme does not contain other HSPs because their role in the negative regulation of heat shock gene expression is not as evident as that of HSP70; however, their possible participation in cytoprotection under energy deprivation or ischemia/reperfusion should be taken into consideration as well.

The outstanding questions that need answers in this field include:

1. What is the role of cofactors in the HSP70-mediated cytoprotection?

Although we discussed in chapters 3 and 7 the importance of ADP, ATP, Hip and HSP40 for chaperoning intracellular proteins by HSP70 during energy deprivation and recovery, no direct evidence is available to prove that they do assist HSP70 in the protection. Probably, a comparison of the effects of the overexpressed intact HSP70 and its deletion mutant form lacking the nucleotide-binding domain on protein aggregation in ATP-deprived cells and their survival might, at least in part, clarify this issue (see chapter 7).

2. By what means does HSP70 or other HSPs protect the cytoskeleton?

Despite abundant speculations on this problem and its evident importance, the mechanism of HSP-mediated protection of cytoskeletal structures upon ATP depletion or ischemia/reperfusion virtually remains a mystery. When mammalian cell lines constitutively overexpressing various HSPs are available, the HSP-cytoskeleton interactions in energy-deprived cells may be investigated more successfully.

3. How does excess HSP70 or other HSPs protect mitochondria during anoxia/reoxygenation?

Indeed, better preservation of mitochondria during ischemia/reperfusion is a well-documented attribute of the tolerant cells. Furthermore, the improved mitochondrial function appears to mainly define the

post-stress cell recovery (see chapters 5-7). Taking into account that HSP60, a mitochondrial stress protein, does not maintain the viability of energy-deprived cells,[22] one may connect the tolerance of mitochondria with other induced HSPs and, most of all, with HSP70. The latter molecule participates in protein import into mitochondria (see chapter 1). However, a molecular basis of the mitochondrial protection mediated by excess cytoplasmic HSP70 remains enigmatic.

4. How do other HSPs contribute to cell resistance during ATP deprivation?

While a direct involvement in cytoprotection during ATP-depleting conditions was precisely established for HSP70, it was shown to be unlikely for HSP60 and HSP90 (see chapters 5-7 and refs. 22, 23 this chapter). At the same time, no one has thoroughly studied the possible role of HSP100, GRPs, HSP40, HSP27, αB-crystallin, ubiquitin, etc. in the tolerance of mammalian cells to energy starvation. Besides HSP70, we consider the small HSPs as the most likely candidates for cytoprotection against ischemia and its consequences. This might be easily examined on cell lines constitutively overexpressing HSP27 or αB-crystallin, since several research groups already possess such cell lines. Furthermore, transgenic mice with heart-specific overexpression of αB-crystallin have been created,[24] which provides a remarkable possibility to evaluate an implication of this chaperone in cardioprotection against ischemia/reperfusion injury. Intriguingly, the proposed protection of ATP-deprived cells by HSP27 may be determined not only by its amount but also by the degree of its phosphorylation as was revealed in Landry's lab for heat shock and oxidative stress.[25-27]

9.1.3. HSPs AND RECOVERY FROM ISCHEMIC INSULT: THE PROBLEMS OF REPERFUSION INJURY AND POST-STRESS APOPTOSIS

For most cellular stresses, cell recovery begins immediately after returning to normal conditions. Contrary to that, the recovery from ischemia or anoxia or ischemia-mimicking exposures represents an additional danger for affected cells. In fact, the cells remaining viable during ischemia can be killed by necrosis during subsequent reperfusion through the so-called reperfusion injury that comprises oxidative stress and Ca^{2+} overload (see chapters 2, 5 and 6). Moreover, some cells sustaining ischemia/reperfusion undergo apoptosis afterward (reviewed in chapters 2, 6 and 8). The resistance of tolerant cells to reperfusion calcium repletion[28] and to apoptosis following transient energy starvation[29] is associated with an increase in HSP70; however, molecular mechanisms of the cytoprotection are poorly understood in both cases.

The most important questions in this field include:

1. Which of the HSPs protect a cell against reactive oxygen species and calcium overload during post-ischemic reperfusion and how?

This remains enigmatic, although the HSP-mediated cell resistance to oxidative stress and Ca^{2+} repletion is often discussed as determinative of the phenomenon of ischemic tolerance (see for example chapter 5).

2. Is cellular protein damage a prerequisite for apoptosis following ischemia/reperfusion?

In order to answer this question, the main efforts should be focused on the interrelation between HSP70, p53 and ubiquitin. Another intriguing point is interelation between HSP70, G-actin and DNase I (see chapter 8).

3. Which of chaperones are involved in cytoprotection during post-ischemic apoptosis and what are the ways to regulate their anti-apoptotic capacity?

9.2. HEAT SHOCK PROTEINS AND ISCHEMIA: SIGNIFICANCE FOR MEDICINE

Acute ischemia, the main cause of such fatal human pathologies as myocardial infarction and brain stroke, remains an unresolved medical problem. Usually, ischemia develops as the result of progressing atherosclerosis: diminution of the arterial lumens through the expansion of atherosclerotic plaques as well as the active thromb formation at atheromas result in the limitation of blood supply (ischemic insult). However, necrosis in ischemic tissue does not occur instantaneously and thrombolytic therapy may be effective within the first few hours after a stroke. Therefore, the resistance of heart or brain cells to transient ischemia followed by reperfusion defines the outcome of the ischemic attack for a patient. In this respect, attempts to develop any compensatory reactions that might retain cell viability during ischemia/reperfusion are of paramount importance.

As we wrote in chapters 5-7, the HSP expression is that adaptive response which may really augment cell survival under ischemia or ischemic-like conditions. What can this knowledge give to practical medicine? Actually, nonabusive HSP induction in affected patients' organs could be a good device for protecting them from ischemic injury. Contrary to that, the HSP-mediated adaptation to ischemia intrinsic to quickly growing tumors may represent a serious oncological problem, since increased HSPs in tumor cells confer them resistance to some kinds of anticancer therapy. Below we review both of these issues.

9.2.1. HSPs, ISCHEMIA AND SALVAGE OF MYOCARDIUM

Perhaps cardiology is the field of medicine where the HSP-mediated defense from ischemia might be applied with a maximal effect. On the one hand, ischemic heart disease is the leading cause of death and disability of men in industrialized countries. On the other hand, it was directly demonstrated on experimental animals that enhanced expression of cardiac HSP70 correlates with reduction of the infarct size and better recovery of the myocardium after ischemia/reperfusion (see chapter 5). The clinical context in which these data are discussed is an important one. However, the most popular ways to achieve the beneficial effect of excess HSP70 (whole body hyperthermia, ischemic or hypoxic preconditioning, transgene overexpression) cannot be adopted by cardiologists for clinical use. Indeed, heating or hypoxia seem to be too harmful for patients with heart pathology. In regard to the "*HSP70* gene cardiotherapy," it is a long way in the future. It seems likely though that the heat preconditioning might be employed in cardiac surgery using hot perfusion of a patient's heart immediately prior to cardiopulmonary bypass or angioplasty. Such a HSP-inducing procedure may be performed on the heart of an open-chest patient scheduled for operation in order to improve preservation of the stunned myocardium during cardioplegia. In support of this approach is a study by Mccully et al[30] who showed the HSP72 accumulation and reliable cardioprotective effect after 15 minutes retrograde hyperthermic (42°C) perfusion of an isolated rat heart only 5 minutes prior to global ischemia. An analogous procedure might be recommended for heat preconditioning of an isolated donor's heart prior to its transplantation into a recipient. We believe that a series of brief ischemic/reperfusion episodes as the preconditioning stimulus may also be used in cardiac surgery instead of the above described hyperthermic perfusion.[30]

A more intriguing question exists: is the HSP70 induction in human hearts a suitable remedy for effective prophylaxis of myocardial infarction among patients with ischemic heart disease? Except for the *HSP70* transgene overexpression, it is difficult to imagine how the level of cardiac HSP70 could be increased to provide constant or at least prolonged cardiotolerance to ischemia. Nevertheless, the transient enhancement of HSP70 expression in a patient's heart appears to be quite beneficial when a risk of infarction arises (e.g., at the beginning of acute ischemic attack). By what means may this enhancement be attained? Evidently, induction of the classic stress response by hyperthermia or by any other proteotoxic exposure is impossible here. A far more likely method of increasing HSP70 in a patient's heart will be through its pharmacological induction. Indeed, a number of drugs induce HSP70 in mammalian cells and tissues. For instance, cardiac HSP70 can be augmented by injected amphetamine, and the cardioprotective effect of the drug injection on post-ischemic recovery of porcine myocardium is observed (see chapter 5 and refs. 31, 32 this chapter). Based on their encouraging results achieved with amphetamine in pigs[31,32] Maulik et al discuss the possibility of using this drug in clinical applications; however, its application in the therapy of serious patients is doubtful, since amphetamine can elevate body temperature to an extreme by enhancing endogenous lipolysis.[32] Perhaps the more promising inducer of HSP70 is herbimycin A. This benzoquinoid ansamycin antibiotic is known to be a tyrosine kinase inhibitor and a nonstressful inducer of HSP70 in mammalian cell lines.[33,34] Morris and colleagues showed that herbimycin A causes the selective accumulation of HSP70 in cultured rat cardiomyocytes, thus conferring them tolerance to simulated ischemia.[35] Although the effect of herbimycin A on experimental models of ischemia in animals is not yet examined, the in vitro data obtained[35] already characterize this drug as a potential tool of anti-ischemic cardiotherapy. In the same context, indomethacin should be of interest to cardiologists, since this anti-inflammatory drug activates human HSF1 and thus considerably decreases the temperature threshold in the heat shock response.[36] Further screening of new pharmaceuticals could result in the discovery of the most suitable drug to safely manipulate HSP expression in patients with heart pathology.

From a molecular biologist's point of view, transgenic methods have the best prospects for cardiotherapy in the future. As a matter of fact, this technology is still very far from clinical application. One of the main problems in gene therapy strategy lies in the targeted delivery of transgene vectors to cardiomyocytes or endothelial cells within the human heart.[37] However, when the technical problems are successfully resolved, *HSP* transgene therapy will open new ways to prolong life and decrease death-rate in patients with ischemic heart disease. Besides the overexpression of *HSP* transgenes themselves, ischemia-inducible promoter elements from heat shock genes may be used in gene cardiotherapy as primers to overexpress other gene products possessing cytoprotective potential (e.g., bcl-2, antioxidative enzymes, thrombolytic and angiogenic factors, etc.).

9.2.2. HSPs AND ISCHEMIC TOLERANCE IN THE BRAIN

In contrast to myocardial ischemia, brain ischemia has not yet been studied in transgenic animals overexpressing HSPs. It was shown though that in vitro transfection with *HSP70* transgene protects cultured neurons and astrocytes from necrosis induced by ischemia-mimicking exposures.[38,39] Likewise, in vivo experiments and ischemic preconditioning of gerbil brain suggest an important role of the induced HSP(s) in tolerance of the central nervous system to ischemic insult (see for review chapter 6). The clinical relevance of the findings lies in the possibility that cell viability would be preserved in the ischemic brain before blood reflow, as in the case of successful thrombolytic therapy.

Importantly, excess HSPs appear to protect ischemic neurons from excitotoxic injury and apoptosis-like delayed death (reviewed in chapter 6 and ref. 40 this chapter). Clinical benefit is expected if HSPs can be induced in the brain safely as prophylactic procedures for patients in situations of known high risk of stroke. Apparently, the same strategy may be undertaken

to save patients experiencing severe brain insult. Thus, HSPs have great potential for clinical application, if we can learn to manipulate their expression. As we already wrote in the previous subsection, the most realistic approach herein is the pharmacological induction or amplification of the heat shock response in affected human tissues (see refs. 33-36).

9.2.3. ISCHEMIA, HSPs AND TUMOR RESISTANCE TO ANTICANCER THERAPY

Some cancer cells can also undergo ischemia or ischemic-like stress during in vivo growth (see for review chapters 4, 6 and refs. 20, 21 this chapter). In large solid tumors, ischemia occurs as a result of disproportion between the size of the malignancies and the degree of their vascularization; in ascites forms of cancer, the tumor cells achieving a high density can experience ischemia-like energy deprivation on account of nutrient and oxygen deficiency in ascites fluid.[20,21] In both cases, tumor cells adapting to ischemia can overexpress HSPs. It seems important that the ischemia-induced HSP accumulation in tumor cells confers not only tolerance to chronic energy starvation but also resistance to those cytotoxic exposures which are tools of combined anticancer therapy. This supposition is based on numerous data showing the HSP70- and/or HSP27-mediated protection of tumor cells against hyperthermia and various chemotherapeutic agents (briefly summarized in Table 9.1). Moreover, the overexpression of HSP70 in malignant cells appears to enhance their tumorigenicity.[49] From this it follows that ischemia may generate a certain tumor phenotype which is characterized by a high level of HSP expression, high tumorigenic potential and high resistance to many cytotoxic exposures. It is not excluded that these resistant cells, remaining viable after combined therapeutic treatments, result in tumor relapse (also discussed in refs. 20 and 21).

It seems likely that the artificial decrease in intratumor levels of HSPs (first

Table 9.1. HSP-mediated tumor resistance to various cytotoxic treatments

Cytotoxic Agent	HSPs Associated with Protection	Tumor Cell Origin	Ref.
Tumor Necrosis	HSP70	fibrosarcoma	41
Factor (TNF)	HSP27	carcinoma	42
Doxorubicine	HSP70	fibrosarcoma	43
	HSP27	testis	44
		breast	45
Cis-platinum	HSP60	head	46
	HSP27	testis	44
Etoposide	HSP70	breast	47
	HSP27	breast	47
Actinomycin D	HSP70	breast	47
	HSP27	breast	47
Vincristine	HSP70	brain	48

of all, HSP70 and HSP27) will lead to the reverse effect, namely diminution of tumorigenicity of malignant cells and their sensitization to therapeutic treatments. In this respect, the inhibitory effect of quercetin on the HSP70 expression in various tumor cell lines[50] and the suppression of heat shock gene transcription in human colon carcinoma HT-29 cells by staurosporine[51] should attract oncologists' attention.

We hope that the questions and approaches outlined in our book will motivate researchers to investigate more actively the role of HSPs in ischemic tolerance.

CONCLUDING REMARKS

In our book, we have intended to uncover a new and rather unexpected function of molecular chaperones, namely the maintenance of cell viability under energy deprivation. We believe that this activity of HSPs has a much greater importance for physiology of mammals, which have a rather constant body temperature, than protection from heat shock. It was only by chance that Ritossa who observed in 1962 chromosomal puffs in *Drosophila* salivary glands treated with a blocker of ATP generation (2,4-dinitrophenol), also subjected

the glands to heat shock and designated the reaction as heat shock response. Otherwise this response might be called metabolic stress response or something else and to date we could know more detailed information about HSPs and energy deprivation, at least as obtained for heat-protective capacity of chaperones. Despite some similar aspects between energy deprivation and heat shock, the main likeness being the apparent proteotoxicity of both exposures, a few features of the stresses are very diverse, and this idea we have also tried to underline in our book.

When we began our book some time ago, evidence for HSP-mediated protection from ischemic stress was rather scarce. To date, upon finishing the book, the missing facts have been obtained by several independent groups of researchers who are using the most advanced techniques of molecular biology. There is no doubt that HSP70 does protect various cells and tissues from ischemic injury. In the near future, however, much work must be done in order to determine the mechanism of this phenomena. But this will be the topic of other books...

REFERENCES

1. Abravaya K, Myers MP, Murphy SP, Morimoto RI. The human heat shock protein hsp70 interacts with HSF, the transcription factor that regulates heat shock gene expression. Genes Dev 1992; 6: 1153-1164.

2. Baler R, Welch WJ, Voelmy R. Heat shock gene regulation by nascent polypeptides and denatured proteins: hsp70 as a potential autoregulatory factor. J Cell Biol 1992; 117: 1151-1159.

3. Mosser DD, Duchaine J, Massie B. The DNA-binding activity of the human heat shock transcription factor is regulated in vivo by hsp70. Mol Cell Biol 1993; 13: 5427-5438.

4. Mifflin LC, Cohen RE. hsc70 moderates the heat shock (stress) response in *Xenopus laevis* oocytes and binds to denatured protein inducers. J Biol Chem 1994; 269; 15718-15723.

5. Anathan J, Goldberg AL, Voellmy R. Abnormal proteins serve as eukaryotic stress signals and trigger the acivation of heat shock genes. Science 1986; 232: 252-254.

6. Mifflin LC, Cohen RE. Characterization of denatured protein inducers of the heat shock (stress) response in *Xenopus laevis* oocytes. J Biol Chem 1994; 269: 15710-15717,

7. Kabakov AE, Gabai VL. Protein aggregation as primary and characteristic cell reaction to various stresses. Experientia 1993; 49: 706-710.

8. Kabakov AE, Gabai VL. Stress-induced insolubility of certain proteins in ascites tumor cells. Arch Biochem Biophys 1994; 309: 247-253.

9. Nguyen VT, Bensaude O. Increased thermal protein aggregation in ATP-depleted mammalian cells. Eur J Biochem 1994; 220: 239-246.

10. Gabai VL, Kabakov AE. Induction of heat-shock protein synthesis and thermotolerance in EL-4 ascites tumor cells by transient ATP depletion after ischemic stress. Exp Mol Pathol 1994; 60: 88-99.

11. Benjamin IJ, Horie S, Greenberg ML et al. Induction of stress proteins in cultured myogenic cells. J Clin Invest 1992; 89: 1685-1689.

12. Beckmann RP, Lovett M, Welch WJ. Examining the function and regulation of hsp70 in cells subjected to metabolic stress. J Cell Biol 1992; 117: 1137-1150.

13. Kabakov AE, Gabai VL. Heat shock proteins maintain the viability of ATP-deprived cells: what is the mechanism? Trends Cell Biol 1994; 4: 193-196.

14. Palleros DR, Welch WJ, Fink AL. Interaction of hsp70 with unfolded proteins: Effects of temperature and nucleotides on the kinetics of binding. Proc Natl Acad Sci USA 1991; 88: 5719-5723.

15. Palleros DR, Reid KL, Shi L et al. ATP-induced protein-Hsp70 complex dissociation requires K^+ but not ATP hydrolysis. Nature 1993; 365: 664-666.

16. Tsukiyama T, Becker PB, Wu C. ATP-dependent nucleosome disruption at a heat-shock promoter mediated by binding of GAGA transcription factor. Nature 1994; 367: 525-532.

17. Gabai VL, Kabakov AE. Tumor cell resistance to energy deprivation and hyperthermia can be determined by the actin skeleton stability. Cancer Lett 1993; 52: 3648-3654.

18. Gabai VL, Kabakov AE. Rise in heat-shock protein level confers tolerance to energy deprivation. FEBS Lett 1993; 327: 247-250.

19. Kabakov AE, Gabai VL. Heat shock-induced accumulation of 70-kDa stress protein (HSP70) can protect ATP-depleted tumor cells from necrosis. Exp Cell Res 1995; 217: 15-21.

20. Gabai VL, Mosina VA, Budagova KR, Kabakov AE. Spontaneous over-expression of heat-shock proteins in Ehrlich ascites carcinoma cells during in vivo growth. Biochem Mol Biol Int 1995; 35: 95-102.

21. Kabakov AE, Molotkov AO, Budagova KR et al. Adaptation of Ehrlich ascites carcinoma cells to energy deprivation in vivo can be associated with heat shock protein accumulation. J Cell Physiol 1995; 165: 1-6.

22. Heads RJ, Yellon DM, Latchman DS. Differential cytoprotection against heat stress or hypoxia following expression of specific stress protein genes in myogenic cells. J Mol Cell Cardiol 1995; 27: 1669-1678.

23. Amin V, Cumming DV, Latchman DS. Over-expression of heat shock protein 70 protects neuronal cells against both thermal and ischaemic stress but with different efficiences. Neurosci Lett 1996; 206: 45-48.

24. Gopalsrivastava R, Haynes JI, Piatigorsky J. Regulation of the murine alpha B-crystallin/small heat shock protein gene in cardiac muscle. Mol Cell Biol 1995; 15: 7081-7090.

25. Lavoie JN, Gingras-Breton G, Tanguay RM, Landry J. Induction of Chinese hamster HSP27 gene expression in mouse cells confers resistance to heat shock. HSP27 stabilization of the microfilament organization. J Biol Chem 1993; 268: 3420-3429.

26. Lavoie JN, Lambert H, Hickey E et al. Modulation of cellular thermoresistance and actin filament stability accompanies phosphorylation-induced changes in the oligomeric structure of heat shock protein 27. Mol Cell Biol 1995; 15: 505-516.

27. Huot J, Houle F, Spitz DR, Landry J. HSP27 phosphorylation-mediated resistance against actin fragmentation and cell death induced by oxidative stress. Cancer Res 1996; 56: 273-279.

28. Marber MS, Walker JM, Latchman DS, Yellon DM. Attenuation by heat stress of a submaximal calcium paradox in the rabbit heart. J Mol Cell Cardiol 1993; 25: 1119-1126.

29. Gabai VL, Zamulaeva IV, Mosin AF et al. Resistance of Ehrlich tumor cells to apoptosis can be due to accumulation of heat shock proteins. FEBS Lett 1995; 375: 21-26.

30. Mccully JD, Lotz MM, Krukenkamp IB et al. A brief period of retrograde hyperthermic perfusion enhances myocardial protection from global ischemia: Association with accumulation of Hsp 70 mRNA and protein. J Mol Cell Cardiol 1996; 28: 231-241.

31. Maulik N, Wei Z, Engelman RM et al. Improved post-ischemic ventricular recovery by amphetamine is linked with its ability to induce heat shock. Mol Cell Biochem 1994; 137: 17-24.

32. Maulik N, Engelman RM, Wei Z et al. Drug-induced heat-shock preconditioning improves post-ischemic ventricular recovery after cardiopulmonary bypass. Circulation 1995; 92 [suppl II]: II-381-II-388.

33. Murakami Y, Uehara Y, Yamamoto C et al. Induction of Hsp72/73 by Herbimycin A, an inhibitor of tyrosine kinase oncogenes. Exp Cell Res 1991; 195: 338-344.

34. Hegde RS, Zuo J, Voellmy R, Welch WJ. Short circuiting stress protein expression via a tyrosine kinase inhibitor, Herbimycin A. J Cell Physiol 1995; 165: 186-200.

35. Morris SD, Cumming DVE, Latchman DS, Yellon DM. Specific induction of the 70-kD heat stress proteins by the tyrosine kinase inhibitor herbimycin-A protects rat neonatal cardiomyocytes—A new pharmacological route to stress protein expression? J Clin Invest 1996; 97: 706-712.

36. Lee BS, Chen J, Angelidis C et al. Pharmacological modulation of heat shock factor 1 by antiinflammatory drugs results in protection against stress-induced cellular damage. Proc Natl Acad Sci USA 1995; 92: 7207-7211.

37. Benjamin IJ, Williams RS. Expression and function of stress proteins in the ischemic heart. In: Morimoto RI, Tissieres A, Georgopoulos C, eds. The Biology of Stress Proteins and Molecular Chaperones. Cold Spring Harbor, NY: Cold Spring Harbor Laboratory Press, 1994: 533-552.

38. Papadopoulos MC, Sun XY, Cao JM et al. Over-expression of HSP-70 protects astrocytes from combined oxygen-glucose deprivation. Neuroreport 1996; 7: 429-432.

39. Amin V, Cumming DV, Latchman DS. Over-expression of heat shock protein 70 protects neuronal cells against both thermal and ischaemic stress but with different efficiences. Neurosci Lett 1996; 206: 45-48.

40. Koroshetz WJ, Bonventre JV. Heat shock response in the central nervous system. Experientia 1994; 50: 1085-1091.

41. Jaattela M. Overexpression of major heat shock protein hsp70 inhibits tumor necrosis factor-induced activation of phospholipase A_2. J Immunol 1993; 151: 4286-4294.

42. Mehlen P, Kretzremy C, Briolay J. Intracellular reactive oxygen species as apparent modulators of heat-shock protein 27 (hsp27) structural organization and phosphorylation in basal and tumour necrosis factor alpha-treated T47D human carcinoma cells. Biochem J 1995; 312: 367-375.

43. Karlseder J, Wissing D, Holzer G et al. HSP70 overexpression mediates the escape of a doxorubicin-induced G2 cell cycle arrest. Biochem Biophys Res Comm 1996; 220: 153-159.

44. Richards EH, Hickey E, Weber L et al. Effect of overexpression of the small heat shock protein HSP27 on the heat and drug sensitivities of human testis tumor cells. Cancer Res 1996; 56: 2446-2451.

45. Oesterreich S, Weng C-N, Qiu M et al. The small heat shock protein hsp27 is correlated with growth and drug resistance in human breast cancer cell lines. Cancer Res 1993; 53: 4443-4448.

46. Nakata B, Hom DK, Barton RM et al. mRNA levels of molecular chaperones hsp27, hsp60 and hsp70 in cisplatin resistant squamous cell carcinomas. Int J Oncol 1996; 8: 1229-1235.

47. Samali A, Cotter TG. Heat shock proteins increase resistance to apoptosis. Exp Cell Res 1996; 223: 163-170.

48. Lee WC, Lin KY, Chen KD et al. Induction of HSP70 is associated with vincristine resistance in heat-shocked 9L rat brain tumour cells. Br J Cancer 1992; 66: 653-659.

49. Jaattela M. Over-expression of hsp70 confers tumorigenicity to mouse fibrosarcoma cells. Int J Cancer 1995; 60: 689-693.

50. Wei Y, Zhao X, Kariya Y et al. Induction of apoptosis by quercetin: Involvement of heat shock protein. Cancer Res 1994; 54: 4952-4957.

51. Kim SH, Kim JH, Erdos G, Lee YJ. Effect of staurosporine on heat shock gene expression and thermotolerance development in HT-29 cells. Biochem Biophys Res Commun 1993; 193: 759-763.

INDEX

A

α-actinin, 50, 53-55, 65
α-crystallin, 7, 65, 72-73, 126, 191, 194-195
α1-adrenoreceptor, 126
αA-crystallin, 7
αB-crystallin, 7, 63, 134, 136, 159, 194, 197, 215, 225
abnormal protein, 13, 64, 77
acidification, 31-33, 36, 40, 58, 72-73, 87, 102, 105, 134, 194, 214-215, 222
actin, 3-4, 7, 40, 49-58, 60-61, 63, 65, 67-68, 72-73, 76, 78, 105, 134, 149, 159, 162, 187, 190-194, 196-199, 212, 214, 224
actin aggregate, 50
actin dynamics, 50
actin microfilament(s), 7, 49-50, 53, 68, 187, 190, 192, 194
actin phosphorylation, 61
actin polymerization, 50-51, 53, 61, 76, 192, 197-198, 214
actin-myosin, 3, 50, 53, 56, 60, 190
actinomycin D, 148, 151, 214
adenine nucleotide, 38, 99, 101
adenocarcinoma, 86, 164
adenosine A1 receptor, 126, 128, 130
ADP, 4, 24-26, 31, 50, 56, 60, 67, 70, 78, 101, 105-106, 180-183, 186, 222, 224-225
3-amino-1,2,4-triazole, 122, 124
AMP-activated protein kinase, 24
amphetamine, 128-129, 135, 227
angioplasty, 227
ankyrin, 55-56, 78
antioxidant, 8, 128, 130, 135, 149, 194-195
AP-1, 107
APO-1/FAS, 210-211, 215
apoptosis, 149-157, 163-166, 177-178, 188-191, 195, 198, 205-220, 225-226, 228
arachidonic acid, 108, 111, 154
arrhythmia, 122
ascites tumor cells, 51-53, 55-56, 60-61, 64-65, 178
assembly, 2-3, 6, 8, 54, 60, 190-191, 196, 200
astrocyte(s), 94-95, 102, 105, 108, 145-147, 149-150, 166, 228
ATP hydrolysis, 4, 26, 32, 41, 67, 70, 186
ATP production, 24, 161
ATP sensor, 60-61, 65, 72
ATP stabilization, 22, 24, 27
ATP-binding, 2-3, 60-61, 73, 78, 185-187
ATP-binding domain, 185-186
ATP/ADP ratio, 39, 50-51, 56, 61, 67-68, 70, 76, 78, 99, 182, 184, 186, 189, 222, 224

ATPase, 2-4, 6-8, 25-27, 30, 33-34, 38, 41, 51, 56, 60-61, 67, 85, 157, 180, 185-186, 188-189, 191, 198
azide, 50, 85, 111, 211

B

β-galactosidase, 61-62, 64, 180-181
Bax, 151, 167, 210-211, 215
Bcl-2, 167, 215
brain stroke, 226

C

c-fos, 104, 106-107, 208
c-fos protein, 208
c-jun, 104, 106-107, 208
C2C12 mouse myogenic cell, 86
CA1 pyramidal neuron, 93-94
Ca2+ homeostasis, 33-34, 38, 111, 151
Ca2+-ATPase, 24, 33-34, 36
Ca2+-channels, 36, 96
Ca2+-overload, 34
Ca2+-paradox, 34
CA3 region, 94-95, 142, 146
calpain, 34, 36, 56, 154-155, 162, 198-199, 212-213
calpastatin, 212
cAMP, 25, 198, 210
camptotecin, 214
cap32/34, 67, 197-198
capping activity, 192, 197-98
carcinoma, 21-22, 29, 53, 76, 87-88, 98, 103-104, 159, 162, 164, 211, 229
cardiac α-crystallin, 72, 194
cardiac HSP27, 191
cardiac surgery, 125, 227
cardioplegia, 125, 135, 227
cardiopulmonary bypass, 124, 227
catalase, 92, 107, 122, 124, 128-129, 153-154, 189
cell shrinkage, 29, 164
cell survival, 2, 64, 86, 130, 147, 151, 178, 188, 194, 200, 209, 226
cell viability, 21-47, 49, 121, 130, 134, 180, 189, 196, 226, 228-229
chaperone-protein complexes, 63, 70, 182
chaperone-target complex, 186
chemical ischemia, 31
Chinese hamster ovary, 27, 66
Chinese hamster V79 cell, 86
chromatin, 3, 27, 29, 39, 58, 60, 63, 68, 164, 188, 190, 212, 224
chromatin condensation, 29, 39, 164, 212